P9-CDA-968

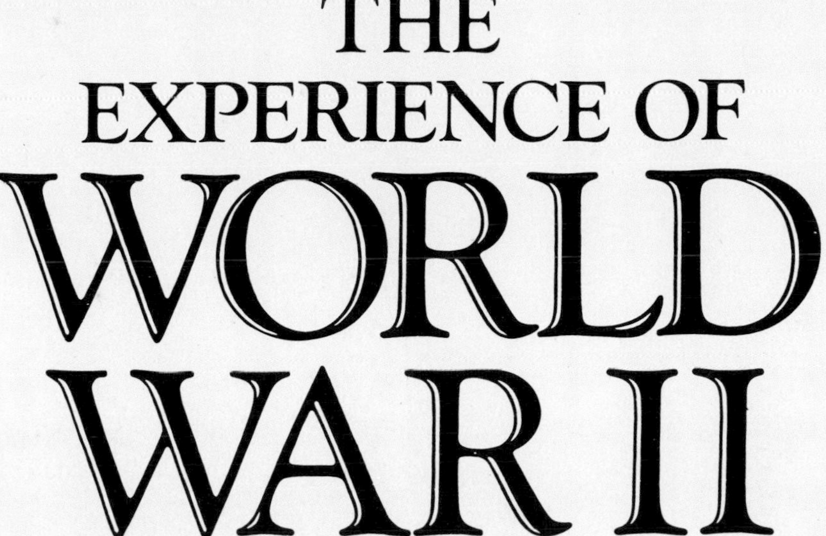

THE EXPERIENCE OF WORLD WAR II

EDITED BY
JOHN CAMPBELL

New York
OXFORD UNIVERSITY PRESS
1989

Volume editor Robert Peberdy
Art editor Ayala Kingsley
Designers Frankie Macmillan, Wolfgang Mezger, Tony de Saulles
Picture researchers Jan Croot, Diane Hamilton (USA)
Cartographic editor Olive Pearson
Project editor Peter Furtado

AN EQUINOX BOOK

Planned and produced by
Equinox (Oxford) Ltd,
Musterlin House, Jordan Hill Road,
Oxford, England OX2 8DP

Copyright © Equinox (Oxford) Ltd 1989

Published in the United States of America by Oxford University Press, Inc., 200 Madison Avenue, New York, NY 10016

Oxford is a registered trademark of Oxford University Press

All rights reserved. No part of this publication may be reproduced, stored in a retrieval system, or transmitted, in any form or by any means, electronic, mechanical, photocopying, recording or otherwise, without the permission of the publisher.

Library of Congress
Cataloging-in-Publication Data

The Experience of World War II
/edited by John Campbell.
p. cm.
Includes bibliographical references.
ISBN 0-19-520792-0
1. World War, 1939–1945.
I. Campbell, John, 1947–
II. Title: Experience of World War 2.
III. Title: Experience of World War two.
D743.E86 1989
940.53–dc20 89–16293

Printing (first digit) 9 8 7 6 5 4 3 2 1

Printed in Yugoslavia by
Gorenjski Tisk, Kranj,
by arrangement with Papirografika

CONTRIBUTORS

Dr Duncan Anderson Royal Military Academy, Sandhurst 20–27, 36–43, 82–87

Professor Geoffrey Best Formerly of University of Sussex 210–217, 230–231

Dr Kathleen Burk Imperial College, London 134–139, 166–171, 224–229, 240–245

Dr Angus Calder Open University 44–51, 52–53, 152–157, 174–181, 196–209

Dr Malcolm Cooper Formerly of University of Newfoundland, Canada 92–101, 102–103, 104–109, 110–111, 112–117, 118–119, 122–125, 182–185

Dr Martin Dean Formerly University of Cambridge 219–219

Dr Anne Deighton University of Reading 246–247

Professor John Erickson University of Edinburgh 54–63, 158–163, 164–165

Dr Julian Jackson University College of Swansea 194–195

Dr Keith Jeffery University of Ulster 126–131

Dr Peter Lowe University of Manchester 140–145, 234–239

Dr Brian Holden Reid King's College, London 28–35, 64–71, 74–81

Denis Ridgeway Formerly of Royal Naval Scientific Service 120–121

Dr Keith Sainsbury University of Reading 72–73

Dr Jill Stephenson University of Edinburgh 146–151, 186–193

Dr J. M. Winter Pembroke College, Cambridge 232–233

ADVISORY EDITORS

Dr Alan Borg Imperial War Museum, London

Lt-Gen. Sir Napier Crookenden Formerly Lieutenant of Her Majesty's Tower of London

General Andrew J. Goodpaster US Army (retired)

Dr Wolfgang Krieger Forschungsinstitut für Internationale Politik und Sicherheit, Ebenhausen, West Germany

Professor Geoffrey Warner Open University

CONTENTS

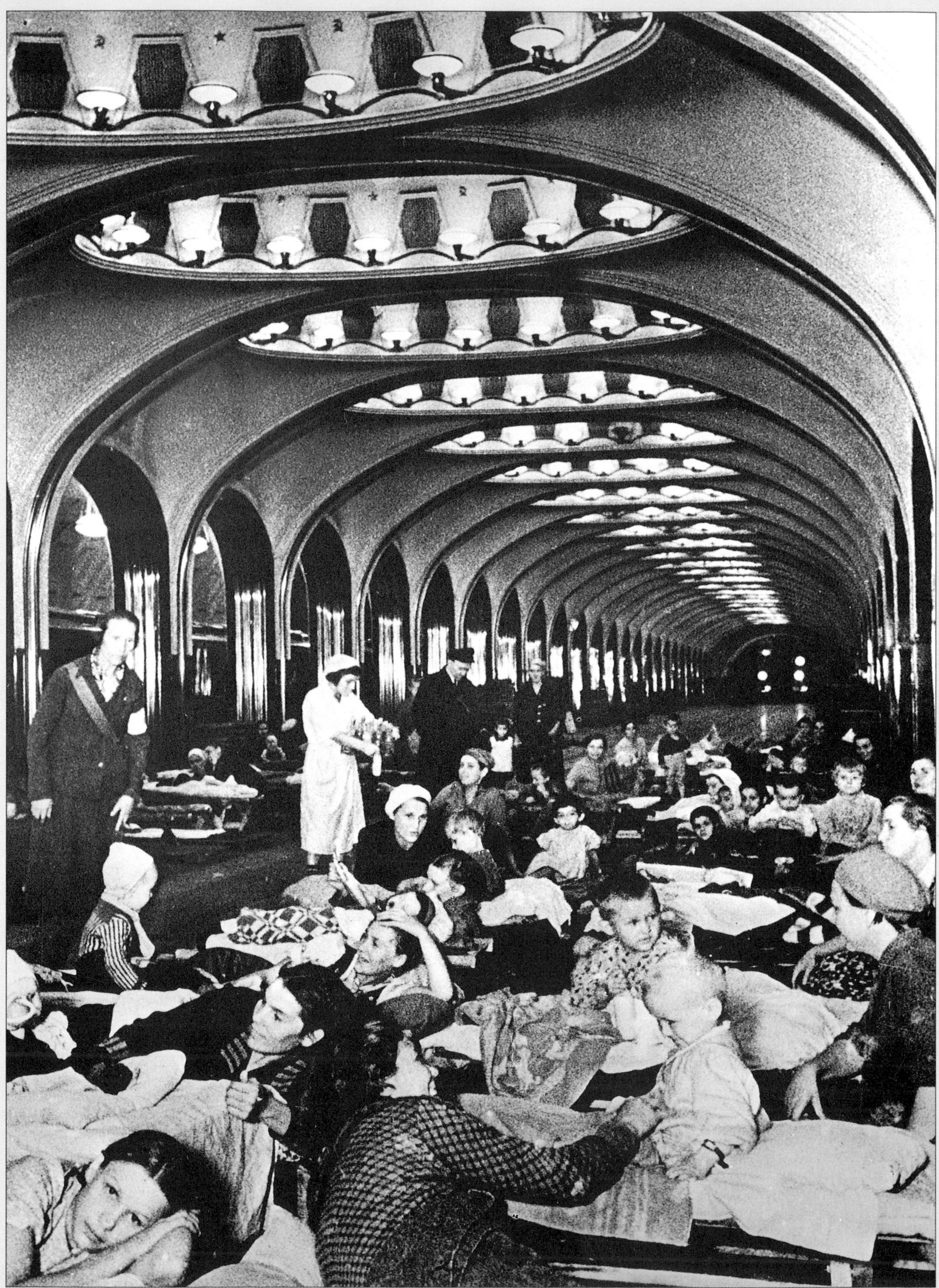

PREFACE

Though commonly known as the *Second* World War, the war of 1937–45 was the first truly worldwide conflict. The Great War of 1914–18 was essentially a European struggle, in which overseas colonies and dominions were only marginally, and the United States only belatedly – though decisively – involved. The war of 1937–45 originated as two distinct conflicts, provoked by the ambitions of Japan in Asia and of Germany in Europe, which from 1941 merged into one. Thereafter it was fought simultaneously over the land of three continents (Europe, Asia and North Africa) and the seas of a fourth (Australasia), with heavy commitment in all theaters of a fifth (North America). In the whole world only South America and southern Africa were not directly involved; but even they could not escape from the disruption which convulsed the whole of the rest of the globe.

It was a war in which the entire populations of all the affected countries were involved to an unprecedented degree, whether as combatants in the fighting services, as workers deployed in war-related industries, as civilians suffering invasion, occupation and aerial bombardment, or as the victims of persecution and mass extermination. It was the first truly "total" war in which the morale of whole peoples was put to the test.

The aim of this book is to encompass in one volume all these different aspects of the worldwide struggle, from its origins to its consequences, trying to hold in simultaneous perspective both Asian and European theaters and giving equal weight to both the military course of the war and its social and economic impact on the nations and individuals caught up in it. Accordingly, the book has five sections.

The first, *The Theaters of War*, presents the military and political narrative, from the war's two separate points of origin in China and Poland to the ultimate defeat of Germany and Japan, with chapters on each of the major geographic regions in which the fighting was concentrated. The second, *The Fighting Services*, describes the main types of military organization and the tactics and weaponry of the different forms of warfare – by land, sea and air – and under cover. The third, *The Mobilization of Peoples*, describes the social and economic impact of "total" war in the major belligerent countries. The fourth, *The Frontline Civilians*, discusses the impact of the war on nonmilitary populations: the effect of bombing, the experience of occupation and resistance, the propaganda war and atrocities and war crimes committed against prisoners, conquered peoples and – most horrifically – the Chinese and the Jews. Finally, *The Aftermath*, looks briefly at the postwar settlement that emerged from the conclusion of the war and the abrupt reversal of alliances that froze into the ideological "Cold War".

The book has been written in collaboration by a team of historians, each a specialist in his or her own field. By the pooling of their expertise we have attempted to comprehend and convey, in a single volume, something of the diversity of worldwide human experience that made up World War II.

John Campbell London

INTRODUCTION

What we call World War II began as two separate wars, which had quite different origins and were fought in different continents. The war in Asia arose from the slow disintegration of the ancient empire of China and the ambition of a dynamic new power, Japan, to move into the resulting vacuum. The war in Europe was essentially a resumption of the Great War of 1914–18, caused by a revival of the military power and territorial ambition of Germany. The story of World War II is the way that these two regional conflicts merged into a single global conflict, resulting in a new world order dominated by two victorious – but very soon rival – superpowers.

The rise of Japan

The emergence of Japan as a world power had been bewilderingly rapid. Within a single lifetime many Japanese had witnessed the self-contained agrarian Japan of their childhood transformed into the world's seventh largest industrial producer. Western penetration in the 1860s had provided the catalyst: in 1868 young Samurai noblemen, smarting from the humiliation, centralized power in the hands of the Emperor Meiji and subsequently engineered a political, economic and military revolution. By the 1870s Japan had a modern army and navy which defeated China in 1895, the Russian Empire in 1905 and in 1914 overran Germany's eastern and Pacific colonies. By 1933 Japan's new empire covered Formosa, Korea, Manchuria and the Marshall, Caroline and Mariana islands in the Pacific.

But by the late 1920s the strain of rapid modernization was beginning to tell. Japan's population explosion – from 30 million in 1870 to 55 million by 1920 and 80 million by 1937 – led to the fragmentation of the land into smaller and smaller holdings, the migration of the population to huge new industrial cities and phenomenal economic growth, concentrated in heavy industry (shipbuilding in particular) and textiles. Export of the latter provided some 70 percent of Japan's foreign earnings, which went to feed its population and pay for industrial raw materials. But in 1930 its overseas textile markets vanished as western nations tried to protect their own hard-hit industries from the depression behind high tariff walls. The effect on Japan was devastating.

Hitherto industrial growth had gone hand-in-hand with the development of political democracy. Seventy years earlier Japan had been as feudal as medieval England. In 1889, however, the Samurai responsible for the Meiji restoration introduced a constitution based on that of imperial Germany. Power remained in the hands of the emperor and his cabinet, but a diet was established, elected on a limited franchise. Universal (male) suffrage came in 1924. Mass political parties began to evolve and in normal circumstances the elected diet should have exerted steadily increasing influence over the executive.

But the crisis of the 1930s put a halt to democratization. Disillusion with western ideals, and the social tensions of rapid modernization, gave birth to ultra-conservative patriotic societies. In a more advanced political system these might have evolved into a national socialist party. Instead they formed alliances with the army and navy, which were developing their own rival ambitions – the army for expansion on the mainland at the expense of China and Russia, the navy to make Japan self-sufficient in a raw materials by the conquest of British Malaya and the Dutch East Indies. In 1935 Japan repudiated the Washington Naval Treaty, negotiated in 1922 by a civilian government, and began building a modern fleet which should challenge British and American domination of the Pacific. In 1936 an army revolt was put down, but henceforth the government was effectively under the thumb of the armed forces and an attempt to solve Japan's problems by war became increasingly inevitable.

The recovery of Germany

The growing power of imperial Germany had made its first bid to dominate the European continent in 1914–18. Following the surrender of Wilhelm II's armies in November 1918, the victorious western Allies – France, the UK and the USA – had tried to prevent any renewal of the attempt by means of the punitive Treaty of Versailles. The German army was to be limited to 100,000 men, with no general staff, no heavy artillery and no aircraft; the navy to ships under 10,000 tons, with no submarines. The territory of Alsace-Lorraine was returned to France, the Rhineland demilitarized, and the union of Germany with Austria (*Anschluss*) was forbidden. The leaders of the new German republic signed the Treaty, under protest, on 28 June 1919.

But, apart from the loss of Alsace-Lorraine and the temporary occupation of the Rhineland, Germany was not dismembered and many Germans did not believe that they had been militarily defeated. Over the next ten years a powerful myth of national betrayal was fed by economic depression. In 1932, following spectacular inflation and high unemployment, the National Socialist ("Nazi") party, led by Adolf Hitler, won 37 percent of the popular vote, becoming by far the largest party in the German parliament, the Reichstag. In 1933 Hitler was appointed chancellor. Once in power, he quickly swept aside the discredited democracy of the Weimar Republic and set about overturning the provisions of Versailles, taking advantage of the reluctance or inability of the other powers to prevent him. France and the UK were exhausted by the last war and unwilling to contemplate another, preferring to believe that Hitler's legitimate desire to restore German pride could be met by reasonable concessions; the old Austro-Hungarian Empire had fragmented in 1918, leaving only a string of vulnerable successor states on Germany's eastern borders; Soviet Russia had turned in on itself following defeat and revolution in 1917; and the United States, after its decisive contribution to the defeat of Germany in 1917–18, had withdrawn back into its traditional isolationism. There was thus a power vacuum in the heart of Europe which Hitler was able to exploit to create the "living space" (*Lebensraum*) he claimed Germany needed. He rearmed, remilitarized the Rhineland in 1936, engineered *Anschluss* with Austria in 1938, and bluffed and bullied the western democracies into letting him take over much of Czechoslovakia in 1938–39; then – still calculating that France and Britain would stand aside – he invaded Poland. By the time the democracies were finally driven to declare war in September 1939 they were helpless to prevent Germany overrunning practically the whole of continental Europe.

From regional conflicts to world war

Fighting in Asia had actually begun in July 1937 when – to the fury of the Japanese admirals – Japanese troops invaded northern China, where they soon became bogged down. But the wider war in Asia really stemmed from Japanese naval expansionism, which quickly ran into conflict not only with the old European empires in Asia – the British in India, Burma and Malaya, the French in Indochina and the Dutch in Indonesia – which offered little resistance, but more importantly with the other rising power in the Pacific – the USA. Japan's military leaders sought to preempt the American response by striking first at the US navy at Pearl Harbor in December 1941. But this was a desperate gamble whose initial success only served to ensure the inevitability of Japan's ultimate destruction when the whole industrial and military power of the United States was eventually mobilized against it.

The aggressive ambitions of both Japan and Germany were motivated by a similar mixture of resurgent national pride, contempt for inferior races and glorification of militarism, exacerbated by the effects of economic recession. The crucial element which linked the two regional conflicts into a world war, however, was not ideology but the military involvement of the United States – and to a lesser extent the UK – in both Asia and Europe. This is not to underrate the immense part played in the defeat of Hitler by the Soviet Union. In 1940 the UK, fighting for its life in lone resistance to Germany in Europe, had successfully resisted German invasion but could do nothing by itself to encompass Hitler's defeat. In June 1941 Hitler took the first step to ensuring his own defeat by invading the Soviet Union. In August 1939 he had been careful to neutralize the latent power of the Soviet Union by means of the Soviet-German Nonaggression Pact before he launched his

▲ **The rise of German fascism: Nazi supporters at a Nuremberg rally, 1936.**

attack on Poland; but now he wantonly provoked it. Like the French emperor Napoleon's *Grande armée* in 1812, his armies penetrated deep into Russia and then foundered; more men and materiel were expended in Russia than on any other front in World War II before the Red Army rolled westward in 1944–45 to occupy half of what had been Hitler's Reich. The USSR unquestionably made the major contribution to the Allied victory over Germany. But from 1943 – as in 1917–18 – the equal and opposite power of the United States made the decisive contribution in western Europe. This too Hitler largely brought upon himself by quite unnecessarily declaring war on the United States following the Japanese attack on Pearl Harbor – an act of suicidally insane bravado which made it enormously easier for President Roosevelt not merely to bring the US into the Euro-pean war, overcoming the still strong isolationist instinct of much American opinion, but also to agree with his British and Soviet allies to make the defeat of Germany, rather than Japan, America's first priority. It was the simultaneous global involvement of the United States in both theaters that made 1941–45 a single war and truly a world war.

There was no necessity for Hitler to declare war on the United States. His action was no more than a gesture of support for an ally who he hoped would help defeat the British for him in their empire while keeping the Americans occupied in the Pacific. Beyond a superficial identity of interest between two militaristic regimes with complementary territorial ambitions in their own spheres, there was no genuine alliance between Germany and Japan. There was little in practice they could do to help each other and they made no attempt to coordinate their strategy. The British, Soviets and Americans, on the other hand, did establish a Grand Alliance with regular summit meetings between the leaders. Between the American and British forces in Europe there was a complete coordination of strategic planning under a single joint command. For reasons of both geography and politics this was not possible between the Red Army and the western democracies. But the UK and America sent vast quantities of material supplies to the USSR; and with the Soviet Union posing as a democratic nation they fought supposedly for shared values against a common enemy, loosely defined as Fascism. Behind the high-flown rhetoric, naturally, they fought at the same time for their own national security and economic interest. In places the interests even of the English-speaking democracies conflicted, notably in Asia where American anticolonialist rhetoric and economic imperialism clashed with Britain's vain concern to preserve the British Empire.

To this extent the war remained two wars, with the Americans and British pursuing somewhat different objectives in each theater. Though the Soviets formally and belatedly joined in the war against Japan just before Hiroshima, there was no serious territorial overlap between the two conflicts. The Soviets and Americans (with the British), converging on Berlin simultaneously from east and west, jointly won the war in Europe and divided the continent between their respective spheres of influence. But the Americans won the war in Asia essentially by themselves and were left – until China and Japan recovered – the predominant power in Asia as well. The conclusive event of the entire war was the American development and use of the atomic bomb. This achievement – matched within a few years by the Soviet Union – changed the nature of the postwar world. Where before the wars in both Asia and Europe there had been power-vacuums waiting to be filled, after 1945 the two nuclear superpowers were omnipresent, locked in worldwide ideological competition and maintaining the peace only through a balance of terror.

The Course of World War II

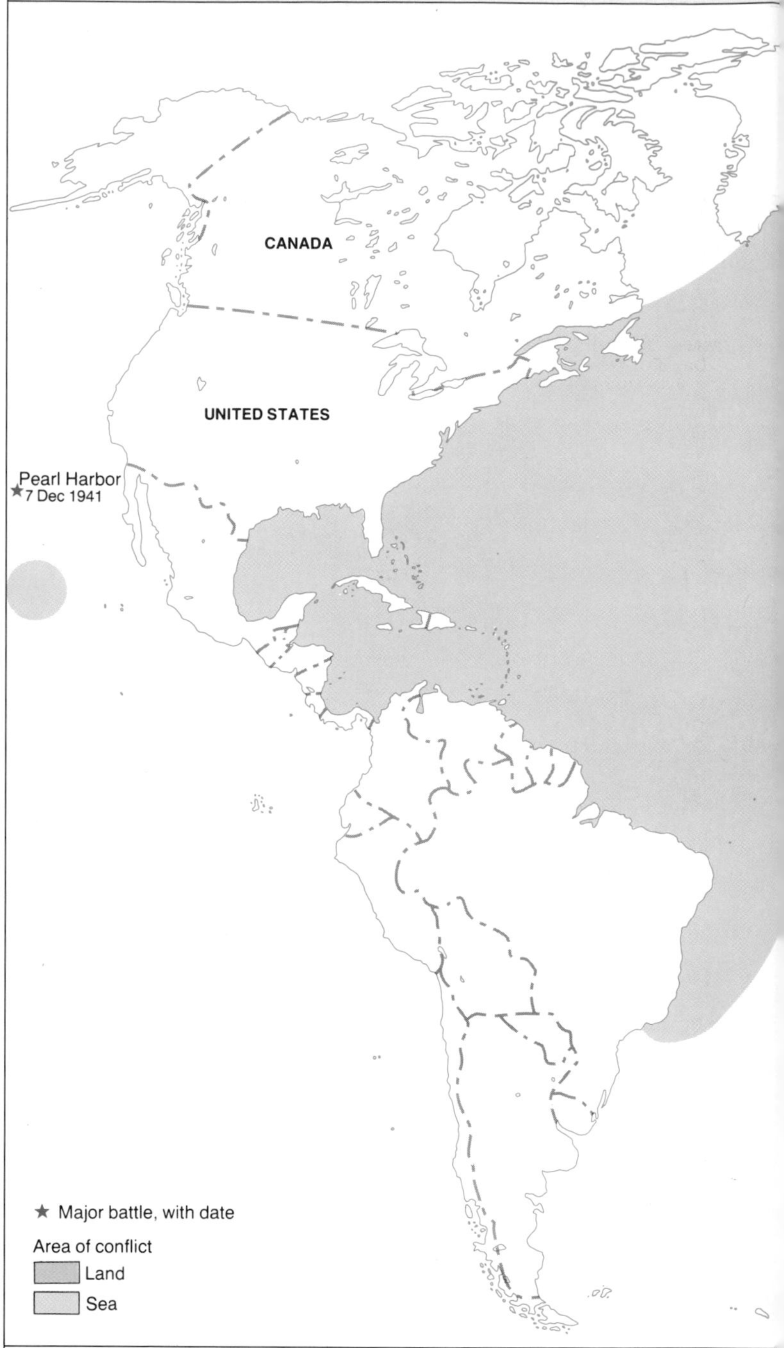

CANADA

UNITED STATES

Pearl Harbor
★7 Dec 1941

★ Major battle, with date

Area of conflict

Land

Sea

ASIA/PACIFIC

1937

On 7 July 1937, Japanese troops invade China. From August to November they also fight for and take Shanghai.

1938

From March to October the Japanese advance south and take Hankou. In the autumn (October-November) they systematically seize Chinese ports.

1939

The seizure of Chinese ports continues. In September a Japanese offensive against Changsha fails, bringing Japanese expansion to an end. In November the Japanese cut the railroad from Hanoi.

1940

In June the USA begins a massive expansion of the navy. In September Vichy France allows Japanese troops to land in Indochina. The USA responds with a trade embargo.

1941

In July Japan declares a Franco-Japanese protectorate over Indochina and occupies the country: the USA, UK and Netherlands stop exports of oil to Japan. On 7 December Japan attacks the US Pacific Fleet in Pearl Harbor. It also starts a conquest of SE Asia.

1942

Japanese expansion in SE Asia continues. Singapore falls in February, Mandalay in April. On the Philippines organized resistance ceases in May. But Japan begins to suffer reverses. US success in the Battle of the Coral Sea in May saves Australia. In June, in the Battle of Midway, the USA destroys four of Japan's carriers. In August US marines capture and hold the Japanese airfield on Guadalcanal in the Solomons.

1943

The Japanese evacuate Guadalcanal in

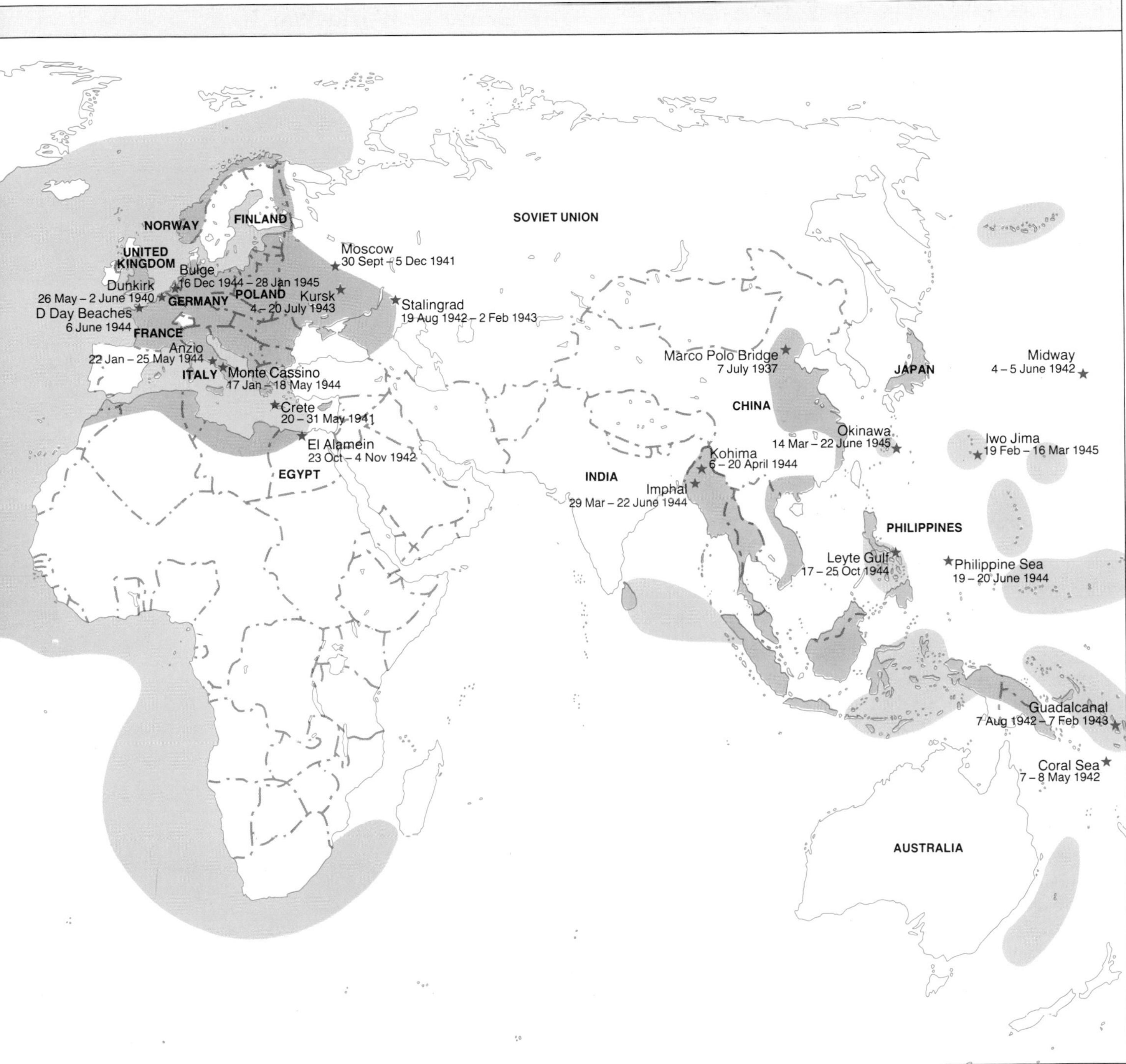

February. In the summer the Allies begin a two-pronged offensive to take the Philippines, cross the Pacific and invade Japan.

1944

In March Japanese troops besiege Imphal and Kohima in India. British forces relieve the towns in April and June. In May Japan mounts an offensive in China. In the Pacific Japan is defeated in the Battles of the Philippine Sea in June and Leyte Gulf in September. In October the Allies invade Burma.

1945

The Allies return to the Philippines and take Manila in March. But the conquests of Iwo Jima (February-March) and Okinawa (March-June) cost huge casualties. Japan prepares to defend the "home islands" but in August Hiroshima and Nagasaki are destroyed by atomic bombs and the USSR declares war. On August 15 Hirohito announces the Japanese surrender.

EUROPE/MEDITERRANEAN

1936–39

The early stages of German expansion. In March 1936 the Rhineland is remilitarized. In March 1938 a union with Austria is engineered. In September the UK and France agree to the transfer of German-speaking areas of Czechoslovakia to Germany. In March 1939 Germany annexes the rest of Czechoslovakia. France and the UK respond by guaranteeing Poland.

1939

On 1 September Germany invades Poland (as does the USSR on 17 September). France and the UK declare war on Germany.

1940

In April Germans invade Denmark and Norway; in May the Low Countries and France. British troops escape from the Continent in a massive evacuation from Dunkirk (May-June). The German-French armistice signed in June divides France into two zones. In July German air attacks are launched (Battle of Britain) in preparation for an invasion of England. RAF resistance causes Hitler to postpone his planned invasion. In Africa Italians invade Egypt in September but are thrown back. Their invasion of Greece in October is also repulsed.

1941

In North Africa in March Rommel's Axis army retreats across Cyrenaica. In April Germans invade Yugoslavia and Greece and in May force the British to evacuate Crete. On 22 June Germans invade the USSR, but are halted outside Moscow in December.

1942

In North Africa the British are forced to retreat. In the USSR Germans make a major drive into the Ukraine and in August launch an attack against Stalingrad. In October, at El Alamein, British troops force the Germans into a retreat across North Africa.

1943

On 31 January the Germans at Stalingrad surrender. The Red Army counterattacks. In North Africa the Axis is defeated in May. In the USSR, in July, Germans attack the Kursk salient, and are defeated. In the Mediterranean the Allies invade Sicily in July, the Italian mainland in September.

1944

In March the Red Army crosses the Dniester. In June the Western Allies invade Normandy and the Red Army attacks Germans in central Russia. But in December the Germans launch an offensive in the Ardennes (Battle of the Bulge).

1945

After victory in the Ardennes, US troops cross the Rhine in March. The Red Army is advancing from the east. Berlin falls on 1 May; Germany surrenders on 8 May.

The social upheaval of war

World War II was a "total" war to an extent that no previous war – not even World War I – comes close to matching. By whatever measurement we use – geographic extent, the scale of military and economic mobilization in the belligerent countries, the direct impact on noncombatants in belligerent and occupied countries, the toll of casualties among both combatants and noncombatants, the disruption of civilian lives or the dislocation of the world economy – the war of 1937–45 affected more people around the world more directly than any other war before, or since. For all who lived through it, it was the overwhelming experience of the 20th century.

The actual fighting was concentrated in China, around the west and southwest Pacific, in Europe and along the Mediterranean coast of Africa. There were also occasional forays into West Africa, India and the south Atlantic. In Europe neutral countries were in a minority: in Asia an even smaller one. The United States, though itself virtually untouched by military action, was a major combatant in both Asia and Europe. Even southern Africa and Brazil supplied the Allied armies with troops. Almost every part of the world was affected, if only economically, by the reverberations of the fighting.

Five major countries were principally involved: Germany, Japan, the UK, the Soviet Union and the United States. (Among others France was knocked out in 1940, Italy in 1943; Australia and Canada played important but subordinate roles on the Allied side; and a number of other countries, like Greece and Yugoslavia, were the scene of significant guerrilla activity.) The scale of military and economic mobilization in the main countries was unprecedented – almost literally total, with virtually all industrial production and manpower, including women, conscripted to serve the needs of the war. (Hitler's reluctance to press German women into war work was a curious exception.) This human and material mobilization can be measured statistically and compared from one belligerent country to another. So, more or less, can the scale of combatant casualties: the Russians lost at least 7 million men killed, the Germans 3.5 million, the Japanese 1.2 million, the UK 250,000, and the USA 300,000. More difficult to quantify are the civilian casualties. Perhaps 300,000 died as a result of aerial bombing in Germany, 60,500 in the UK and 500,000 in Japan, including the 106,000 victims of the atomic bombs on Hiroshima and Nagasaki. But then there are the incalculable numbers of those – men, women and children – killed in the course of invasion and occupation, resistance and reprisals: those who perished in labor and concentration camps, by slave labor, forced marches, torture and all the other atrocities by which this war was so gruesomely characterized – a roll-call of bestiality led by the Nazi's systematic extermination of some 5–6 million European Jews; not forgetting those who simply died, as refugees or displaced persons, in bewildering conditions of social disruption, deprivation and distress.

The lesser effects of the social upheaval caused by the war are likewise unquantifiable. They affected, in different ways, countries far from the front line, like the United States and Canada, and even neutral countries, as well as those threatened or overrun by invasion or bombing. But they were not all bad. War shook up social rigidities in many countries, broke down sexual and moral inhibitions, gave many individuals – particularly women – opportunities of independence and self-discovery which they would not have had in peacetime. The human experience of the war, mixing excitement with tragedy, heroism with cruelty, and an intoxicating sense of purpose with long stretches of tedium and futility, was infinitely varied but, one way or another, universal.

The war was characterized, too, by new levels of mass communication: between belligerent governments and their own peoples, whose morale they were concerned to keep up; between governments and enemy populations, whose morale they hoped to break down; between Allied peoples, and even between opposed peoples, by means of the worldwide diffusion of news and popular entertainment, particularly music and film. It was a war of the air waves: it was a war of propaganda, true and false, of words and images. Like any intense, widely shared experience, it laid down a rich seedbed of memory which – for the most part after the war was over – bore a crop of powerful works of art, in all media, which are a permanent legacy of the conflict, with enduring universal resonance. In addition the war left an indelible imprint on the collective mind of all the participant countries: on the defeated Japanese and Germans a deep scar of national shame and guilt, to be expiated by hard work and exemplary devotion to the profitable pursuits of peace; on the victorious Americans and British a delusive sense of limitless power, righteousness and responsibility for protecting freedom and democracy throughout the world; on the Soviet peoples, who had suffered the heaviest casualties of all, the determination at any cost that the soil of the motherland must never be invaded again.

Above all the other consequences of the war, however – military, social, economic and political – there stand out two features in particular, immense in their implications and unique to this war: the Jewish holocaust and the American dropping of the atomic bomb. Both belong in categories already mentioned, but both events are on a scale which lifts them above generalizations about routine atrocities or conventional bombing. The discovery at the end of the war of Hitler's attempted extermination of the Jews was a shocking revelation of the depth of inhumanity to which a civilized nation could sink under a fanatical regime. It led to a widespread sense of collective guilt in Europe and America, and to the establishment in 1948 of Israel as a Jewish national home in Palestine. Sadly Hitler's monstrous example has had its imitators in the postwar world; yet the holocaust remains the benchmark against which the attempted genocide of other crazed rulers – in Cambodia, Uganda and elsewhere – is measured. More positively, the unprecedented wartime crimes against humanity have led to a worldwide awareness of human rights, enshrined and monitored – however inadequately – by the United Nations.

The mushroom cloud of the atom bombs at Hiroshima and Nagasaki has hung over the world since 1945 like a Sword of Damocles, threatening – as ever more powerful and sophisticated weapons have succeeded one another – the destruction of the world itself if they should ever be deployed. The appalling power placed in the hands of the United States and, within a few years of the war's end, of the Soviet Union as well has had the effect of making war between them unwinnable and therefore rationally unthinkable. Atomic power helped to freeze into seeming perpetuity the crude political settlement arrived at by the final position of the Allied armies in May 1945.

Nuclear weapons have not prevented smaller wars all over the world since the end of World War II. But they have placed an effective limitation on these local wars. The world since Hiroshima has been dominated by the ideological "Cold War" between the Soviet Union and the United States, but no global conflict on the scale of 1937–45. If the outcome of World War I sowed at least some of the seeds of World War II, the conclusion of the Second World War has served for nearly half a century to avert the danger of a third.

▶ Total war: victims of bombing are pulled from the rubble.

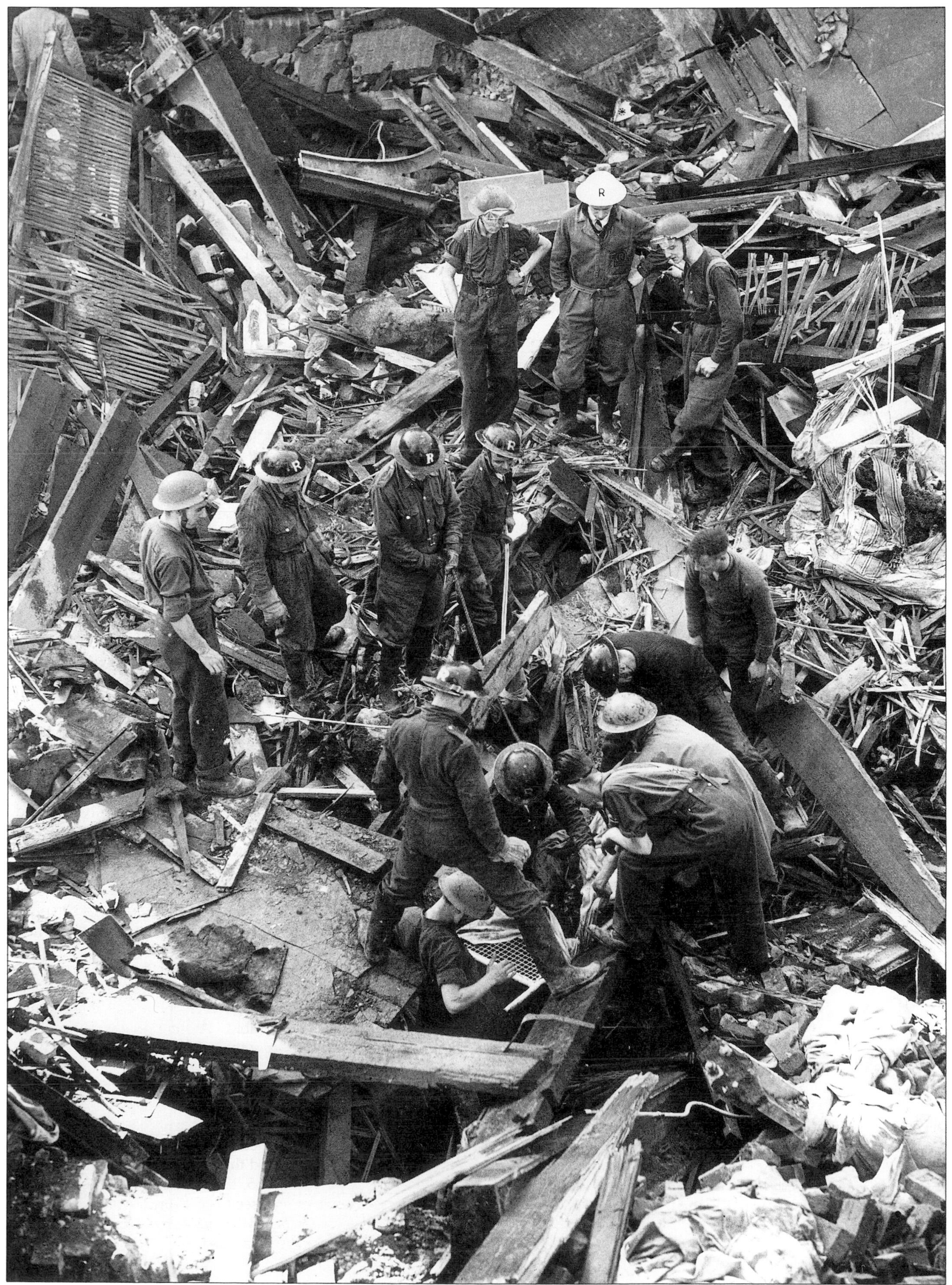
R

Time Chart

	1936–38	1939	1940	1941
Asia/Pacific	• 7 July 1937: Marco Polo Bridge incident • 7 July–23 Nov 1937: Japanese conquest of NE and E China		• Sep 22: Japanese occupation of Indochina begins	• July 25: Occupation of Indochina complete • Dec 7: Japan bombs US naval base at Pearl Harbor • Dec 8: Japanese occupation of SE Asia begins
Europe/Med	• March 1936: Germany remilitarizes the Rhineland • 13 March 1938: Union of Germany and Austria • 12–30 Sep 1938: Czech crisis	• March 16: Germany annexes Czechoslovakia • Apr 7: Italy invades Albania • Sep 1: German invasion of Poland • Sep 17: Soviet invasion of Poland	• Apr 9–June 13: German conquest of Norway • May 10–June 22: German conquest of Low Countries and France • July 10–Nov: Battle of Britain • Sep 15: War in N Africa begins	• May 20–31: German invasion of Greece
Eastern Front		• Nov 30: Soviet–Finnish war begins	• March 12: Soviet–Finnish war ends	• June 22–Dec 5: German invasion of USSR • Dec–Mar: Soviet counteroffensive
Japan	• Oct 1937: Planning agency prepares ground for full-scale war economy • June 1938: Formation of patriotic industry association, state-sponsored • Nov 1938: "New Order in East Asia" proclaimed	• July: First labor mobilization plan • July: National conscription ordinance: workers given designated employment • Oct: "Stop Order" freezes prices and wages	• Voluntary dissolution of political parties; formation of Imperial Rule Assistance Association • Sep: US embargo on war–trade with Japan • Sep: Neighborhood associations revived	• Apr: Rice-rationing begins • Oct: Hideki Tojo becomes prime minister • Dec: Conscription expanded • Dec: Agricultural Control Law tries to check migration from farm to factory
Germany	• 13 March 1938: Union of Germany and Austria • 12–30 Sep 1938: Czech crisis • 9–10 Nov 1938: "Night of Broken Glass" – anti-Jewish pogrom	• Mar: Germany annexes Czechoslovakia • Aug 23: Soviet–German Nonaggression Pact signed • Aug 27: Food rationing introduced • Sep: Introduction of war economy regulations • Nov: Formation of war economy and armaments office of the armed forces		• Mar: "Women for Victory" campaign launched • Aug: T4 euthanasia scheme halted, after protests • Dec: Hitler assumes personal command of the German army
United Kingdom		• June: Women's Land Army formed • Aug: Emergency Powers Act passed • Sep: Internment of Nazi sympathizers and aliens begins • Sep: First wave of evacuation	• Jan: Rationing begins • May: Chamberlain resigns as prime minister; succeeded by Churchill • May: Home secretary given power of arbitrary internment • July: Local Defence Volunteers reformed as Home Guard • Sep: London Blitz begins	• Mar: Mobilization of labor by Registration and Essential Work Order • May: House of Commons destroyed by bombs • Dec: National Service Act
Soviet Union		• Apr: Third Five Year Plan approved by party congress • Aug 23: Soviet–German Nonaggression Pact signed	• June: Working day increased from seven hours to eight • July: Agricultural reforms confiscate land from farmers and impose stricter discipline • July: Large price rises introduced • Oct: Tighter regulation of industrial labor and establishment of compulsory "labor reserve schools" for the young	• June: Committee on evacuation set up • July: Stalin calls for loyalty to fatherland • July: Ration-cards introduced in Leningrad and Moscow • Aug: Emergency War Plan for economy approved • Nov: Evacuation of industrial plant completed
United States			• July: US embargo on trade with Japan • Sep: Conscription introduced	• Mar: Lend-Lease Act • June: Committee on Fair Employment Practises established to eliminate racial discrimination in labor

1942	1943	1944	1945
• May 7–8: Battle of Coral Sea • May 18: Japanese occupation of SE Asia complete • June 4–5: Battle of Midway • Aug 7: Battles of Guadalcanal begin	• Feb: Battles of Guadalcanal end • July 1: US offensives in SW Pacific begin (to March 1945) • Nov 13: US offensives in central Pacific begin (to 10 August 1945)	• Oct 17–25: Battle of Leyte Gulf	• Feb 19–Mar 16: US conquest of Iwo Jima • Mar 14–June 22: US conquest of Okinawa • Aug 6 and 9: Atom bombs dropped on Japan • Aug 15: Japan surrenders
• Oct 24–Nov 4: 2nd Battle of El Alamein	• May 7: War in N Africa ends with defeat for Germany • July 10–June 4: Main Allied campaign in Italy	• June 6–Aug 25: Allied liberation of northern France and Paris • Dec 6: Battle of the Bulge begins	• Jan 28: Battle of Bulge ends in German defeat • March 7: US forces cross Rhine • May 8: Unconditional surrender of German forces
• Soviet offensive in Ukraine • Aug 19: Battle of Stalingrad begins	• Jan 31: Germans defeated at Stalingrad • July 4–20: Battle of Kursk • July 20–Nov 30: Soviet counteroffensive • Dec 24: Soviet winter campaign begins	• May: Soviets retake Leningrad and Ukraine • June 22–Aug 23: Operation Bagration	• Jan 12–May 8: Soviet invasion of Eastern Europe • May 1: Berlin surrenders to Red Army
• Oct: Neighborhood associations take over rationing at local level	• Mar: Wartime Special Administrations Act gives Tojo supreme control over economy • Nov: Munitions ministry established	• July: Resignation of Tojo	• Mar: US B-29s begin incendiary campaign against Japanese cities • July: 10 percent cut in all staple rations imposed • Sep 2: Formal Japanese signature of surrender document
• Jan. 20: Wannsee Conference proposes extermination as the Final Solution of the "Jewish Problem" • Feb: Albert Speer becomes armaments minister • Apr: Central planning committee formed to coordinate war production • May: RAF "Thousand bomber" raid on Cologne	• Jan: Labor conscription introduced for women aged 17–45	• June: Qualitative weapons' superiority abandoned, and the economy mobilized to concentrate on quantitative production • July 20: Assassination attempt on Hitler • July: Goebbels becomes plenipotentiary-general for total war • Oct: Volkssturm created • Nov: Speer creates Ruhr staff to maintain production in the face of devastating Allied bombing raids	• Apr 30: Hitler commits suicide
• Jan: First American GIs arrive in UK • Dec: Beveridge Plan published	• Dec: Balloting for "Bevin Boys" begins		• July: Churchill resigns after electoral defeat; Attlee forms Labour government
• Apr: Nonagricultural labor conscripted for harvest • Apr: First war loan floated • May: Central partisan staff set up to organize guerrilla warfare in occupied territory	• Apr: Peak of prices on free market		
• Jan: Establishment of War Production Board • Feb: War Relocation Authority established to intern Japanese-Americans • Feb: Prohibition of non-military automobile production • June: Office of War Information established to facilitate propaganda • Nov: Secretary of Agriculture calls for 18 million "Victory Gardens"	• Smith–Connally Act • Jan: Army invites Japanese-American men in concentration camps to enlist • Feb: Allied shipping losses reach peak • May: Office of War Mobilization established to coordinate production • Nov: War production at peak	• June: Servicemen's Readjustment Act ("GI Bill of Rights")	• Apr: Death of Roosevelt; Truman becomes president

92
92

PART 1

THE THEATERS OF WAR

WARTIME LEADERS AND GLOSSARY

AUSTRALIA

Governor General
Baron Gowrie 1936–44
Duke of Gloucester 1945–47

High Commissioner
Sir G. Whiskand 1936–41
Sir R. Cross 1941–46

Prime Minister
J. Lyons 1932–39
E. Page 1939
R. Menzies 1939–41
A. Fadden 1941
J. Curtin 1941–45
F. Forde 1945
J. Chifley 1945–49

Foreign Minister
G. Pearce 1934–37
W. Hughes 1937–39
H. Gullet 1939–40
J. McEwen 1940
F. Stewart 1940–41
H. Evatt 1941–49

BELGIUM
(ministers in exile 1940–44)

Head of State
Leopold III, King 1934–44
Regency 1944–50

Prime Minister
P. van Zeeland 1935–37
P. Janson 1937–38
P. Spaak 1938–39
H. Pierlot 1939–45
A. van Acker 1945–46

Foreign Minister
P. Spaak 1936–39
P. Janson 1939
E. Soudan 1939
H. Pierlot 1939
P. Spaak 1939–49

CANADA

Governor General
Baron Tweedsmuir 1935–40
Lord Athlone 1940–46

Prime Minister
W. Mackenzie King 1930–48

Foreign Minister
W. Mackenzie King 1935–46

CHINA

Head of State
Lin Sen, President 1932–43
Jiang Jieshi, President 1943–49

President of the Executive Council (Yuan)
Jiang Jieshi 1935–45

Minister of Foreign Affairs
Wang Zhonghui 1937–41
Kuo Daiji 1941
T. V. Soong 1941–45
Wang Shijie 1945–48

Minister of War
General He Yingqin 1930–44
General Chen Cheng 1944–45

Chief of Staff
General He Yingqin 1938–45
General Chen Cheng 1945–48

DENMARK

Head of State
Christian X, King 1912–47

Prime Minister
T. Stauning 1924–42
V. Buhl 1942
E. Scavenius 1942–45

Foreign Minister
E. Scavenius 1935–45

FINLAND

Head of State
Dr P. Svinhufrud, President 1931–37
K. Kallio 1937–40
Dr R. Ryti 1940–44
Field-Marshal C.G. Mannerheim 1944–45
J. Paasikivi 1945–46

Prime Minister
T. Kivimaki 1932–37
A. Kajander 1937–39
Dr R. Ryti 1939–41
J. Rangell 1941–43
E. Linkomies 1943–44
A. Hackzell 1944
U. Castren 1944
J. Paasikivi 1944–46

Foreign Minister
A. Hackzell 1932–37
E. Holsti 1937–38
V. Voionmaa 1938
E. Erkko 1938–39
V. Tanner 1939–40
R. Witting 1940–43
H. Ramsay 1943–44
C. Enckell 1944–50

FRANCE

Head of State
A. Lebrun, President 1932–40
C. de Gaulle, Head of State 1945–46

Prime Minister
L. Blum 1936–37
C. Chautemps 1937–38
L. Blum 1938
E. Daladier 1938–40
P. Reynaud 1940
Marshal P. Pétain 1940

Foreign Minister
Y. Delbos 1936–38
J. Paul-Boncour 1938
G. Bonnet 1938–40
P. Reynaud 1940
E. Daladier 1940
P. Reynaud 1940
G. Bidault 1944–46

Minister of National Defense and War
E. Daladier 1936–40
P. Reynaud 1940
A. Diethelm 1944–45

Chief of the General Staff
General M. Gamelin 1931–40
General M. Weygand 1940

FRENCH STATE

Head of State
Marshal P. Pétain, President 1940–41
Admiral F. Darlan, Chief of State 1941–42
P. Laval, Chief of State 1942–44

Prime Minister
Marshal P. Pétain 1940–42
P. Laval 1942–44

GERMANY

Head of State
Adolf Hitler, Chancellor 1933–45
and President 1934–45
Admiral Karl Dönitz, President 1945

Foreign Minister
K. von Neurath 1932–38
J. von Ribbentrop 1938–45

Minister of War
Field Marshal W. von Blomberg 1933–38

Head of the Oberkommando der Wehrmacht (OKW)
Field Marshal W. Keitel 1938–45

Commander-in-Chief of the Army
General W. von Fritsch 1934–38
Field Marshal W. von Brauchitsch 1938–41
Adolf Hitler 1941–45

Chief of Staff of the Army
General L. Beck 1933–38
General F. Halder 1938–42
General K. Zeitzler 1942–44
General H. Guderian 1944–45
General H. Krebs 1945

GREECE (govt in exile 1941–44)

Head of State
Georgious II, King 1935–47

Prime Minister
C. Demerdjis 1935–37
J. Metaxas 1937–41
A. Korizis 1941
E. Tsouderos 1941–44
S. Venizelis 1944
G. Papandreou 1944–45
General Plastiras 1945
Admiral Voulgaris 1945–46

Foreign Minister
C. Demerdjis 1935–37
J. Metaxas 1937–41
E. Tsouderos 1941–44
S. Venozelis 1944
G. Papandreou 1944–45
J. Sophianopoulous 1945
I. Politis 1945

INDIA

Viceroy
Lord Linlithgow 1936–43
Lord Wavell 1943–47
Lord Mountbatten 1947

ITALY

Head of State
Victor Emmanuel III, King 1900–46

Prime Minister
Benito Mussolini 1922–43
Marshal P. Badoglio 1943–44
I. Bonomi 1944–46

Foreign Minister
Count Galeazzo Ciano 1936–43
Benito Mussolini 1943
Baron Guariglea 1943–45

Chief of the General Staff
Marshal P. Badoglio 1925–40
Marshal U. Cavallero 1940–43
Marshal V. Ambrosio 1943

JAPAN

Head of State
Hirohito, Emperor 1926–89

Prime Minister
Senjuro Hayashi 1937
Funimaro Konoe 1937–39
Kiichiro Hiranuma 1939
Nobuyuki Abe 1939–40
Mitsumasa Yonai 1940
Funimaro Konoe 1940–41
General Hideki Tojo 1941–44
Kuniaki Koiso 1944–45
Kantaro Suzuki 1945

Foreign Minister
Senjuro Hayashi 1937
Naotake Sato 1937
Koki Hirota 1937–38
Kazushiga Ugaki 1938
Funimaro Konoe 1938
Hachiro Arita 1938–39
Nobuyuki Abe 1939
Kichisaburo Nomura 1939–40
Hachiro Arita 1940
Yosuke Matsuoka 1940–41
Teijiro Toyoda 1941
Snigenori Togo 1941–42
General Hideki Tojo 1942
Masayuki Tani 1942–43
Mamoru Shigemitsu 1943–45
Kantaro Suzuki 1945
Shigenori Togo 1945

Minister of War
Kotaro Nakamura 1937
General H. Sugiyama 1937–38
Seishiro Itagaki 1938–39
Shunroku Hata 1939–40
General Hideki Tojo 1940–44
General H. Sugiyama 1944–45
Korechika Anami 1945

Army Chief of Staff
Prince Kanin 1931–40
General Hajime Sugiyama 1940–44
General Hideki Tojo 1944
General Yoshijiro Umezo 1944–45

NETHERLANDS
(govt in exile 1940–45)

Head of State
Wilhelmina, Queen 1890–1948

Prime Minister
H. Colijn 1933–39
D. de Geer 1939–40
P. Gerbrandy 1940–45
W. Schermerhorn 1945–46

Foreign Minister
A. de Graeff 1933–37
H. Colijn 1937
J. Patijn 1937–39
E. van Kleffens 1939–46

NEW ZEALAND

Governor General
Lord Galway 1935–41
Lord Newall 1941–45
B. Freyberg 1945–52

High Commissioner
Sir H. Batterbee 1939–45
Sir P. Duff 1945–49

Prime Minister
M. Savage 1935–40
P. Fraser 1940–49

Foreign Minister
M. Savage 1935–40
F. Langstone 1940–42
P. Fraser 1943–49

NORWAY (govt in exile 1940–45)

Head of State

Haakon VI, King	1905–57

Prime Minister

J. Nygaardsvold	1935–45
E. Gerhardsen	1945–51

Foreign Minister

H. Kont	1935–41
T. Lie	1941–45

SOUTH AFRICA

Governor General

Lord Clarendon	1931–37
P. Duncan	1937–43
N. de Wet	1943–46

Prime Minister

General J. Hertzog	1924–39
General J. Smuts	1939–48

Foreign Minister

General J. Hertzog	1927–39
General J. Smuts	1939–48

UK OF GREAT BRITAIN & N. IRELAND

Head of State

George VI, King	1936–52

Prime Minister

N. Chamberlain	1937–40
W. Churchill	1940–45
C. Attlee	1945–51

Foreign Minister

A. Eden	1935–38
Viscount Halifax	1938–40
A. Eden	1940–45
E. Bevin	1945–51

Defense Minister

Sir T. Inskip	1936–39
Lord Chatfield	1939–40
W. Churchill	1940–45
C. Attlee	1945–46

Minister of War

L. Hore-Belisha	1937–40
O. Stanley	1940
A. Eden	1940
Captain D. Margesson	1940–42
Sir P.J. Grigg	1942–45
J. Lawson	1945–46

Chief of the Imperial General Staff

Field Marshal Lord Gort	1937–39
Field Marshal Sir E. Ironside	1939–40
Field Marshal Sir J. Dill	1940–41
Field Marshal Sir A. Brooke	1941–46

UNITED STATES OF AMERICA

Head of State

Franklin D. Roosevelt, President	1933–45
Harry S. Truman, President	1945–52

Secretary of State

C. Hull	1933–44
E. Stettinius	1944–45
J. Byrnes	1945–46

Secretary of the Army

H. Woodring	1936–40
H. Stimson	1940–45

Chief of Staff of the Army

General M. Craig	1935–39

General of the Army

George C. Marshall	1939–45

UNION OF SOVIET SOCIALIST REPUBLICS

Head of State

M.I. Kalinin, Chairman	1936–46

Secretary General of the Communist Party

J.V. Stalin	1922–53

Prime Minister

V.M. Molotov	1931–41
J.V. Stalin	1941–53

Foreign Minister

M.M. Litvinov	1931–40
V.M. Molotov	1939–49

Commissar for Defense

Marshal K. Voroshilov	1934–40
Marshal S. Timoshenko	1940–41
Josef Stalin (and Commander in Chief)	1941–46

Chief of Staff of Red Army

Marshal B. Shaposhnikov	1937–40
Marshal K. Meretskov	1940–41
Marshal G. Zhukov	1941
Marshal B. Shaposhnikov	1941–42
Marshal A. Vassilevskii	1942–48
Marshal A. Antonov	1945–46

YUGOSLAVIA (govt in exile 1941–45)

Head of State

Peter II, King	1934–45
Prince Paul, Regent	1934–41
Dr I. Ribar, President	1945–53

Prime Minister

M. Stojadinovic	1935–39
D. Cvetkovic	1939–41
General D. Simovic	1941–42
S. Jovanoic	1942–43
M. Trifimovic	1943
B. Puric	1943–45
Marshal J. Tito	1945–63

Foreign Minister

M. Stojadinovic	1935–39
A. Markovic	1939–41
M. Nintchic	1941–45
I. Subasic	1945–46

Amphibious operation
Military operation for landing troops from the sea, usually in enemy-held territory.

Appeasement
Prewar policy of political and territorial concessions to Nazi Germany to avoid war.

Attrition
Combat to defeat the enemy by wearing down his military resources.

Axis
The military alliance of Germany, Italy and Japan.

Barbarossa
The code name for the German invasion of Russia on 22 June 1941.

Battleaxe
Code name for General Wavell's unsuccessful offensive in North Africa in June 1941.

Bolshevism
Communist government as practiced in the USSR under Lenin and Stalin.

British Expeditionary Force (BEF)
British army of ten divisions deployed in France in 1940.

Commandos
Elite British force for raiding and sabotage in enemy-occupied territory.

Communist
A follower of the political doctrines of Karl Marx, believing in the people's right to own the means of production.

Dynamo
Code name for the evacuation of the British Expeditionary Force from Dunkirk in May and June 1940.

Fascist
A follower of right-wing authoritarian doctrines.

Fortress Europe
The area of German-occupied western Europe fortified for defense against an Allied amphibious landing.

Free French
French forces, led by General Charles de Gaulle, fighting for the Allies after the defeat of France in 1940.

Guerrilla warfare
Clandestine military action by irregular forces.

Guomindang
The Chinese Nationalist party led by General Jiang Jieshi.

Imperialism
The imposed rule of a foreign nation over indigenous native populations.

July Plot
Conspiracy by high-ranking German generals to assassinate Hitler in July 1944.

Kriegsmarine
The German navy.

Lend-lease
Arrangement by which the USA provided weapons and munitions for delayed payment.

Luftwaffe
The German air force.

Market Garden
Code name for ambitious airborne operation to capture the bridges at Eindhoven, Nijmegen and Arnhem in September 1944.

Mobilization
The calling-up of a country's armed forces. Also the harnessing of its economic resources for war.

Nationalist
A supporter of a nation's or people's right to self-government.

Nazis
The German national socialist party, led by Adolf Hitler. Also supporters of its political doctrines.

Overlord
Code name for the Allied invasion of occupied Europe on 6 June 1944.

Panzergruppe
(Ger: "tank group") Battle group of armored and mechanized forces.

Partisans
Irregular guerrilla forces operating behind enemy lines in occupied national territory.

Plan D
French plan to counter a German invasion through Belgium by advancing to meet it on Belgian territory.

Rationing
Governmental control of the supply of food and other resources.

Sealion
Code name for the planned German invasion of the UK in the autumn of 1940.

Torch
Code name for the Anglo-American amphibious landings in North Africa on 8 November 1942.

Unconditional surrender
Capitulation by a nation decisively defeated, allowing the victor freedom to impose conditions of peace.

Vichy government/France
The pro-German regime established in France under Marshal Pétain after France's defeat in 1940, named after its capital Spa town in the center of the country.

Victory disease
Japanese overconfidence in the conduct of the Pacific war, brought on by their string of easy early victories.

Wehrmacht
The regular German army.

Winter war
The war between Russia and Finland in the winter of 1939–40.

Datafile

Japan's invasion of China in 1937 was reckless in the extreme, for it placed enormous burdens on its empire's military resources. Japan's army was relatively small. Fewer than 1 million of the empire's 100 million served in it at any one time. Although the Japanese Empire had a population of 100 million, labor-intensive agriculture and industry, combined with a primitive conscription apparatus, meant that mobilization was inefficient. Between 1937 and 1945 7.4 million Japanese served in the armed forces, a very low proportion when compared to the UK where 6.2 million of a population of 48 million served in the armed forces between 1939 and 1945.

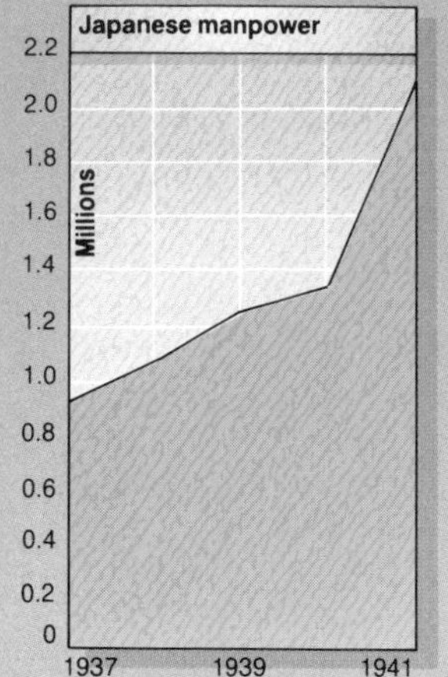

▲ Between 1937 and 1941 the Japanese army more than doubled in size placing enormous strains on the economy. The Japanese began the Pacific war in 1941 with armed forces which could not be expanded easily and which would consequently be very vulnerable to attritional warfare.

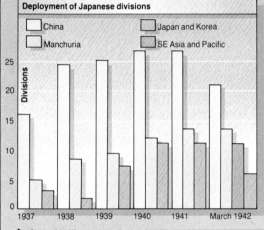

Deployment of Japanese divisions

China
Manchuria
Japan and Korea
SE Asia and Pacific
Divisions
25
20
15
10
5
0
1937
1938
1939
1940
1941
March 1942

◀ To the Japanese army the main war was being fought in China where most of its men were deployed. The army regarded the Pacific conflict as very much the navy's war.

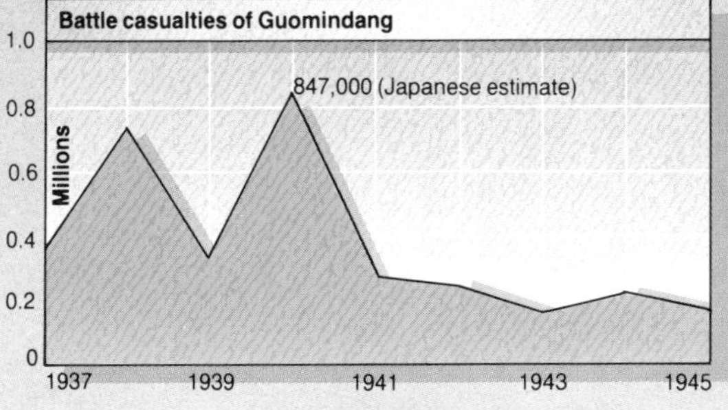

▶ A graphic illustration of the steadily diminishing role of the Guomindang armies after December 1941. Even before this date it is likely that some of the number are civilian rather than military casualties.

In the mid 1930s the officer corps of Japan's army and navy were riven by competing factions, most of which had schemes for the expansion of Japan's empire. Within the army the *Koda ha* (Imperial Way) faction, a group of low-ranking officers allied with the extreme nationalist Black Dragon Society, pressed for expansion into Siberia. For the *Tosei ha* "control" faction (composed largely of senior officers and the general staff), invasion of China with its population of 400 million was a tempting possibility, but time was running out. Jiang Jieshi's Guomindang army seemd to be on the verge of reunifying the country after the anarchy of the 1920s.

Feuding between the *Tosei ha* and *Koda ha* came to a head in 1935. *Tosei ha* officers managed to transfer their opponents to remote and unpopular locations: but when in February 1936 the 1st Division, staffed by *Koda ha* officers, was ordered to go from Tokyo to the Mongolian–Manchurian frontier, the entire dividion rose in revolt, seized part of Tokyo and murdered several government members. *Tosei ha* troops suppressed the rebellion and the ringleaders were executed, but the government had a price to pay. From 1936 onward army and navy ministers had to be serving officers, subject to dismissal for behavior deemed inimical to military interests. According to the constitution of 1889, if any minister resigned the entire cabinet had to follow. Consequently even if civilian ministers united against their service colleagues, the latter could still threaten their

JAPAN'S WAR IN ASIA

The outbreak of the war in Asia

Japan's conquest of north China

The struggle for Shanghai

The Japanese occupation of Chinese ports

US resistance to Japan's expansion

resignation and bring down the entire government. The armed forces effectively held the government under their thumb.

The navy also had plans for expansion, but in yet another direction. Naval officers argued that Japan needed to secure control over supplies of raw materials – oil, rubber, tin, iron ore and bauxite – materials which the Dutch East Indies and British-ruled Malaya contained in limitless quantities. The first step would be the construction of a fleet capable of challenging Anglo-American domination of the Pacific – an aim supported by Japan's industrial combines, the *zaibatsu*, and much of Japanese society. In 1935 Admiral Yamamoto, Japan's delegate to the London Disarmament Conference, announced that Japan would no longer be bound by the Washington Treaty, which imposed limits on fleet sizes. He returned to a hero's welcome, and as vice-minister of the navy was soon overseeing a massive naval expansion program. It would be many years before Japan could challenge the British and the Americans. Domination of the Pacific could be achieved by about 1950.

Japan attempts a quick campaign in China

Japan's admirals were furious when, on 7 July 1937, large-scale fighting grew out of a minor clash between Japanese and Guomindang forces near the Marco Polo Bridge half-way between Tianjin and Beijing. The Guandong army, now firmly under the *Tosei ha*'s control, advanced south in force and occupied the Beijing-Tianjin region as well as striking west in Outer Mongolia. For a few days in early August it looked as if the conflict might quickly be settled, but Jiang Jieshi (who a few months earlier had entered into an anti-Japanese alliance with Mao's communists) showed unexpected resolve. He concentrated his best units in the Shanghai area and on 8 August launched a heavy attack on the Japanese concession, jeopardizing Japan's Shanghai garrison. Even the most moderate civilians in the Tokyo government wanted to avoid the humiliation of evacuating the city: on 17 August general mobilization was ordered in Japan and reinforcements rushed to Shanghai. As punishment, Japan occupied territory in northern China down to the Huang (or Yellow) river – intended to be used as a bargaining counter when Jiang Jieshi was

▼ Shanghai-bound reinforcements march through Tokyo streets lined with crowds, 11 October 1937.

Emperor Hirohito 1901–89

Once portrayed as an unworldly pacifist ineffectually urging restraint on Japan's military, it is now clear that from his accession in 1926 Hirohito was more than a figurehead. He did not actively promote the military's plans2 for expansion but he did little to discourage them and took an intense interest in the conduct of the war. His power within the state system was revealed when he advised capitulation in 1945. His subsequent surrender broadcast on 15 August 1945 caused all but a few diehards to lay down their arms.

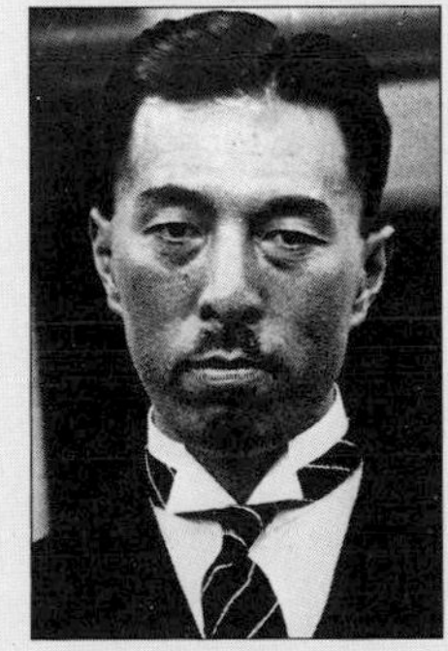

Prince Fumimaro Konoye 1891–1945

A member of the prestigious Fujiwara clan, Konoye entered Japan's house of peers in 1916, and became president of this house in 1933. Between 1937 and 1941 Konoye served three times as prime minister, but proved unable to control either the military or the naval members of his cabinets. He retired from politics in 1941 and although he remained inactive throughout the war, when he learned that the Americans were planning to place him on trial for war crimes he committed suicide.

Kanji Ishiwara 1889–1949

A pan-Asian nationalist who saw Japan's mission as the liberation of its fellow Asians from European domination, Ishiwara used his position on the staff of the Guandong army to engineer the seizure of Manchuria in 1931 and to provoke the "China incident" six years later. Though many high-ranking officers shared his views, Ishiwara fell out with Tojo and the dominant control faction, and in 1941, on the eve of Pearl Harbor, he was cashiered. He spent the rest of the war engaged in propaganda work.

AND FOREIGN DISPENSARY

forced to come to the conference table. Things worked out very differently: eight years later much of Japan's army was still tied down in China.

In early September 150,000 Japanese, organized in two mutually supporting army groups, struck south along the three main railroad lines toward the Huang. Guomindang forces, despite outnumbering their opponents three to one, lacked armor, artillery and air support, and melted away before the Japanese advance. By the first week in November some Japanese units had reached the Huang. Others drove southwest through the mountainous Shanxi province but met unexpectedly determined resistance from Mao's communists and suffered two disturbing reverses; but by November they had achieved most of their objectives.

The struggle for Shanghai was far more difficult. More than 450,000 of Jiang Jieshi's best troops assailed the Japanese: only when the landing of a fresh Japanese army group south of Shanghai in Hangzhou Bay on 5 November threatened them with envelopment did they withdraw to Nanjing. Jiang Jieshi's forces were shattered. He had lost 270,000 men. Japanese casualties, 40,000 of the 200,000 engaged, had not been light either. Tokyo thought the campaign had ended with the fall of Shanghai, but local commanders had already decided to advance on Nanjing, the Guomindang capital, unaware that Jiang Jieshi had moved his government to Hankou in the southwest. Nanjing fell on 13 December: for a month Japanese troops indulged in an orgy of looting, rape and murder. They burnt down a third of the city and butchered between 200 and 300 thousand Chinese.

Japanese atrocities failed to dent the Guomindang's resolve. In mid-December Tokyo used the German ambassador to Hankou to propose terms to Jiang Jieshi. All military activity would cease if he acknowledged the authority of the army's puppet governments in northern China and Mongolia, agreed to further the economic integration of China and Japan, accepted responsibility for the war and agreed to pay reparations. On 26 December Jiang Jieshi rejected these proposals out of hand.

Jiang Jieshi 1887–1975

Leader of China in 1937, Jiang had established his authority in the late 1920s: in 1927 he had ousted the communists from the nationalist Guomindang party and in 1928 had established his government in Beijing. The Japanese forced him out, to Nanjing, Hankou, and finally to Chongqing in summer 1938. He rejected their terms, but in spite of receiving lavish supplies from the USA could not expel them. On Japan's surrender in 1945 he found communists embedded in former Japanese territory. Civil war followed; the Guomindang lost; Jiang fled to Formosa.

Mao Zedong 1893–1976

A member of China's communist party from its foundation in 1921, Mao became its leading figure in the late 1920s. Narrowly avoiding destruction at the hands of the Guomindang, Mao led the communists in the "Long March" of 13,000km (8,000mi) to safety in Shaanxi (1934–36). After the Japanese invasion there was a truce between the communists and the Guomindang but in practice the communists waged guerrilla war against the Guomindang as well as the invaders. By the time of the Japanese surrender they were strong enough to defeat the Guomindang in the civil war of 1946–49. Mao then dominated China for 27 years.

Japan attempts to defeat the Chinese government

Japan's only option was to continue the war. On 16 January 1938 the prime minister, Prince Konoye, announced that Japan would have no further dealings with the Guomindang, and would instead "rely upon the establishment and growth of a new Chinese regime for cooperation" – in effect, create a new government. Two months later in a wave of patriotic zeal the diet voted a national mobilization law which extended conscription and introduced rationing: Japan settled down for a long war.

By March 1938 nearly 500,000 Japanese troops had been sent to China; the army in north China now launched a southward offensive to link up with Japanese forces in the Shanghai-Nanjing area. It went disastrously wrong. Only half of the 60,000 men who found themselves surrounded by 200,000 Chinese at Tai'erzhuang managed to fight their way back north. So one month later the Japanese tried again, their reinforced armies this time advancing simultaneously from north and south; they sliced through Chinese defenses and linked up a Xuzhou on 20 May for what should have been a decisive victory. More than 100,000 Chinese were trapped between Xuzhou and the Yellow Sea, but as the Japanese closed in they melted into the countryside, merging imperceptibly with the peasantry.

Even if Japan could not destroy the Chinese armies, at least it could drive Jiang Jieshi from Hankou. Forces headed west toward the rail junction at Zhengzhou, planning to swing south down to the Guomindang's new capital. They advanced

◀ After the fighting in July 1937, Japanese troops staged a formal triumphal entrance into Beijing in early August.

▼ From August to November 1937 the Japanese attempted to capture Shanghai. Half a million troops were engaged in bloody combat. Japanese aircraft bombed the city repeatedly, and civilian casualties were heavy. This photograph was taken in the wake of a raid in early November on Shanghai's eastern (Nanjing) railroad station.

along the southern bank of the Huang; by early July they were less than 80km (50mi) from Zhengzhou when a wall of water engulfed them. Thousands drowned, most equipment was lost, and the advance halted. The Chinese, unable to stop the Japanese by military means, had broken the Huang dykes.

By September the Japanese had regrouped: reinforced to 12 divisions (about 250,000 men) they drove southwest toward Hankou. On 10 October, after heavy fighting, they entered the now-ruined city only to find that Jiang Jieshi had again moved his capital, this time to Chongqing. Although the Japanese launched limited offensives south of the Yanzi the following year, the capture of Hankou marked the limit of their territorial expansion. In 15 months and at the cost of 70,000 dead they had overrun an area of 1.8 million sq km (700,000 sq mi) with a population of 170 million, the most economically productive area of China.

By this time the Japanese, who physically occupied only the major cities and railroad lines of this vast area, faced a new threat – sporadic guerrilla activity from the communist 8th Route Army in Shaanxi which by September 1937 had started to infiltrate north China. The Japanese handled the guerrilla problem more subtly than western nations did after 1945. They realized that Mao's revolutionary war was essentially political in nature so they created a variety of civilian agencies to win the population's loyalty. After two and a half years of consolidation, while they raised and trained a pro-Japanese Chinese army, Tokyo established, on 30 March 1940, a government in Beijing under a former prominent Guomindang member, Weng Jing-wei. But the communists had also been busy and on 20 August 1940 Mao launched his "Hundred Regiments" offensive, an intensive series of small-scale guerrilla raids all over northern China. Overnight Japanese hopes of peaceful consolidation evaporated. Their response was a program of calculated terrorism, the "Three All Policy" – take all, burn all, kill all. Japanese columns moved through the most heavily infiltrated areas creating a wasteland, but for every communist suspect killed they created ten new guerrillas. Japan was still fighting the insurgents when it was forced to surrender in 1945.

By autumn 1938 Japan had realized the painful truth that there could never be a decisive battle to end the war with China. But they could render the Guomindang forces impotent by cutting off Jiang Jieshi's supply lines from the outside world. As early as 3 May 1938 forces from Formosa, acting without authorization from Tokyo, had occupied the southern Chinese port of Xiamen. In September Tokyo ordered the systematic seizure of China's remaining ports. Guangzhou was occupied on 25 October, Fuzhou on 23 November, Hainan on 10 February 1939; nine months later a force landed at the head of the Gulf of Tongking, advanced inland and captured Nanning, thereby cutting the major rail and road link between the French Indochinese port of Haiphong and Chongqing.

But Japan could do little about the supplies which China continued to receive via the old silk route from the USSR through Xinjiang. On 11 July 1938 elements of the Guandong army, claiming that Russian forces had violated Manchurian territory, had attacked Soviet positions on Chengkaifeng Hill, where the borders of Korea, Manchuria and Siberia met. After 30 days of bitter fighting they were bloodily repulsed. Nine months later much more serious fighting had broken out between the Guandong army and Soviet forces along the Mongolian border; by September 1939, after suffering more than 18,000 casualties, the Japanese had withdrawn, defeated. Although in the summer of 1940 diplomatic efforts improved relations with the Soviet Union, until the Germans invaded the western USSR on 22 June 1941, Tokyo was acutely aware of a major threat to the northwest.

China was also receiving supplies from the south. The Haiphong–Nanning railway line was now defunct but supplies could be trucked to

▶ After their initial onslaught, the Japanese attempted to bring the Guomindang to battle by launching drives up the Huang Ho and Yangzi river valleys, and attempted to starve them of supplies by occupying ports along China's coast (1938–39). It was all to no avail. The Chinese continued to withdraw out of reach and established new supply lines via French Indochina and British Burma.

▲ The bandy-legged, buck-toothed Jap of Western cartoons of the 1930s. This German cartoon is a reminder that until Japan's entry into the Axis on 27 September 1940 relations with Germany were ambivalent.

◀ After a grueling four-month campaign, Japanese forces from northern China linked up with their forces in central China to the west of Xuzhou. This photograph, taken on 19 May 1938, the day the northern and southern thrusts met, shows troops of Japan's 2nd Army streaming westward from Xuzhou, driving the Guomindang ever deeper into south central China.

▼ In the wake of the retreating Guomindang came hordes of refugees.

The Japanese Invasion of China and Indochina 1937–41

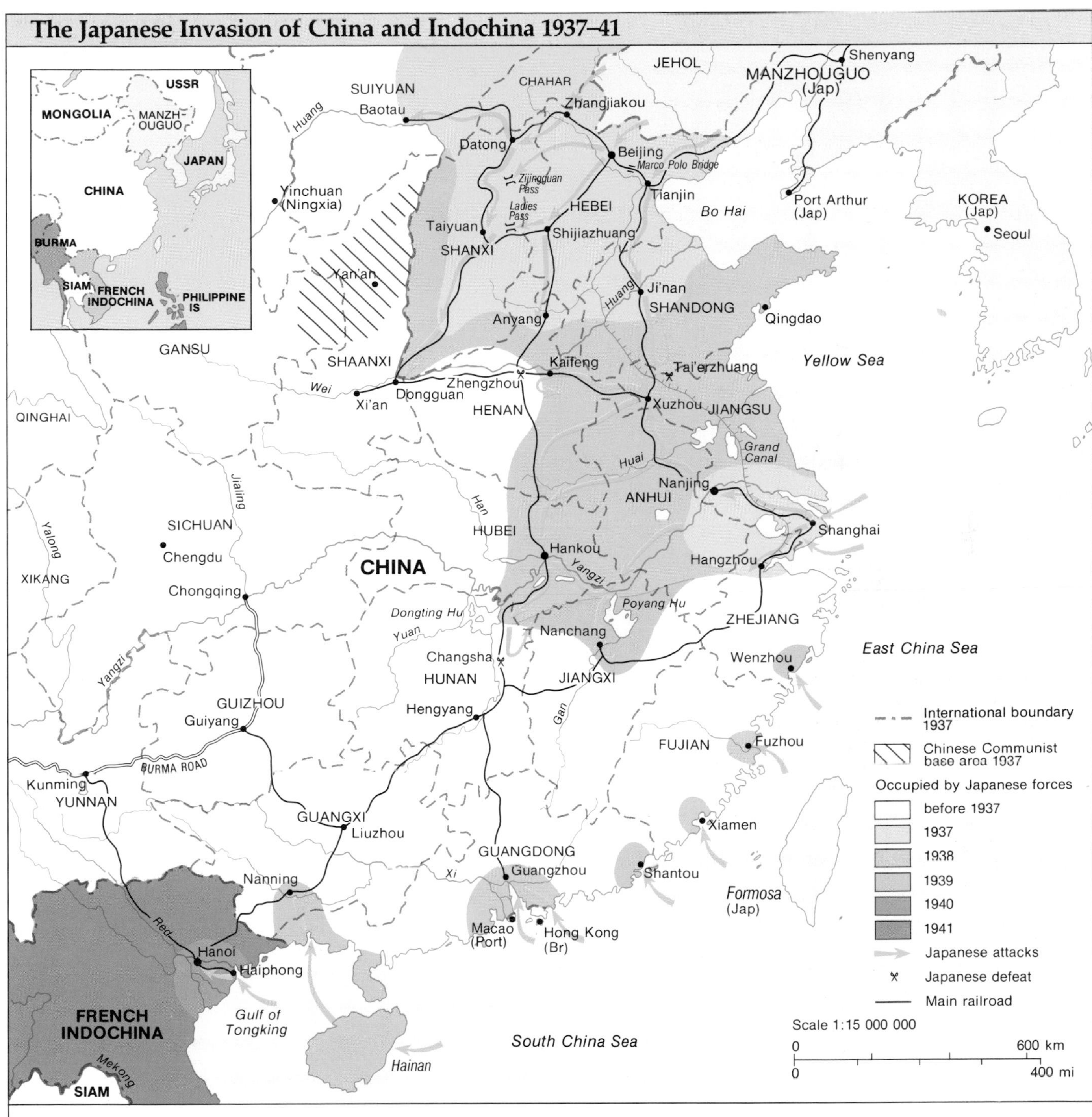

NORTHERN AND CENTRAL THEATERS

1937

July 28
(to Dec. 27). Following clashes at Marco Polo Bridge (July 7) the Japanese strike south into Hebei province along three main railroad lines. Beijing and Tianjin are quickly occupied: by the end of 1937 Japan controls eastern China north of Shanxi province.

August 8
(to Dec. 13) A Japanese amphibious assault against Shanghai runs into Jiang Jieshi's best armies: the city does not fall until Nov. 11. The Japanese drive west and capture Nanjing (Dec. 13).

1938

March 1
(to April 6) A Japanese army advances south along the Ji'nan–Xuzhou railroad and is surrounded by 200,000 Chinese at Tai'erzhuang (March 24). The Japanese fight their way north (April 6).

May 1
(to May 20) Reinforced, the Japanese launch converging attacks from Ji'nan in the north and Nanjing in the south and capture Xuzhou on May 20. Their northern and central forces are now linked.

June 1
(to July 1) The Japanese strike west along the Xuzhou–Xi'an railroad aiming to capture the vital Zhengzhou railway junction. Kaifeng falls on June 6 but as the Japanese approach Zhengzhou the Chinese break the Huang dykes, engulfing the Japanese.

August 1
(to Oct. 10) Shifting their axis of advance, Japanese converge on Jiang's capital Hankou and take the city on Oct. 10. But Jiang has removed his government to Chongqing.

1939

March 27
Japanese capture Nanchang.

September 14
(to Oct. 16) Five Japanese divisions converge on Changsha, and in a month's fighting are driven northward – the end of Japanese expansion in central China until 1944.

SOUTH AND INDOCHINA

1938

May 11
(to Nov. 23) To cut off supply lines, Japan begins occupying China's ports: Xiamen on May 31, Guangzhou on Oct. 25, and Fuzhou on Nov. 23.

1939

February 10
(to Nov. 24) Still trying to isolate China, Japan lands on Hainan Island on Feb. 10 and occupies the remaining ports. A force lands on the south China coast on Nov. 15, drives north and captures Nanning, cutting the Hanoi–Changsha rail link (Nov. 24). China now has only two supply routes: the railroad from Haiphong and a road from Lashio in Burma.

1940

June 25
(to Sept. 22) After France's defeat in Europe, Japan by agreement lands forces in Indochina. Warships in Haiphong harbor effectively close the Haiphong–Kunming railroad. On Sept. 22 Japan begins a military occupation of the Haiphong-Hanoi area.

July 18
(to Oct. 18) The UK capitulates to Japan's demand to close the Burma road. Three months later, the Luftwaffe defeated, the UK reopens it (Oct. 18).

1941

July 24
Japanese forces occupy central and southern Indochina.

Chongqing via a single-track line running northwest from Hanoi to Kunming. Moreover, since early 1938 the Chinese had been constructing a road southwest of Kunming to Lashio in Burma. The Japanese army now saw a solution to the China war in a drive south to occupy French Indochina and British Burma, thereby cutting Jiang Jieshi's supply lines. Thus it was that late in 1939 army policy began to converge with the navy's long-standing plans for southern expansion. It was a dangerous move which might result in conflict with the European colonial powers; since September both army and navy had been observing the progress of war in Europe, hoping that a reverse for the western Allies would afford them an opportunity for expansion.

Friction between Japan and the USA

That opportunity came, almost miraculously, in May and June 1940: Germany's occupation of the Low Countries and France and the apparent imminence of its invasion of Britain suddenly rendered the Southeast Asian colonies vulnerable. Japan's army and navy urged action but the civilians in Yonai's cabinet, wary of US reaction, were more cautious. Relations between the two nations had been deteriorating since 1937. As early as 5 October 1937 US President Roosevelt, influenced by the powerful China lobby, had suggested a quarantine of Japan and other aggressor nations. Two months later Japanese attacks on British and American gunboats on the Yangzi prompted his proposal (which horrified the UK cabinet) for an Anglo-American blockade of Japan. Japan's abject apology on Christmas Day 1937 obviated the need for such action, but during the next 18 months the American press whipped up strong anti-Japanese sentiments within the USA. Private American citizens sent millions of dollars to the Guomindang. In July 1939, the government, responding to public and congressional pressure, gave Japan six month's notice of the termination of a commercial treaty of 1911 which had given favored status to certain categories of Japanese imports. Worse was to come after the fall of France in June 1940; the US congress rushed through an act to expand the US fleet by 17 battleships, 11 aircraft carriers, 40 cruisers, nearly 70 submarines, and over 100 destroyers. Japan's admirals were aghast. Their own naval expansion program was tiny by comparison; by 1943, with the launching of the American program's first ships, they would be consigned to inferiority for ever.

The time window was small; Japan had to act while it still dominated the Pacific. Civilians in the government prevaricated; on 16 July the navy, acting in concert with the army, engineered the fall of the Yonai government. They would only allow Prince Konoye's new cabinet to take office on the condition that Japan would commit itself to southward expansion. Negotiations would settle border differences with the USSR: US intervention would be discouraged by Japan's entry into the Axis alliance with Germany and Italy. This settled, two days later the new government secured a diplomatic triumph by persuading the

◀ **After summer 1938 pitched battles were a rarity but guerrilla attacks took an increasingly heavy toll on the Japanese. This picture taken in 1939 shows the result of a guerrilla attack on the railroad system of north China and the efforts of Japanese engineers to overcome the disruption.**

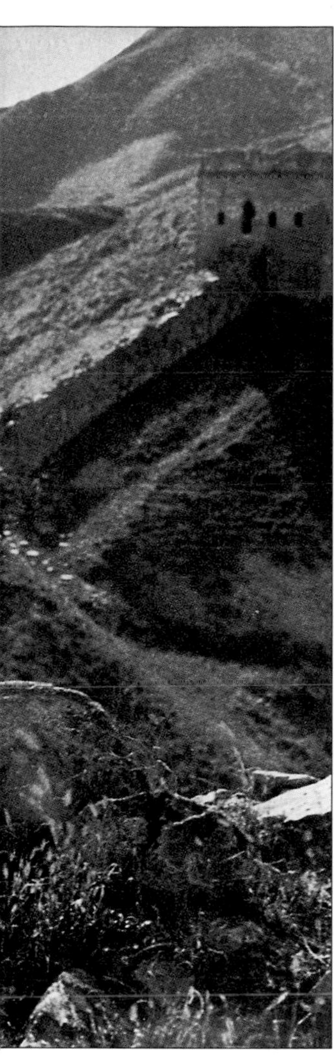

▲ Troops of the communist 8th Route Army pose with the Great Wall as a dramatic backdrop for a propaganda photograph. Mao's forces inflicted a humiliating defeat upon Japanese forces in the Battle of Pingdiquan in the Wutai Mountains on 25 September 1937.

UK to close the Lashio-Kunming road; on 22 September, having demanded and received from the Vichy government the right to land troops, Japan began to occupy northern Indochina.

The Americans responded with vigor. On 26 September they placed an embargo on steel and scrap-iron exports to Japan, which cut its supply of construction materials by 80 percent. The Japanese navy prepared for the worst; on 15 October it ordered full war mobilization, to be completed in 18 months. The USA also geared up. In December Roosevelt gave full approval to a scheme for supplying Jiang Jieshi with new B-17 bombers on condition they were used to attack Tokyo; but delays in their construction forced the scheme to be abandoned. In January 1941 Roosevelt ordered a large proportion of America's battle fleet to move to Pearl Harbor in Hawaii and increased the flow of supplies to General Douglas MacArthur's forces in the Philippines. On 15 April he signed an executive order authorizing American pilots to volunteer to serve in China.

By summer 1941 the Japanese military were alarmed by the strength of the American reaction; the risk of war now seemed a certainty. Though the navy wanted to continue southern expansion, it wished to do so cautiously. On 22 June they had an unpleasant surprise. Elements within the army responded to the German invasion of the USSR by demanding an end to risky southern expansion and a move against Siberia. The navy thought Japan's chance would be lost foreover; it demanded the immediate occupation of southern Indochina and on 25 July Tokyo proclaimed a joint Franco-Japanese protectorate over the whole of Indochina. The American reaction the following day was dramatic; all Japanese assets in the United States were frozen, meaning that trade between the United States and Japan, including the export of oil, ceased. Within days the UK and the Netherlands had followed suit. Although Japan had a two-year stockpile of oil, it was now left with two choices: to negotiate or to fight.

The Konoye cabinet determined to follow both courses simultaneously – to prepare for war but to hope that the Americans would agree to lift the embargo in return for Japanese concessions. But what did the Americans want? On three occasions during August and September Konoye tried to arrange a summit meeting with Roosevelt, offering to withdraw all Japanese troops from Indochina as soon as the "China Incident" had been settled, but received no reply. On 3 September an army–navy liaison conference set a deadline of 10 October (later extended to 15 October) for negotiations to be successful. If not, war was the only recourse. On 2 October the US secretary of state, Cordell Hull, finally replied. America would not resume normal relations unless Japan evacuated not only Indochina but China as well – a demand the army, after four years of war, and 185,000 ahead, could not possibly concede.

On October 15 the deadline expired. Determined not to plunge Japan into war with the USA, Konoye resigned and the war minister, General Hideki Tojo, formed a government. Although in cabinet meetings Tojo had previously argued for war, as prime minister he had grave doubts. The navy refused him assurances that victory was possible; the best they could promise was a year's domination of the Pacific after which the Americans would inevitably gain supremacy. Tojo now extended the deadline for a negotiated settlement until 1 December and on 5 November despatched Admiral Kurusu as a special envoy to Washington with a new set of proposals; Japan would withdraw immediately from southern Indochina and would agree to a phased withdrawal from China over the next 25 years. Roosevelt's administration had learnt about Japan's simultaneous military preparations via intercepted Japanese diplomatic messages and treated the proposals with contempt. On 26 November Hull again insisted that normal relations could be restored only if Japan withdrew from Indochina and China immediately, recognized no government in China other than that of Jiang Jieshi, and withdrew from the Axis alliance. When these demands were received in Tokyo on 27 November they were treated as an ultimatum. Ten days later Japan launched a new war.

To EMPEROR HIROHITO

The people of the United States ... have eagerly watched the conversations between our two Governments during these past months ... We have hoped that a peace of the Pacific could be consummated in such a way that nationalities of many diverse peoples could exist side by side without fear of invasion... and that all peoples would resume commerce without discrimination against or in favour of any nation.

FRANKLIN D. ROOSEVELT
6 DECEMBER 1941

Franklin Delano Roosevelt 1882–1945

President of the USA from March 1933 until his death on 12 April 1945, Roosevelt steered America through the crises of the Great Depression and World War II. He was inclined to intervene in Europe after the outbreak of war there in 1939, but isolationist sentiment prohibited it. The Japanese attack on Pearl Harbor on 7 December 1941 caught Roosevelt and the American people completely by surprise, though it should have been evident that America's economic embargos on Japan would result in conflict. Isolationist sentiment vanished overnight and though Roosevelt remained unpopular in many quarters, for the first time in history a president commanded an America united in a common purpose. Unlike Churchill or Stalin, Roosevelt made few important strategic decisions, but he did offer America inspiring and visionary leadership.

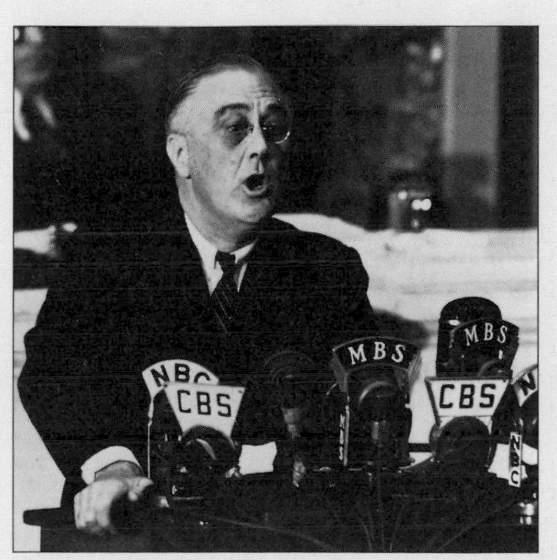

Datafile

In spring 1940 Germany invaded and conquered Norway in two months and defeated the Low Countries and France in seven weeks. These victories at unprecedented speed were achieved through the use of a new style of mobile warfare, Blitzkrieg or "lightning war". It depended for success on organization and technique rather than superiority in numbers. As these statistics show, Germany was outnumbered in manpower and tanks and had superiority only in aircraft. The French and British thought they could hold the Franco-German Front and even sent an expeditionary force to oppose the German invasion.

The most startling feature of Blitzkrieg is that dead and wounded were only a small proportion of the forces engaged. The secret lay not in killing enemies, but in confusing and demoralizing them, so that they surrendered in droves. By the standards of World War I, Germany's victories were cheap.

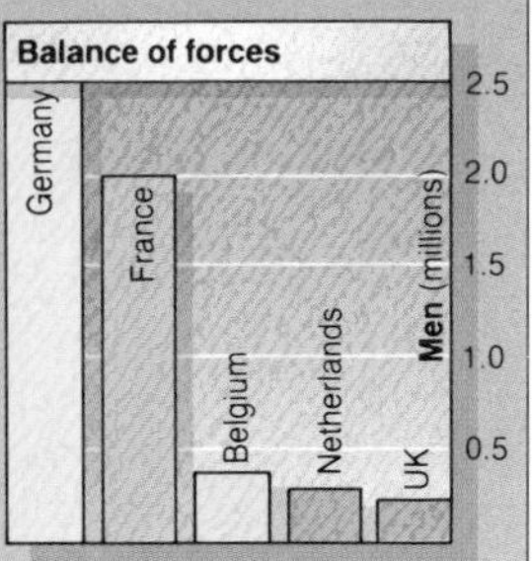

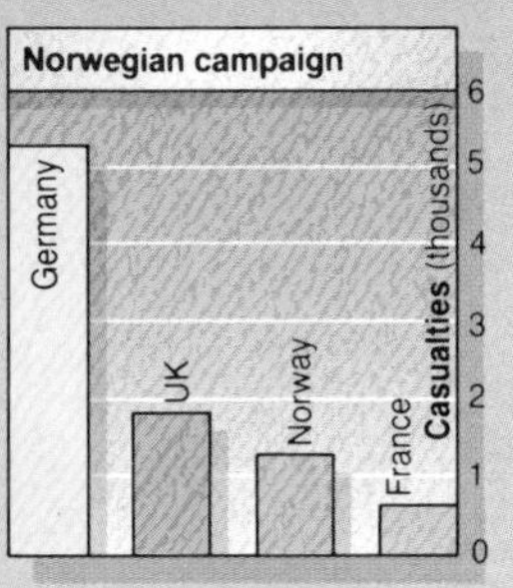

▲ In the late spring of 1940 the UK and France had near parity with Germany in manpower and if Belgian and Dutch forces were included they outnumbered German forces by a healthy margin. The casualty statistics suggest a drawn battle, but this was a quirk of Blitzkrieg; the reality was a runaway German victory.

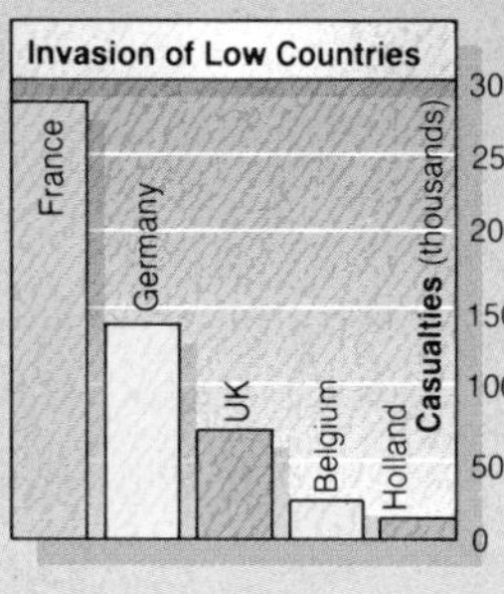

◀ The casualty figures for the German campaign of May–June 1940 underline the lesson of Norway. France lost only 90,000 dead and 200,000 wounded, but more than 1.9 million were taken prisoner. But for the Dunkirk evacuation the British Expeditionary Force would have suffered the same fate.

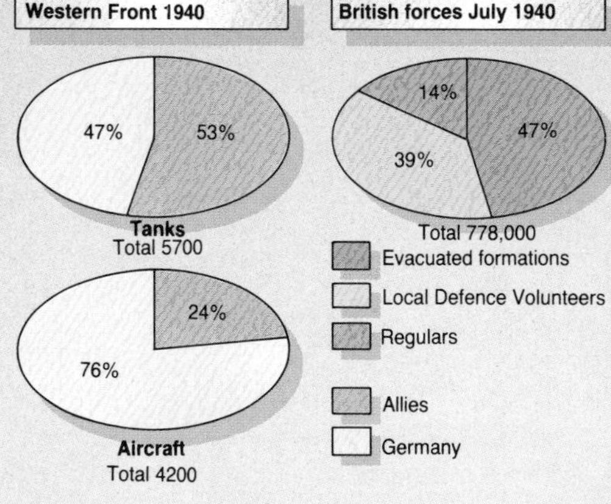

◀ The outcome of the German invasion appeared extraordinary on the basis of a comparison of Allied and German armor. The Allies had more and better tanks. Germany had nearly a 2:1 superiority in aircraft but this should not in itself have been decisive. Had the Germans crossed the Channel the British could have done little to resist. Most forces consisted of untrained and ill-equipped LDVs and the remnants of the BEF which had lost all its heavy equipment (600 tanks and 2,350 guns) in France.

War broke out in Europe in September 1939 in response to Hitler's renewed attempt to establish German dominion over Europe. At first he extended German power with a series of bluffs and diplomatic triumphs. In March 1936, in defiance of the Treaty of Versailles, he successfully remilitarized the Rhineland. In April 1938 he engineered a union (*Anschluss*) between Germany and Austria. So far Britain and France stood by and did nothing. Then Hitler pressed what he claimed was his last territorial demand in Europe: for the German-speaking area of Czechoslovakia, the Sudetenland, to be transferred to the Reich. The two allies were unprepared for war and thought they could take Hitler at his word. At a conference held at Munich in September 1938, they ceded Hitler's demand and believed that the

Hitler extends German power in Europe 1936–39

The German invasion of Norway, April 1940

The German invasion of the Low Countries and France, May–June 1940

The Battle of Britain

Churchill seeks to persuade the USA to enter the war

Munich agreement had averted another major European war.

The ease of this success convinced Hitler that Britain and France would not resist further advances; but when he proceeded in March 1939 to annex the Czech provinces of Bohemia and Moravia they finally concluded that his word was worthless and committed themselves to defend the integrity of Poland. Without an alliance with the Soviet Union, however, the Western democracies had no means of honoring their commitment. On 23 August 1939 Hitler achieved another diplomatic coup by concluding a pact with the USSR. Eastern Europe was divided into Soviet and German spheres of influence and Poland was partitioned. Eight days later Hitler invaded Poland, and France and Britain declared war.

▼ Hitler's triumphal entry into the Sudetenland, 3 October 1938.

Adolf Hitler 1889–1945

If the blame for the outbreak of war in Europe is to be ascribed to a single man, it is to Hitler. A rootless Austrian adventurer, he had picked up in his youth a rag-bag of second-rate ideas and prejudices which had figured on the European right for a century. But through his energy and formidable gifts as an orator, he transformed them into a terrible and nihilistic philosophy. The profound error committed when he was appointed chancellor of Germany in 1933 and became *Führer* (leader) in 1934 was that nobody expected him to act upon what he believed. Racism, and antisemitism in particular, lay at the heart of his creed, and was to dominate his policy, culminating in the attempted destruction of European Jewry.

As a war leader, Hitler's greatest gift was his eagerness to take risks. He continually caught his opponents off their guard. His grasp of detail was phenomenal – he knew more about weapons than the commanders fielding them. His hypnotic hold over his subordinates ultimately worked against him.

He lived in a world of dreams, and lost all touch with reality. Believing himself to be "the greatest conqueror of all time", he refused to sanction any retreats, and the overextended Wehrmacht was overwhelmed by numbers in 1944–45. After Germany's defeat in the Battle of the Bulge, he withdrew to Berlin and committed suicide in the bunker of the chancellery on 30 April.

Wilhelm Keitel 1882–1946

Chief of the high command of the armed forces (OKW) from 1938 to 1945, Keitel was an officer of moderate ability who was completely held in thrall by Hitler. Ironically Hitler held him in low regard, and dismissed him as "a man with the brains of a cinema usher". But these were the attributes that Hitler required. He had surrounded himself with sycophants and nonentities who would not question his genius. Keitel had held no significant operational commands and his frontline service was confined to that of an artillery regiment in 1914–18. After the war he was convicted for war crimes and hanged.

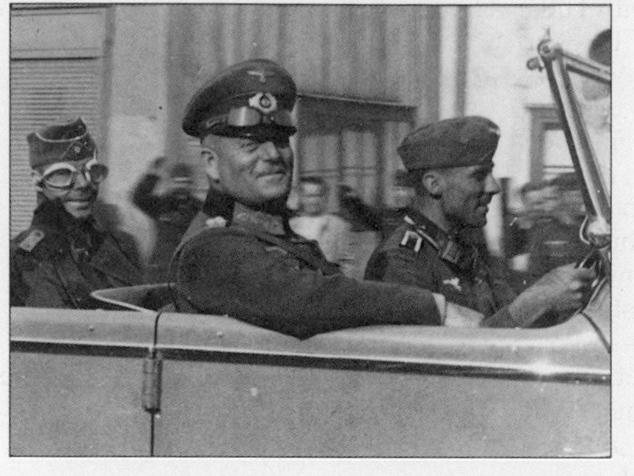

▶ The Germans planned to occupy Norway by an amphibious preemptive strike. This was risky because of the inferior size of the German navy. The Germans relied on surprise, speed and ruse to land and occupy the ports before the British. Their plan worked brilliantly, except at Narvik.

▼ Below left, German troops in a Norwegian village. Movement and initiative were the keynotes of German infantry tactics. Below, ships sunk by German aircraft. Germany's losses at Narvik crippled its surface fleet.

The "phony war"

After the defeat and occupation of Poland, the war settled down to a routine of waiting. Except for the odd dramatic event at sea, like the sinking of the German pocket battleship *Graf Spee* in the Plate estuary in December 1939, little happened, though the British Expeditionary Force (BEF) was sent to France to join the considerable French forces in northern France.

This was the so-called "phony war". It suited Britain very well. The official British view was that Germany could, if necessary, be defeated in a long war: the greater economic strength of Britain and France would eventually wear Germany down, if they were only given time to mobilize it. France was confident that the Maginot Line, stretching along the French-German border from Longuyon to the Swiss frontier, was invulnerable. But the British prime minister, Neville Chamberlain, probably also hoped that stalemate would allow him to renew his efforts at appeasing Hitler, which were so rudely interrupted by Hitler's occupation of Prague in March 1939.

Thanks to the Soviet-Nazi Pact concluded in August 1939, Hitler was able to concentrate on one front at a time. The Soviet government was determined to divert German aggression away from the USSR and build up a defensive cordon around its frontiers. Stalin was not impressed by

what he had seen of the Western democracies in 1939 and threw in his lot with Hitler. He invaded Poland from the east on 17 September – two weeks after Hitler had done so from the west. Two months later he invaded Finland. (See p.55.) The heroic struggle of the Finns excited Western sympathy. There were even those who wanted to send British and French troops through Norway and Sweden to aid them. This would merely have added the USSR, not to mention the neutral Scandinavian countries, to the list of Britain's enemies.

Buoyed up by his success in Poland, Hitler had planned originally to attack the west in November 1939. Severe winter weather forced the abandonment of this plan, much to the relief of the German high command.

The German campaign in the west opened on 9 April 1940, not in France but in Scandinavia. Hitler was determined to secure his right flank and supplies of bauxite ore from Sweden, which the Allies had threatened by laying minefields in Norwegian waters on 8 April.

Denmark was occupied within 24 hours. The Germans then aimed to seize the main Norwegian ports, Oslo, Bergen and Kristiansund, as rapidly as possible. German air power was crucial, and the British Home Fleet of three capital ships, six cruisers and 21 destroyers had no aircraft carrier sailing with it. The fall of these ports within hours made it difficult for the Allies to build up a base. Only at Narvik did they meet with success. Here 10 German destroyers were sunk for the loss of only two. An ill-equipped Allied expeditionary force under the British Lieutenant-General Claude Auchinleck was dispatched, and troops were withdrawn from France to reinforce it. But Auchinleck's task was hopeless and with the German attack on the Low Countries on 10 May, it was withdrawn. This evacuation was the first of a series of amphibious strategic withdrawals which were to mark the British conduct of the war in its first two years. The only major casualty was the loss of the unaccompanied aircraft carrier, *Glorious*, on 8 June. Losses on the ground were slight on both sides. The severest losses were sustained by the German Navy, whose surface fleet was crippled. But the acquisition of naval bases which threatened Britain's hold on the Northern Approaches of the Atlantic Ocean enabled the German navy to mount a much greater threat: a U-boat campaign against British sea lines of communication.

Hitler had not missed the bus, as Chamberlain had claimed. The Norwegian campaign's greatest casualty was Chamberlain's government. The irony was that it raised up in his place the man who, as first lord of the admiralty, had been most responsible for the disaster. On 10 May 1940 Winston Churchill became British prime minister, his decisive recommendation being that the Labour Party agreed to serve under him. Many, including King George VI, were sorry to see Chamberlain go. Chamberlain himself may have calculated that his surrender of power would be only temporary. Churchill had a reputation for reckless misjudgment. Chamberlain thought he would blunder; certainly he lacked a political base in the Conservative Party. But Chamberlain was

The German Invasion of Denmark and Norway 1940

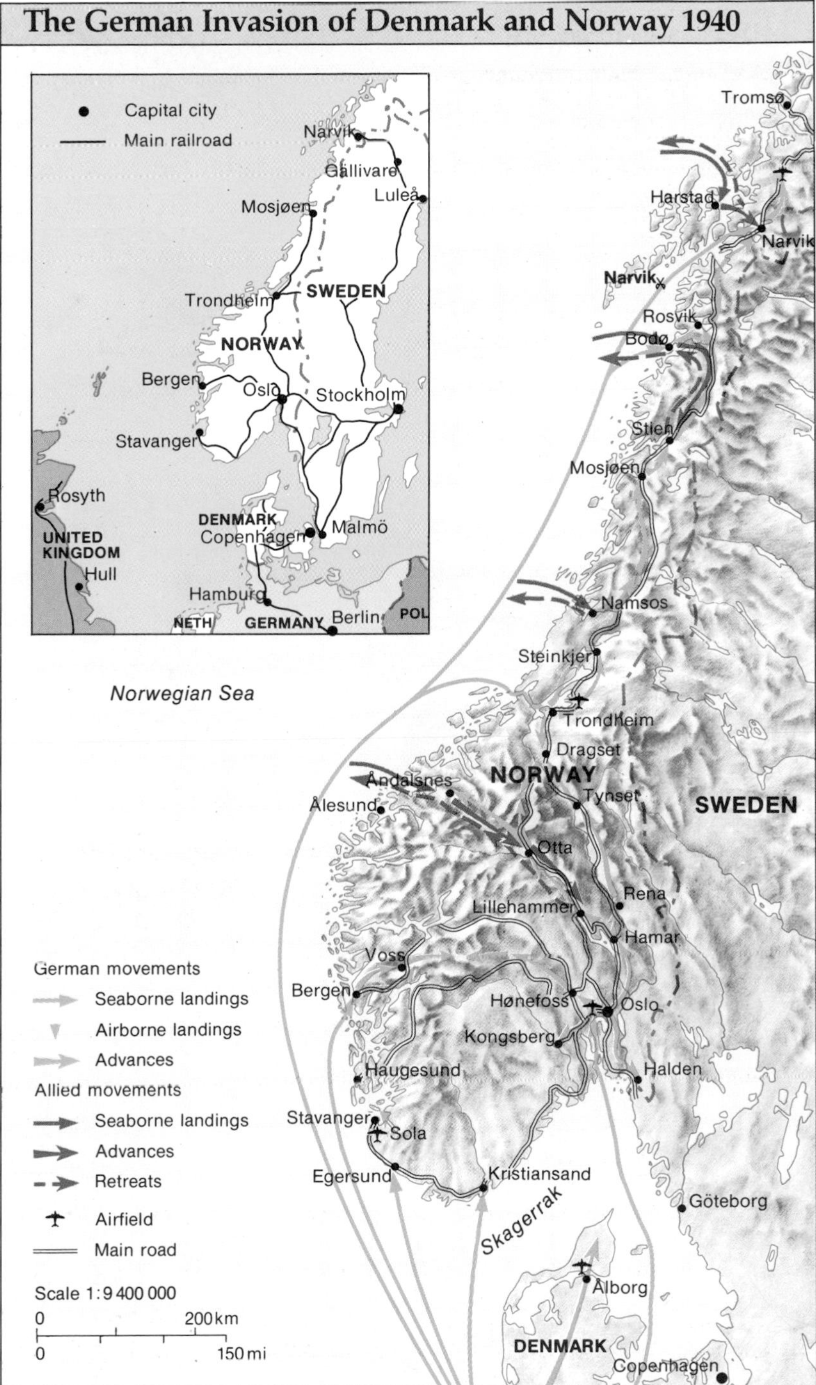

April 9
Operation *Weserübung* ("Exercise Weser"). German troops invade Denmark: it capitulates within four hours, giving Germany control of the Skagerrak. Meanwhile German forces seize Norway's major peripheral ports and cities.

April 10–11
The Norwegian government and King Haakon VII refuse to submit to German terms. In the north a British destroyer fleet attacks German destroyers at Narvik (First Battle of Narvik).

April 12–20
Reinforced German forces push out from Oslo along the major roads. They capture Halden, Kongsberg and Hønefoss (April 13). To the north they move along the two main routes toward Trondheim. By the 20th they are approaching Norwegian positions at Lillehammer and Rena. Meanwhile the British and French have landed troops at Namsos (April 14 and 16) and Åndalsnes (April 17) to reinforce the Norwegians. On April 13, in the Second Battle of Narvik, British destroyers again attack German ships. Meanwhile German territory around Narvik is expanded.

April 21
(to May 3) The Germans on the western road to Trondheim meet strong resistance from the Norwegians and British. They reach Otta on April 28. A final Norwegian and British action on April 30 fails to hold the Germans. British troops are evacuated from Åndalsnes on April 30 and May 1. Germans on the eastern road reach Tynset on the 25th and meet German troops from Trondheim at Dragset on April 29. The British force at Namsos is withdrawn on May 2–3.

May 4–31
German troops advance north along the Arctic Highway. Norwegian, British and French troops attempt to stop the advance (notably at Stien on May 17) but fail. On May 31 the Germans reach Bodø and Rosvik. Meanwhile the Norwegians and their allies capture Narvik on May 28.

June 1–8
As German troops pour into France, the British and French evacuate Narvik.

June 9–13
German troops continue their advance north to Narvik.

The German Invasion of the Low Countries and France 1940

May 10
German Army Groups A and B launch a Blitzkrieg invasion of the Low Countries. The air force bombs Dutch and Belgian airfields while airborne troops land ahead of ground forces.

THE NETHERLANDS

May 10–14
Airborne troops attack The Hague and secure bridges and buildings near Rotterdam. The 18th Army drives Dutch defenders from the border and directs its main weight to Rotterdam. On May 14 the city is heavily bombed and capitulates. The Dutch surrender the next day.

BELGIUM AND LUXEMBOURG

May 10–13
Glider-borne troops capture the Fort of Eben Emael and bridges over the Albert Canal. Armored divisions of the 6th Army cross the canal and advance (May 11). The main Belgian defenders, stationed behind the R. Dyle, begin to retreat. Liège falls on May 13. Meanwhile three armored corps of Group A cross the Ardennes to obtain bridgeheads. German forces also invade Luxembourg.

BELGIUM AND FRANCE

May 14–23
Operation *Sickelschnitt* ("The cut of the sickel"). German forces break the French line at Sedan and the Allies retreat from the Dyle/Meuse defense line. German armored units race for the coast. Units reach the sea on May 20 and thereby split the Allied forces apart. French and British counterattacks (May 17–19 and 20 respectively) fail to reestablish contact between the two halves. German units then swing north, isolating Boulogne and Calais (May 22–23). The French 1st Army, the Belgian army and the British Expeditionary Force (BEF) are trapped between Group A and Group B advancing from the north. On May 23 Hitler and von Runstedt order units to halt.

May 25
(to June 5) The Allies fall back around Dunkirk and are withdrawn in a large-scale evacuation (May 26–June 2). The German advance continues. Calais falls on May 27; Belgium surrenders on May 28; Dunkirk is occupied on June 5. German forces then regroup.

FRANCE

June 5–27
German armored units pour into France. June 16, French premier Reynaud resigns; his successor, Pétain, seeks an armistice. It is signed on June 22. France is divided into an occupied zone and an autonomous state. German forces reach the Pyrenees on June 27.

Henri Philippe Pétain 1856–1951

The savior of France at Verdun in 1916–17, Pétain lived too long for the benefit of his historical reputation. The colossal casualty lists inflicted on France in 1914–18 shocked him and he was determined that it should never again endure that experience. Between the wars he became increasingly pessimistic and defeatist. He believed that "To make union with England was fusion with a corpse". He sued for peace with Germany and Italy in June 1940. He presided over the "unoccupied" Vichy zone as chief of state, a position he held until 1945. He was an elderly self-centered man who took most notice of the last person he had spoken to. After the Liberation he was tried and condemned to death. De Gaulle commuted the sentence.

struck down by cancer and was dead by the end of the year. Churchill immediately had himself elected leader of the Conservative Party and easily routed the appeasers. Though he had some difficult moments in 1941–42, he was not seriously challenged for the remainder of the war.

The fall of France

The campaign to the west of Germany began with a shattering air attack followed by parachute landings in the Netherlands and Belgium. They immediately appealed for Allied help. Plan "D", the Allied advance into Belgium to the River Dyle, was put into operation. But within five days the Dutch army had laid down its arms. Rotterdam was devastated by massive air attacks. The Belgian bastion at Eben Emael on the Albert Canal fell within hours. The Belgian army withdrew to the line Antwerp–Louvain and linked up with the BEF.

Once the Allies were committed in Belgium, the German advance through the Ardennes began. By 14 May Army Group A, moving behind a screen of Stuka dive bombers, crossed the Meuse at Sedan. The French 9th Army, covering this sector of the front, was of poor quality and dissolved before the German assault. Paul von Kleist, commander of the Panzer groups, now wanted to halt and regroup. Guderian was furious. "I neither would nor could agree to these orders", he wrote later, "which would involve the sacrifice of surprise we had gained and of the whole initial success that we have achieved". He was given permission to continue the advance for a further 24 hours, and Guderian advanced 90km (55mi) beyond Sedan. By 17 May a bridgehead had been secured at St Quentin. The French army was paralyzed by this blow.

On 16 May Churchill visited Paris. In conference with the French prime minister, Paul Reynaud, he turned and asked Gamelin, the French commander in chief, "Where are the mobile reserves?" Gamelin replied, "There are none."

On 18 May Gamelin was sacked and replaced by Maxime Weygand who had a formidable reputation dating from his service as chief of staff to Marshal Foch in 1918. But his appointment came too late. Amiens fell on 20 May. Weygand hoped to organize a two pronged counterattack. On 21 May only the northern thrust toward Amiens could be mounted. It caught the Germans off-guard but was too weak for decisive effect.

The German advance on the English Channel continued the next day. May 23 was the critical day for the survival of the BEF. Its commander, Gort, was a courageous and determined fighting soldier. Though not the most intelligent of British generals, he was decisive. On hearing that the Belgian army had capitulated, he gave orders

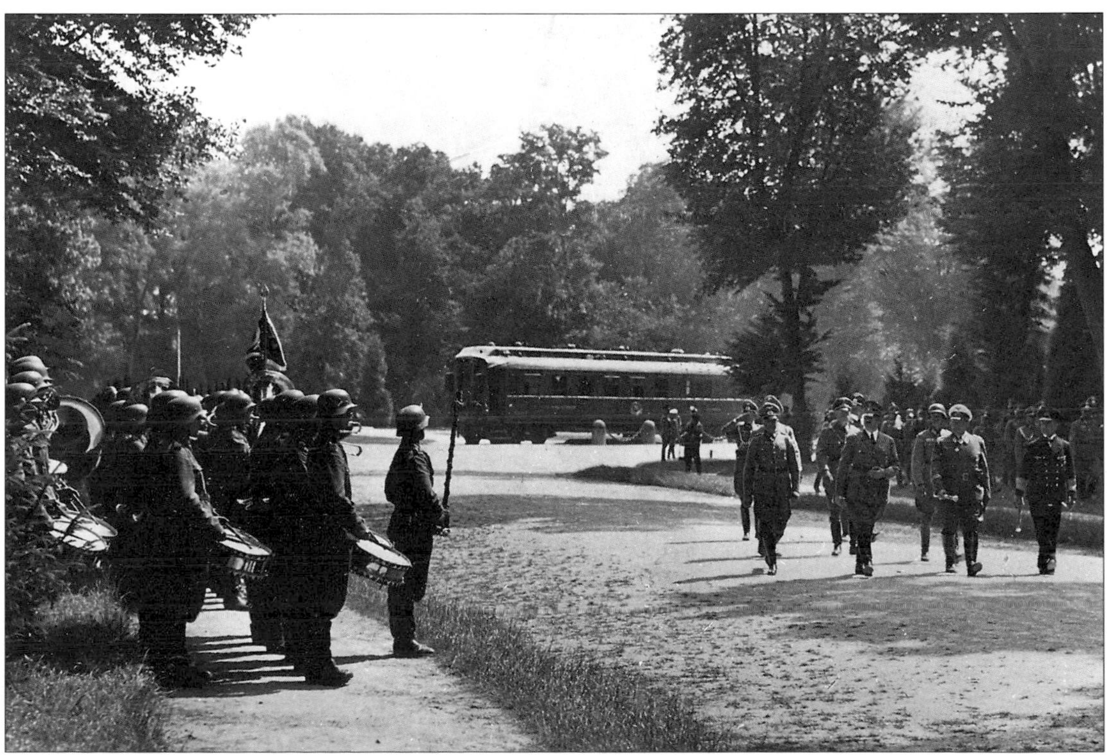

► A staged humiliation, Hitler and his generals after the signing of the French surrender of 22 June 1940. Hitler was determined to avenge the signing of the Armistice in 1918. In the Compiègne forest the railroad car in which Marshal Foch had forced the Germans to surrender was installed, and every effort was made to rub salt into the French wounds.

◄ The original German plan was a repeat performance of the Schlieffen Plan of 1914. But an officer was captured carrying the plans. General von Manstein had developed an alternative. In Guderian's words, this "involved a strong tank thrust through Southern Belgium and Luxembourg towards Sedan, a break through of the prolongation of the Maginot line... and a consequent splitting in two of the whole French front". The success of this strategy was accentuated by the French plan to advance into Belgium.

(fortified by Ultra intelligence) that the BEF should retire to the Channel ports, toward a perimeter around Dunkirk. There was not a moment to be lost, for Calais fell on 26 May.

That day the British admiralty issued orders for the commencement of Operation "Dynamo", the evacuation of the BEF from Dunkirk. This was made possible by the issue of the "halt order" by von Runstedt, with Hitler's agreement, on 24 May. It was rescinded the following day but 24 vital hours were lost. "Dynamo" lasted nine days: it was hoped that 35,000 might be saved; in the event ten times that number were rescued.

The Battle of France now moved rapidly toward its climax. Weygand intended to organize another defensive line along the Somme. But with the Maginot Line outflanked and the morale of the French army shattered this was an empty hope. Although individual fighting units showed courage, defeatism had rotted the morale of the French army and prolonged resistance was impossible. The Germans crossed the Seine on 10 June, and two days later the French government declared Paris an open city. Churchill's offer of a complete Anglo-French Union was ignored. Reynaud resigned and was succeeded by the aged and pessimistic Marshal Philippe Pétain. He asked for an armistice and accepted German terms on 22 June. In Churchill's pithy phrase, the Battle of France was over, the Battle of Britain was about to begin. France was divided into two zones, the occupied north and west and the unoccupied south and east. France was to bear the cost of the German occupation; Pétain set up a puppet capital at Vichy. Fearful that the French navy would be seized by the Germans, the Royal Navy attacked a French squadron at Oran. For a few days it appeared possible that Vichy France would declare war on Britain.

The Battle of Britain

The most celebrated aspect of the Battle of Britain was the war in the air, which is described in another chapter. (See p.114.) But in its entirety the Battle of Britain was the last moment when the British held the destiny of Europe in its hands. A heady spirit of defiance characterized British thought and deed. Despite defeats, Churchill demanded from Germany nothing less than unconditional surrender – or, as he put it, he offered nothing but "blood, toil, tears and sweat". There would be no negotiations with the Axis powers, no compromise peace. After the fall of France, Hitler offered peace terms which were discussed seriously by the British Cabinet: some of the appeasers thought that they should be pursued. In private Churchill was less dismissive than he was in public. He was still in a weak political position. But Churchill not only sensed the public mood, he embodied it, and Hitler's terms were rejected.

Hitler's directive for the invasion of Britain, Operation "Sealion", was issued on 16 July 1940. The main thrust of the invasion would be delivered by Army Group A, which would cross from the Pas de Calais with six divisions. Another thrust based on Le Havre, would move on the

► German bombers over London 1940; impressive but not decisive. In reality German bombers were designed for air–ground cooperation and not strategic bombing.

▼ A miracle of deliverance, the British army escapes at Dunkirk the humiliation heaped on the French army. But in France it was regarded as a betrayal.

Winston Churchill 1874–1965

Churchill had earned the reputation before 1934 as a man of great ability, but also of romantic inclinations and reckless misjudgment. On being appointed prime minister and minister of defence in May 1940, Chamberlain thought he would blunder. Certainly he lacked a political base in the Conservative Party. But Churchill routed his opponents. He was probably the most gifted politician produced by Britain. In the range of his gifts he has been compared with Caesar. Like Caesar he could be wilful, brutal and impulsive. Yet he displayed a streak of noble romanticism that matched the exigencies of the hour. During the war he was able to bolster domestic morale, win American support for the British cause and contribute imaginatively to the formulation of strategy. In A.J.P Taylor's view, Churchill was "the saviour of his country". It is difficult to think of anyone remotely qualified to take his place.

Isle of Wight, Portsmouth and Brighton. A secondary force would land at Lyme Bay and advance on Bristol. But this plan demanded control of the English Channel, and thus air superiority. Also, the German armed forces not only lacked a tradition of carrying out amphibious operations, but lacked specialized equipment.

The commander of the German air force, Hermann Goering, boasted that he could achieve victory over British skies. But Hitler was impatient and his heart was not wholly in the operation. On 15 August came *Adler Tag*, Eagle Day, in which the air force made an all-out assault on the RAF's airfields. Air chief marshal Sir Hugh Dowding, furnished with Ultra intelligence, knew the pattern of the raids in advance. The RAF remained in being; German air superiority was not achieved. The air battle continued for another month, until on 17 September Hitler issued instructions for "Sealion's" indefinite postponement. The air war now changed character. Provoked by an RAF raid on Berlin, the German air force concentrated on the bombing of British cities.

How was Britain to survive, let alone defeat Germany? The clue was provided in Churchill's "we will fight them on the beaches" speech. Here, he explained, Britain would "carry on the struggle, until, in God's good time, the New World, with all its power and might, steps forth to the rescue and liberation of the old". Since his accession to the premiership, Churchill had cultivated the president of the USA, Franklin D. Roosevelt. In December 1940 Churchill visited Washington and the basis for future wartime collaboration with the United States was laid. Though bound by election promises to keep the USA out of the war, Roosevelt was conscious of the threat posed to American security by a Nazi superstate controlling the resources of Europe. On 10 January 1941 the president placed a Bill before the US Congress which permitted the handing over to states at war with the Axis weapons which need not be bought with dollars – "lend-lease". The bill was passed. Thus the first major step was taken toward ultimate Allied victory.

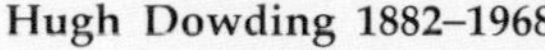

Hermann Goering 1893–1946

On the strength of his repuation as an air ace in World War I, Goering was given responsibility for the conduct of the German aerial war. A man of ability, unlike many of Hitler's Nazi cronies, Goering was also flabby and corrupt – his real political gifts canceled out by a penchant for good living, over-eating and the consolation of drugs. The longer the war went on, the less weight (figuratively speaking) he carried in the inner circles of the *Führer*. He was convicted at Nuremberg but committed suicide on the eve of execution.

Hugh Dowding 1882–1968

Dowding had been an advocate of fighter defense before 1939, and was never persuaded by the "bomber will always get through" school. He had supported the decision to deploy radar in 1935. He resisted Churchill's desire to send Fighter Command piecemeal to France during the crisis in May and June 1940. In summer and fall 1940 he won the Battle of Britain. This was a major achievement, but Dowding was aloof and morose and had many enemies. In 1940 he was dismissed as the result of a shabby cabal.

Datafile

The expansion of the US navy authorized in June 1940 threatened the predominance of the Japanese navy in the Pacific. It attempted to preempt the challenge by eliminating the US Pacific fleet. On 7 December 1941 it bombed the fleet's base at Pearl Harbor: battleships were destroyed but the US carriers were elsewhere. Thereafter Admiral Yamamoto sought to destroy the US fleet in a climactic battle, for fear of a war of attrition. But when engaged in the Battles of the Coral Sea and Midway in 1942 the Japanese suffered irreplaceable losses. The decision to challenge the US landing on Guadalcanal in August 1942 involved them in exactly the kind of conflict they feared.

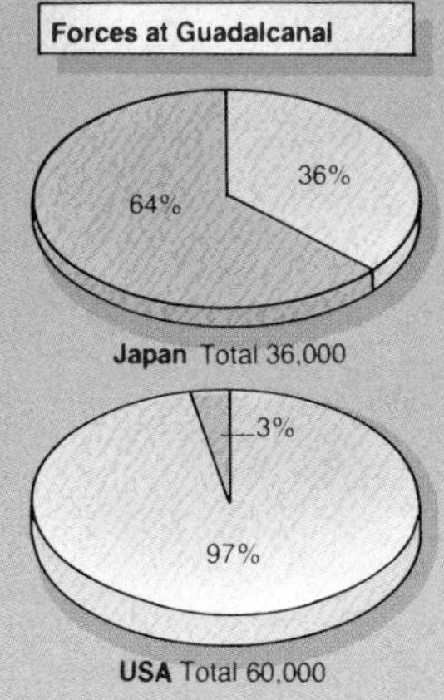

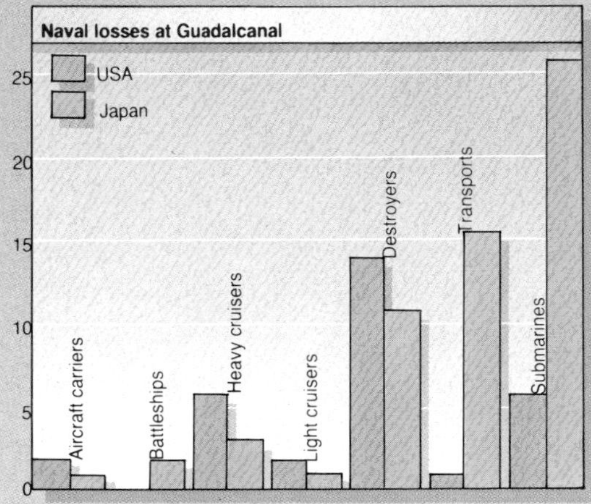

▲◀ The comparison of US and Japanese forces on Guadalcanal and the number of casualties (above) graphically illustrates the superior ability of the US navy to keep supplies moving to the island, while denying supplies to the enemy. In the naval campaign (left) the Japanese did better, their losses being less than those they inflicted on the Americans. But the United States' losses were being rapidly replaced.

▼▶ The Japanese began with a slight numerical advantage over the fleets of the USA, UK and the Netherlands (below), but in reality the advantage was much greater, the Japanese having a common system of training and communication. This advantage disappeared as the USA set in motion a construction program the like of which the world had never seen (right). New American ships were better constructed and fitted with sophisticated equipment.

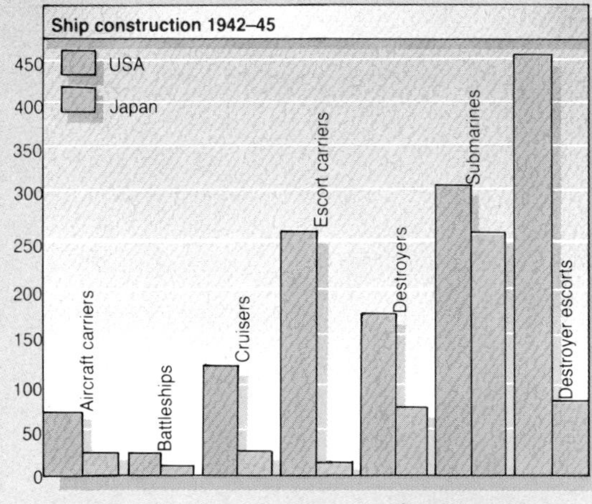

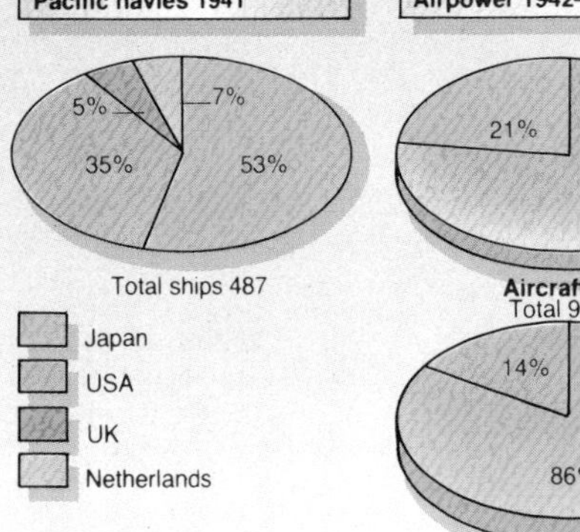

Airpower 1942–45

21%

79%

Aircraft built
Total 95,000

14%

86%

Aircraft lost
Total 19,700

◀ The United States outbuilt Japan in warplanes by a factor of 4:1. Although the Japanese had a marked qualitative advantage at first, by the summer of 1943, the Americans were introducing machines of greater speed, endurance, maneuverability, robustness and firepower than Japanese models. America's quantitative and qualitative superiority was reflected in the comparative air losses.

Since the 1920s the Japanese navy had been preparing for battle with the American fleet. It assumed that the American Pacific fleet would strike against the Japanese mandates. By the summer of 1940, with war now a real possibility, the newly appointed commander of the Combined Fleet, Admiral Yamamoto, began a systematic revision of the navy's plans. The decisive action strategy now seemed very risky: if it failed the American fleet could cut the supply lines of Japanese forces in Malaya and the Indies. Far better, Yamamoto reasoned, to destroy the American fleet before it had time to put to sea. Japan would then have about two years in which to consolidate its conquests and construct a defensive perimeter in the central and south Pacific.

The problem was finding a means of destroying the American fleet in harbor. Midget submarines seemed to be the only means, yet Yamamoto, as one of the few senior officers to have qualified as a pilot, argued vigorously for a simultaneous aerial attack launched from Japan's six fleet carriers. Many officers thought this absurd. No capital ship had been sunk by aerial attack and the US base at Pearl Harbor was too shallow and too narrow for the use of air-dropped torpedoes.

JAPAN'S WAR IN THE PACIFIC

Japan's planning for an attack on the US Pacific Fleet

The bombing of Pearl Harbor, 7 December 1941

MacArthur's escape from the Philippines

The Battles of The Coral Sea and Midway

Japanese operations in the Solomons and on New Guinea

Australian and American successes in New Guinea and at Guadalcanal

The bombing of Pearl Harbor

On 11 November 1940 attitudes changed. That morning 21 obsolete Swordfish torpedo bombers launched from the British carrier HMS *Illustrious* sank three Italian battleships within Taranto harbor – both shallower and smaller than Pearl Harbor. The Japanese navy quickly obtained a full report and began planning a similar but large-scale attack. Technical problems were systematically overcome; by the summer of 1941 new quick-fusing shallow-run torpedoes and massive armor-piercing bombs were being manufactured. Japanese agents in Hawaii had by now given an accurate picture of the operations of the American fleet. The naval commander who planned the attack, Minoru Genda, decided that the best time to attack would be early on a Sunday morning, when the ships would be in harbor and many of the crews on leave in Honolulu. Sunday 7 December was selected: six carriers would launch 360 planes while midget submarines would penetrate the base.

In the early hours of 7 December US anti-submarine patrols reported several contacts with unidentified craft off the entrance to Pearl Harbor and at 7.00 am a radar station on the north coast of Oahu detected the approach of a large flight of aircraft. These reports were ignored. Shortly before 8.00 am the first of 360 torpedo and dive-bombers and fighters swept down on Pearl and adjacent airbases, taking the Americans completely by surprise. During the next two hours all eight battleships in harbor and more than 300 aircraft were either destroyed or badly damaged. Exultant pilots, returning to the carriers, urged the carrier fleet commander, Nagumo, to renew the assault; harbor installations and oil storage tanks were still largely undamaged and the American carriers had still to be located. Nagumo, however, felt he had pushed his luck far enough and turned for home. It was a grievous misjudgment: Pearl Harbor was left sufficiently unscathed to continue functioning as a major base.

The consequences of the attack were ironic. Intended to destroy the fighting capacity of America's Pacific fleet, the loss of the battleships forced the Americans to rely on their carriers and to evolve a new and much more effective means of naval warfare. Moreover Japan's envoys in Washington did not deliver their country's ultimatum until 2.20 pm Washington time, about 50 minutes after the attack had started. As America's radio networks broadcast the news during the afternoon, a wave of rage swept across

▼ **Pearl Harbor, 8.40 am, 7 December 1941. Hit by bombs from Japanese "Vals", a fireball shoots hundreds of meters into the air from USS *Shaw* against the backdrop of the vast cloud of smoke arising from the shambles of "Battleship Row". The destruction at Pearl Harbor initially led both the Japanese and Americans to believe that the power of the US Pacific fleet had been broken.**

Hideki Tojo 1884–1948

Japan's war minister and a leading member of the army's "control faction" in the 1930s, Tojo urged expansion on the Asian continent. He became premier in October 1941 and sought at the eleventh hour to avert what his policies had done so much to bring about: war with the USA and the European colonial powers. As a soldier Tojo paid more attention to the war on the Asian mainland and at times seemed almost to enjoy the discomforture of the navy at the hands of the Americans. The fall of Saipan in summer 1944, which brought Tokyo into US bombing range, led to Tojo's resignation. After Japan's surrender he was tried for war crimes and hanged.

Isorouko Yamamoto 1884–1943

Appointed commander in chief of the Japanese Combined Fleet in 1939, Yamamoto was the architect of the Pearl Harbor attack, though he had severe misgivings about a war with the USA. He was depicted as the "Japanese Nelson" by both the American and Japanese press but his ability was not great and he committed his fleet to the disastrous Midway and Guadalcanal operations. In spring 1943 a now deeply depressed Yamamoto went on tour of forward bases in the Solomons to boost morale, but his aircraft was intercepted and shot down by American fighters, and the admiral was killed.

The War in the Pacific 1941–43

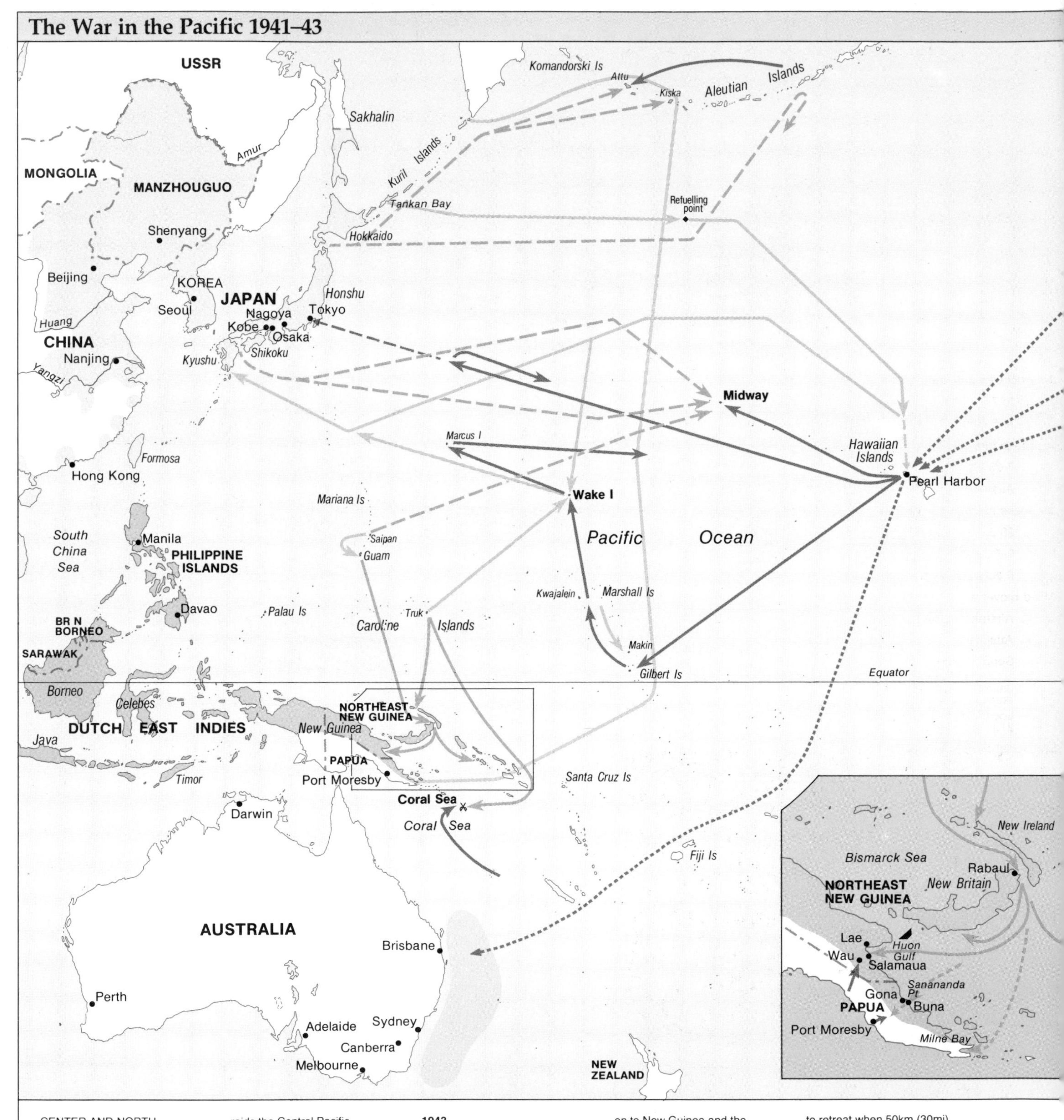

CENTER AND NORTH
1941

December 7
Japanese carrier aircraft attack the US naval base at Pearl Harbor, Hawaii.

December 8
(to Dec. 23) Japanese attack and eventually capture Wake Island (US possession). Japanese landings secure Guam and Makin (Dec. 10).

1942

February 1
(to March 4) A US task force raids the Central Pacific, attacking Japanese bases on the Gilberts, Marshalls and Marcus and Wake Islands.

April 18
B-25 bombers from USS *Hornet* bomb Japan ("Doolittle raid").

June 3
(to June 7) The Japanese seize Kiska (June 6) and Attu (June 7) while the main fleet closes on the US island of Midway; but in the Battle of Midway (June 4–5) US planes sink four Japanese carriers.

1943

March 26
(to Aug. 15) A US naval squadron drives Japanese from the Komandorskis (March 26). Americans retake Attu (May 11–29) and the Japanese evacuate Kiska (July 29).

SOUTHWEST AND SOUTH
1942

January 23
Japan's South Seas force captures Rabaul in New Britain.

March 8
(to July 7) The Japanese move on to New Guinea and the Solomons: they occupy Lae and Salamaua, Tulagi (May 3) and Guadalcanal (July 7).

May 7–8
Battle of the Coral Sea. Both sides withdraw.

July 21
(to Jan. 22, 1943) Japanese troops land on the north Papua coast and advance toward Port Moresby. A subsidiary force lands at Milne Bay but is repulsed by Australians (Aug. 25–Sept.5). Meanwhile Australians force the Japanese to retreat when 50km (30mi) from Port Moresby and then annihilate them (Nov. 20–Jan. 22, 1943).

August 7
(to Feb. 7, 1943) US marines seize Tulagi on Guadalcanal (Henderson Airfield). The Japanese navy counterattacks, sinking US ships off Savo Island (Aug. 9). The next five months see a series of naval battles. On Guadalcanal the Japanese launch unsuccessful attacks (Sept. 12–14, Oct. 23–25, Jan. 10–Feb. 7) and withdraw.

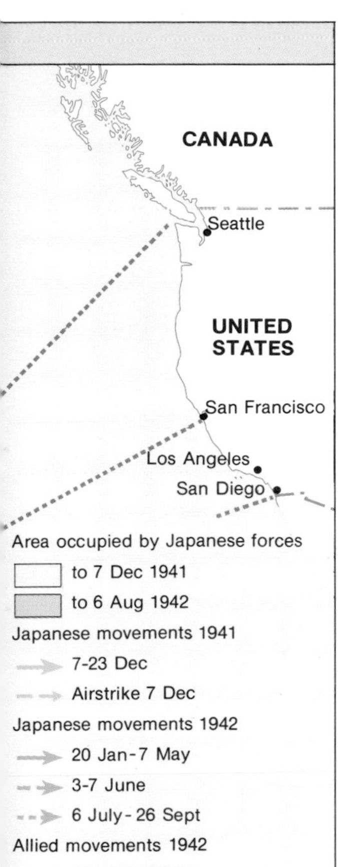

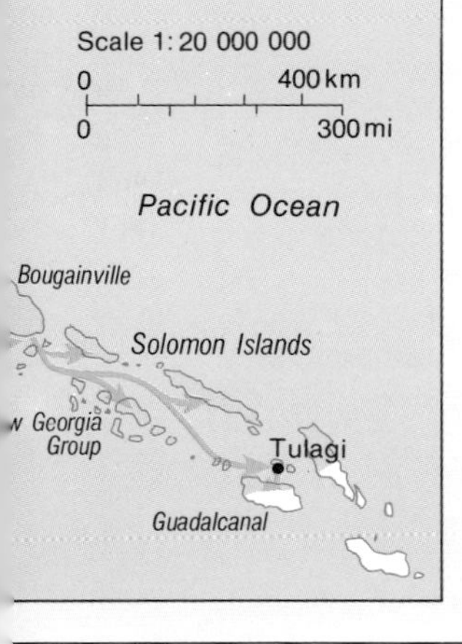

1943

January 18
Japan deploys submarines off the east coast of Australia.

January 20–30
The Japanese reinforce Lae and Salamaua.

March 2–3
The Japanese send a convoy from Rabaul to Lae. But it is destroyed by 200 US planes.

June 16
Japan withdraws its submarines from Australia.

◀ Manila, 31 December 1941. MacArthur declared it an "open city" on 27 December 1941. Four days later it was bombed by the Japanese.

◀◀ After the bombing of Pearl Harbor and the conquest of Southeast Asia, Japan launched a two-pronged offensive. A task force attempted to expand the perimeter in the Southwest Pacific, but was turned back in the Battle of the Coral Sea. A little later the bulk of the Imperial Navy thrust into the Central and North Pacific in an attempt to bring the remnants of the US navy to battle. This operation ended in the disastrous Battle of Midway.

Douglas MacArthur 1880–1964

As US army commander in Southeast Asia in 1941, MacArthur was responsible for a faulty defense of Luzon which led to rapid enemy advance. But his defense of the Bataan Peninsula turned him from a failure into a hero. He escaped to Australia in March 1942. Thereafter he sought to return to the Philippines as soon as possible, an aim that brought him into bitter conflict with the US navy, especially with admirals Nimitz and King. He returned in October 1944 and was preparing for the invasion of Japan when it capitulated in August 1945. He took the formal surrender on 2 September aboard USS *Missouri* in Tokyo Bay..

the country. Overnight isolationist sentiment vanished and when Roosevelt asked Congress for a declaration of war the following day he addressed a nation united as never before. Japan could scarcely have engineered a worse result.

American and Japanese responses

For the moment, however, it seemed in Tokyo that a mighty victory had been won. Japan's conquest of the British and Dutch Empires could now proceed unhindered (see pp. 46–47) and a defensive perimeter could be established in the Pacific. Within the area enclosed by the perimeter the American bases of Guam and Wake were quickly overrun, and the American-protected Philippines isolated. America's commander in the Philippines, General Douglas MacArthur, had long expected a Japanese attack, and had at his disposal 30,000 American and 100,000 Filipino troops, supported by the United States' small Asiatic fleet, and about 200 modern fighters and bombers. Yet although MacArthur learned of the Pearl Harbor attack at 3.00 am Manila time, when Japanese bombers from Taiwan swept over Luzon's airbases at midday they found the American squadrons still sitting on the runways. It was a second Pearl Harbor: within 30 minutes more than half MacArthur's air force had been destroyed. Over the next few days unopposed Japanese air attacks forced the Asiatic fleet to withdraw to Java and on 22 December large Japanese forces landed at Lingayen Gulf on the western coast of Luzon, followed two days later by another landing at Limon Bay in the south of the island. MacArthur's inadequately trained Filipino divisions soon crumbled, and as Japanese columns closed on Manila from north and south, MacArthur withdrew his forces northwest to the rugged Bataan Peninsula which, with the island fortress of Corregidor off its tip, controlled the entrance to Manila Bay. American forces in the Manila Bay area did not surrender until 6 May, and those still at large in the southern Philippines continued organized resistance until 18 May, when many melted into the jungles and started guerrilla campaigns.

The prolonged resistance of MacArthur's forces was embarrassing for the Japanese but even more embarrassing for the Roosevelt administration. Since the first assault MacArthur had kept up a barrage of demands for reinforcements, demands which could not be entirely ignored because he was fast becoming a national hero. His presence in the Philippines was particularly irritating to Admiral King, chairman of the joint chiefs of staff (JCS), and Admiral Nimitz, the commander in chief in the Pacific (Cincpac). After Pearl Harbor, US admirals concentrated on defending Hawaii and maintaining communications with Australia.

MacArthur's position posed a problem. Roosevelt solved it in early March by ordering MacArthur and some of his senior staff to escape

▲ **The flight deck of the carrier USS *Yorktown* during the Battle of the Coral Sea (7–8 May 1942): a Douglas Dauntless dive bomber followed by a row of Douglas Devastator torpedo-bombers. Coral Sea was the first battle in which opposing fleets did not come into direct conflict.**

to Australia, where, shortly after his arrival on 17 March, MacArthur was placed in command of the newly created Southwest Pacific Area. Like Nimitz, MacArthur found he had few forces at his disposal, and would have to remain on the defensive for some months to come. However, without consulting either the president or the JCS MacArthur had already announced his policy to the Australian press:– "I came through and I shall return". For MacArthur the only road to Tokyo ran through the Phillipines; for King and Nimitz it ran through the islands of the Central Pacific. Herein lay the genesis of the bitter disputes over strategy which were to bedevil the conduct of America's war in the Pacific in the years to come.

The Japanese, too, were now having problems deciding on a strategy. With the capture of Rabaul on 21 January, and the occupation of Lae and Salamaua on the north coast of New Guinea in early March, the southeastern part of their Pacific perimeter had been secured. One faction within the army, whose most vociferous spokesman was General Yamashita, the conqueror of Malaya, wanted to invade Australia immediately. The predominant China faction within the army and Yamamoto's "combined fleet" faction within the navy united to oppose this scheme and gradually Yamamoto's plans were accepted by imperial headquarters. The admiral devised two complimentary strategies: the isolation of Australia by the capture of Port Moresby, followed by a drive southeast through the Solomons, the New Hebrides, New Caledonia and Fiji; and a combined fleet offensive into the central Pacific to take the island of Midway and bring the remnants of the American fleet to battle. Japan thus embarked on two enterprises which divided its forces and placed an enormous strain on its already overloaded merchant marine. The astounding successes of the first four months of the war had created an atmosphere in which anything seemed possible and the meticulous planning which preceded the Pearl Harbor attack was now very much less in evidence. Senior officers had been infected with

Ernest Joseph King 1878–1956

King became chief of naval operations in March 1942 and was also a member of the US joint chiefs of staff and the Allied combined chiefs of staff. While agreeing in principle to a "Germany first" strategy, he resisted British pressure for a purely defensive stance against Japan. He also sought to relegate MacArthur's Southwest Pacific Area to the status of a backwater and to keep Britain's navy out of the Pacific. Though unsuccessful in both endeavors, King had engineered the growth of the US navy into the most powerful naval force in the world history.

what Japanese historians would later call the "victory disease", the major symptom of which was continued attempts to achieve difficult objectives with slender resources.

The Battles of Coral Sea and Midway

The American fleet was still very much smaller than Japan's – as yet Nimitz had only four carriers – but intelligence had now broken Japan's naval codes. This coup gave Nimitz an inestimable advantage, for by gaining advance warning of Japan's moves he could concentrate his forces and achieve at least local equality. By 17 April intelligence had discovered Yamamoto's plans to take Port Moresby, and Nimitz concentrated a task force in the south Pacific built around the carriers *Lexington* and *Yorktown*. The three task forces Yamamoto sent south outnumbered the Allied fleet but contained only three carriers, the *Shoho, Shokaku* and *Zuikaku*, which together could launch some 140 aircraft, about the same number as the *Yorktown* and *Lexington*. The ensuing Battle of the Coral Sea was the first in history in which warships of the opposing fleets did not come into direct conflict. It was carrier aircraft against carrier aircraft, a weapon in which the Americans had parity with the Japanese. On 7 May American divebombers scored their first major success of the war by sinking the *Shoho*. The following day Japanese aircraft sank the *Lexington* and damaged the *Yorktown*, but not before American machines had hit and badly damaged the *Shokaku*. Alarmed by the loss of two carriers the Japanese withdrew north, temporarily abandoning their plans to take Port Moresby. It was their first strategic reverse.

The Coral Sea action convinced Yamamoto that the decisive battle had to be fought before the Americans became any stronger. Japan had already suffered a rude shock on 18 April when a flight of B-25 medium bombers raided Tokyo. It had been a morale boosting stunt – the bombers had been launched 1,300km (800mi) off the coast of Japan from the carrier *Hornet* – but many Japanese were convinced that they had flown from Midway. Yamamoto, anxious to press ahead with his Combined Fleet offensive, decided not to wait for the *Shokuku* to be repaired or for the *Zuikaku* to have its severely depleted squadrons rebuilt. His 165 ships, including four large and four light carriers, seemed more than sufficient to deal with the Americans. His plan was complex, the fleet being divided into four widely separated task forces.

Once again naval intelligence had kept Nimitz apprised of Japanese intentions. The damaged *Yorktown* sped back from the south Pacific and was repaired in an astonishing 48 hours (her captain had thought she would be in dry dock for at least three months) then sailed for Midway to rendezvous with the *Enterprise* and *Hornet*. The fleet commander, Admiral Spruance, had some 50 ships in all, and was outnumbered three to one by Yamamoto. But the Coral Sea battle had shown that what now mattered was aircraft, and the 250 aboard his three carriers, supported by 109 machines based on Midway, gave Spruance a slight superiority over the Japanese. Early on 4 June Nagumo's carriers came within air range of Midway and, unaware of the proximity of the American carriers, launched strikes against the island. Convinced they had destroyed American airpower on Midway the aircraft returned to their carriers around 9.00 am and began the lengthy procedure of refuelling and rearming. The decks of the carriers were covered with fuel lines and bombs when American torpedo-bombers, followed by waves of dive-bombers, struck the Japanese. By 10.30 am *Akagi, Kaga,* and *Soryu* were flaming wrecks. Nagumo's surviving carrier, the *Hiryu*, managed to launch an airstrike which located and crippled *Yorktown* (she was later sunk by a Japanese submarine) but the *Hiryu* was in turn discovered by the *Enterprise*'s aircraft and sent to the bottom.

Having failed to secure a decisive victory Yamamoto knew that the navy faced the prospect of unremitting warfare with an enemy whose power was increasing daily. But within imperial headquarters the attitude was less gloomy. Midway had been a setback but the "victory disease" still gripped many planners. Japan still dominated the Pacific and headquarters decreed that operations to isolate Australia (suspended since the Coral Sea) should now proceed. As a first step the perimeter in New Guinea and the Solomons was to be expanded by two concurrent operations. In June 1942 engineers began constructing an airfield on the island of Guadalcanal in the southeast

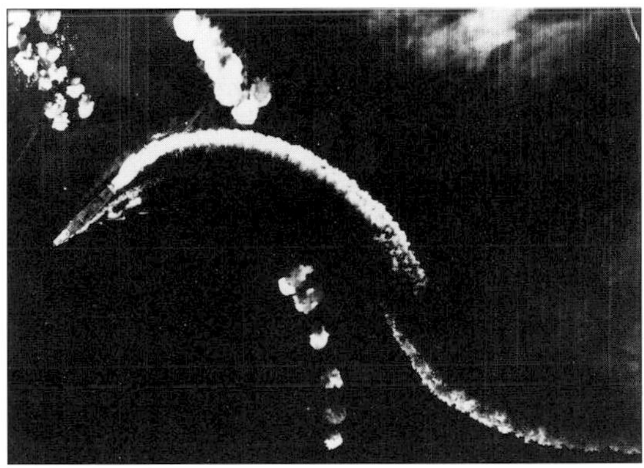

▲◀ At 14.30 on 4 June 1942 "Kate" torpedo bombers from Vice-Admiral Chuichi Nagumo's sole surviving carrier *Hiryu* score a torpedo strike against the damaged USS *Yorktown* (above). At the same time 100km (60mi) to the west Midway-based B-17s make an ineffectual high-level attack on *Hiryu* (left). Within hours of these photos being taken both carriers had been mortally wounded, the *Hiryu* by dive bombers from USS *Enterprise*, and *Yorktown* by torpedoes from the Japanese submarine I-168.

Five minutes! Who would have dreamed that the tide of battle would shift completely in that brief interval? Visibility was good. Clouds were gathering at about three thousand meters, however, and though there were occasional breaks, they afforded good concealment for approaching enemy planes. At 10.24 the order to start launching came from the bridge by voice-tube. The air officer flapped a white flag, and the first Zero fighter gathered speed and whizzed off the deck. At that instant a lookout screamed: "Hell-divers!" I looked up to see black enemy planes plummeting toward our ship.

MITSUO FUCHIDA
MIDWAY, 4 JUNE 1942

Solomons; six weeks later, on 21 July, a division-sized force landed at Buna on the northern coast of Papua. When completed the Guadalcanal airfield was to provide aircover for operations against the New Hebrides. By this time the landing force in Papua would have advanced across the rugged Owen Stanley range via the Kokoda Trail and taken Port Moresby from the north.

But the period of Japan's easy victories was already past. Midway had removed the threat to Hawaii and substantial American forces had been added to those already despatched to the South Pacific. By early July a naval task force, including three carriers and a new battleship, and an army division had arrived in New Caledonia, and the 1st Marine Division had landed in New Zealand. MacArthur's forces in SWPA had also increased substantially. With large forces now available in the Southwest and South Pacific, on 2 July the joint chiefs of staff ordered an Allied offensive – a parallel advance on Rabaul. When the Japanese landed at Buna, MacArthur had been on the point of carrying out the first stage of his part in the operation – the occupation of Papua's north coast – and to this end had already reinforced Port Moresby and established a powerful base at Milne Bay on Papua's eastern tip. South Pacific forces too had been preparing to advance into the Solomons when on 5 July Nimitz received word that the Japanese were constructing an airfield on Guadalcanal, which, if completed, would pose a major threat to American operations. Nimitz responded with characteristic energy. On 7 August 1st Marine Division, transported in a heavily protected convoy, landed on the northern coast of Guadalcanal and the nearby island of Tulagi. They drove surprised Japanese construction troops into the jungle and seized and dug in around the airstrip, renaming it Henderson Field after a hero of Midway.

Reversals for Japan on land and sea

Japanese headquarters in Rabaul now faced a difficult situation. What should they do: close down the Papuan operation and concentrate on the new threat or attempt to contain the marines until Port Moresby had been taken? Either decision would have been sensible. Instead Rabaul HQ decided to continue the advance on Port Moresby and to rush forces to Gaudalcanal to eject the Americans.

The Papuan operation was the first to run into trouble. On 25 August, endeavoring to speed up the capture of Port Moresby, a regiment landed at Milne Bay and, unaware of the strength of the Australian garrison, launched a frontal attack. In ten days of bitter fighting the Japanese lost more than 1,000 dead and by 5 September had been pushed into the sea. Their Guadalcanal counter-offensive became bogged down at the same time. Contrary to Rabaul's expectations, the marines fought tenaciously and in what became known as the Battle of Bloody Ridge killed more than 600 Japanese and smashed their attack.

Having suffered two serious reverses within a week, Rabaul finally accepted the impossibility of simultaneously taking Port Moresby and driving the marines from Guadalcanal. As the marines seemed to represent the greater threat, on 25 September Rabaul ordered the Papuan expeditionary force, which had by then pushed to within 46km (29mi) of Port Moresby, to withdraw to the northern slopes of the Owen Stanleys, and dig in.

It was a decision which should have been made six weeks earlier. Denied adequate supplies, the retreat across the Owen Stanleys quickly turned into a shambles. Veteran Australian brigades pursued them closely, wiped out the Japanese rearguard at Templeton's Crossing near the crest of the range on 16 October, and encircled and annihilated the main body at Gorari on the northern slopes of the range three weeks later. Now only the Buna beachhead remained. By the third week of November the Australians, assisted by the American 32nd Division which had landed on Papua's northern coast a few kilometers east of Buna, closed on the Japanese beachhead. With Japanese airpower completely absorbed by the struggle then raging on Gaudalcanal, Allied aircraft dominated the skies of northern Papua and made resupply and reinforcement of the beachhead virtually impossible. The Japanese fought stubbornly from heavily bunkered positions for nearly eight weeks but on 22 January 1943 the last pockets of resistance were exterminated. It was Japan's first major defeat on land.

Meanwhile the Imperial Japanese Navy had made an all-out effort to destroy Henderson. On the nights of 14 and 15 October two strongly escorted battleships, *Kongo* and *Haruna*, swept down on the beachhead and bombarded the airfield, temporarily putting it out of action. The Japanese had at last reasserted control of the waters around Guadalcanal, an achievement which had far-reaching consequences. Ever since landing on Guadalcanal the marine commander, Major-General Vandegrift, had been complaining bitterly to Cincpac about the relatively poor performance of the navy in supporting his beleagured troops. The fact that Japanese battleships were operating off Guadalcanal was taken as evidence that his complaints were justified and on 18 October Nimitz gave the South Pacific a new commander, Rear Admiral William H. Halsey.

▼ Immaculately clad in their white tropical uniforms, Japanese sailors pose carefully for a photographer on Guadalcanal in August 1942. A few weeks later these men looked very different – filthy, emaciated, with their uniforms in tatters.

▲ **5 October 1942, Port Moresby-based B-17s hit the Japanese airfield at Buna. By the end of the month such raids had given the Allies control of the air over northern Papua.**

Capitalizing on the success of the battleship raid on 25 October, Yamamoto at last committed a large carrier task force to the Solomons. As it plowed south toward the Santa Cruz Islands to the east of the Solomons, Halsey's carriers moved north. The ensuing action was a virtual replay of the Coral Sea, the Americans and Japanese launching simultaneous air strikes. The Japanese sank *Hornet* and badly damaged *Enterprise* but American air strikes also hit home, badly damaged *Zuiho* and *Shokaku*, and forced the Japanese to turn about. Yamamoto was furious. Believing that more aggressive leadership would have produced a major victory he dismissed carrier force commander Nagumo and ordered yet another battleship bombardment of Henderson to cover large-scale landings at Cape Esperance. On the night of 12 November, 18 Japanese warships, including the battleships *Hiei* and *Kirishima*, swept down the northern coast of the island and collided with an American force of 5 cruisers and 8 destroyers. In pitch darkness the fleets opened fire at point-blank range and in the bloody melee which followed all 18 Japanese ships were extensively damaged and the *Hiei* and two cruisers were sunk. Two nights later the *Kirishima* with 4 cruisers and 9 destroyers again raced toward Henderson and this time was intercepted by a powerful American force. Failing to spot the battleship *Washington* the Japanese concentrated their fire on the destroyers and *South Dakota* and within minutes had sunk or disabled the destroyers and silenced the battleship. Meanwhile *Washington* had closed to within 6.5km (4mi) of *Kirishima* and a devastating broadside sank the Japanese battleship. The fighting on 12 and 14 November (Naval Battle of Guadalcanal) gave the Americans control over the waters of the southern Solomons.

The fighting on the island had not gone well. On the night of 23 October the 20,000 Japanese then on Guadalcanal had stormed the American perimeter and had been repulsed with heavy casualties. As American control of the waters of the southern Solomons became tighter their strength on Guadalcanal grew – 58,000 men by December – while that of the Japanese diminished. On 10 January 1943 the Americans attacked and drove the Japanese toward the western end of the island from where the Imperial Navy managed to evacuate 12,000 survivors during the first week in February. The campaign had cost the Japanese 25,000 dead, the Americans fewer than 3,000 casualties. More importantly, Japan had lost 600 aircraft and 24 warships. American aircraft and shipping losses were similar but were being easily replaced – those of Japan were not.

How splendid the first stage of our operations was! But how unsuccessfully we have fought since the defeat at Midway! Our strategy, aimed at invasion of Hawaii, Fiji, Samoa, and New Caledonia... has dissipated like a dream. In addition, the occupation of Port Moresby and Guadalcanal has been frustrated... In war things often do not turn out as we wish. Nevertheless, I cannot stem my feeling of mortification.

ADMIRAL UGAKI
31 DECEMBER 1942

Datafile

World War I had seen the end of European world dominance, as markets were lost to America and Japan and as New York took over from London as the world's financial capital. By the end of World War II the USA had fully realized its potential for world hegemony. In the early years of the war Japan had clearly been boxing above its true weight. In 1942 it conquered a vast area with 200,000 men – yet by the end of that year the USA had 346,000 troops in the Pacific theater. United States gross national product increased from 88.6 billion dollars in 1939 to 198.7 in 1945 and, between 1940 and 1944, US output of manufactured goods grew by 300 percent.

Manufacturing output 1937

5%
4%
35%
22%
14%
11%
9%

USA
Others
USSR
Germany
Britain
France
Japan

▲ H.C. Hillman's figures for manufacturing output closely match, in the ratios shown, those for arms production in 1942 (below center) which were compiled by a different scholar, Dieter Penzina. The rashness of the Japanese attack on the USA is evident. Even with the UK committed on other fronts, and France and the Netherlands impotent, Japan stood no long-term chance in the Pacific. Meanwhile, Australia industrialized rapidly: machine tool factories rose from 3 prewar to over 100.

World merchant fleet

17%
83%
1939

52%
48%
1947

Rest of world (tons)
US fleet (tons)

► The staggering growth of the US share of world merchant shipping was matched in the military sphere. The UK had abandoned the naval equality with the US established by treaty in 1922. By 1944 the US Navy was roughly three times the size of the British Royal Navy. American submarines destroyed almost 90 percent of the shipping Japan had possessed in 1941.

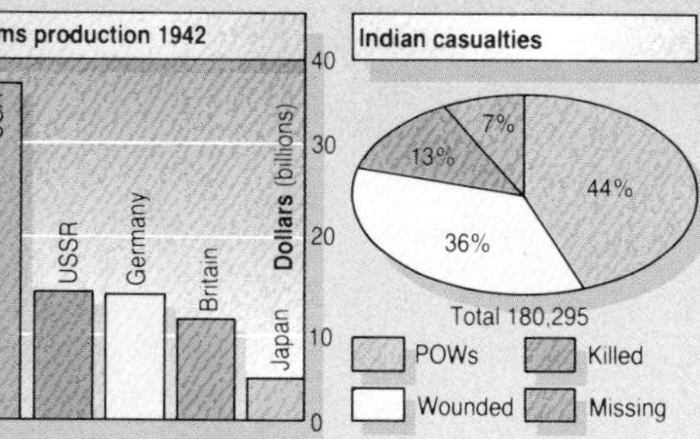

► These figures for arms production underestimate US dominance – US war industry reached top gear only in 1944. Between 1941 and 1944 the rate of Japanese aircraft production increased nearly sixfold to 2,348 per month. But Japanese arms were made by harshly exploited labor at the expense of consumer goods.

Indian casualties

7%
13%
44%
36%
Total 180,295

POWs
Killed
Wounded
Missing

▲ India contributed at peak 2.5 million servicemen to the Allies' war effort – roughly as many as the rest of the British Empire and Commonwealth, bar the UK itself, combined. Indian troops fought bravely in North Africa and Europe, while of 27 Victoria Crosses awarded for the Burma campaign, 20 went to Indians. Among Indians captured, about 4,000 out of 12,000 held in Germany and at most 13,000 out of 60,000 in Malaya-Singapore, joined the Indian National Army formed by Subhas Chandra Bose to fight alongside Japan. INA men went into the Imphal offensive; 400 were killed, 715 deserted and 1,500 died of disease and starvation.

Battle of Imphal

British IV Corps
British XXXI Corps
Japanese 15th Army
Indians
Japanese reinforcements
Thousands
150
100
50
0

◄▼ In the Battle of Imphal from 29 March to 22 June 1944, the Japanese were aiming to strike into India from Burma. But they were always at a numerical disadvantage (left). British forces were heavily hit by disease. The Japanese fared worse: malnutrition accounted for many of their deaths from disease.

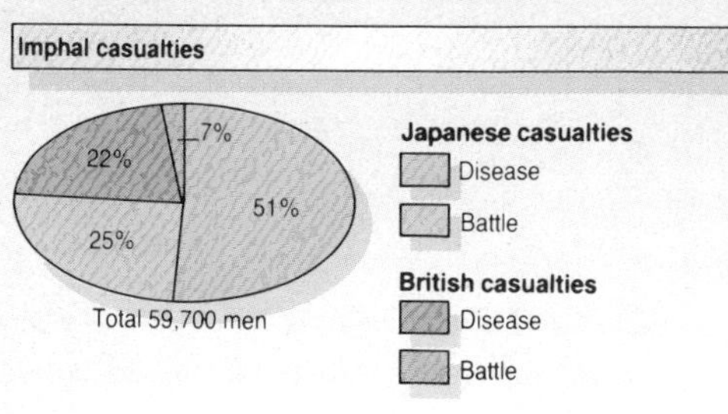

Japanese casualties
Disease
Battle

British casualties
Disease
Battle

The world's longest-lived empire was the Chinese. Since the 3rd century BC one dynasty of emperors had succeeded another. But by the beginning of the 20th century the Manchu dynasty was in decay, undermined by the economic "opening up" of China by the European powers. When it fell in 1911, civil war ensued. The dynamic Japanese, who had proved by their defeat of Russia in 1905 that they could confront a mighty European empire and win, were irresistibly drawn into the vacuum. By the 1930s two leaders had emerged to contend for the honor of restoring the integrity of the empire: Jiang Jieshi, leader of the Nationalist Guomindang, and the communist Mao Zedong. By 1937 Jiang had asserted his authority over the provincial warlords and was presiding over rapid modernization. The communists were reduced to a small rump living in caves at Yan'an in a remote part of northwest China. But Mao was a shrewder strategist than Jiang.

Younger empires from the "barbarian" European world had seized control of Southeast Asia. Three centuries of Spanish rule in the Philippines had ended in 1898, when the USA had taken over. The Portuguese retained only tiny enclaves. But the Dutch had come to govern most of Indonesia. The British, from the 1750s, had acquired piecemeal power over the whole subcontinent of India and moved on in the later 19th century to conquer Burma and control Malaya. Over the same period the French had taken over "Indochina" – Vietnam, Laos and Cambodia.

Contradictions of imperial rule

Briefly these empires had been imposing and profitable, but they were essentially short-lived, based on an anomaly. Four hundred million Indians and 60 million Indonesians outnumbered British and Dutch home populations by eight to one. To govern in the East, thin-spread Europeans had to recruit and train "native" clerks and other subordinates. While Hindu, Buddhist and Islamic traditionalists resented rule by white Christians, young "natives" educated in Western ways acquired ideas which they turned against their white masters. Thus Pandit Nehru welcomed "the Western gift of science" but campaigned against British rule in India, while the Indonesian nationalist Sutan Sjahrir announced that it was only by using "the dynamism of the West" that the East could be released from subjugation. European liberal ideas were in contradiction with colonial tyranny. Nationalism itself was the most potent import, though Europeans often professed to be baffled by Asian adaptations of the idea. How could there be an Indian "nationalism" embracing Hindu, Sikh, Bengali and Punjabi? How could "Indonesia" ever have existed without the

EMPIRES AT BAY

The prewar situation of imperial powers in Southeast Asia

1941–42: the Japanese conquest of Southeast Asia

Pressures on British rule in India

American-Chinese relations

The Japanese offensive in China in 1944

Dutch? Whites failed to understand that the agrarian and urban industries they had introduced had created new social classes detached from tradition, while their roads, telegraphs and railroads had linked hitherto separate communities.

Nevertheless the Dutch, and still more the British, had moved, during World War I and after, to appease nationalism. But every concession of a share of authority to "natives" had reinforced the demand for full independence. The French did not escape by being less generous and eventually reaped the fiercest nationalist whirlwind of all in Vietnam.

During the 1930s the European powers were weakened by economic depression. Imperialist ideology and considerations of prestige, rather than rationality, preserved the European will to dominate politically areas whose assets they could not monopolize.

It was true that the native troops of the Indian army gave the British a reserve of manpower which could be deployed in other parts of the world, but it required 50,000 white soldiers to hold down the country. British exports to India had fallen from 1,175 million rupees in 1914 to 480 million in 1937, and the balance of trade favored India. Britain had lost control of the Indian domestic market to "native" competition. The Dutch East Indies likewise had come to import more from Japan and export more to Europe. The French in Indochina were superficially the most successful. But people in France paid no less for Michelin tyres than the Swiss who had no colonies.

The motor car was becoming a modern necessity. Southeast Asia produced practically all the world's rubber and nearly half its tin. The USA was the chief buyer, Malaya the chief producer, of both commodities. Over the dying empires Japan hovered like a predatory kite, watched from afar by a mighty eagle. The Japanese, modernizing violently and committed to the conquest of China, needed rubber and tin, and the oil and bauxite of the Dutch East Indies, and the rice of Burma and Thailand. In 1942 they showed that they could impose their claim.

▼ A rubber factory in British Malaya before World War II. The fact of white domination is plain enough. Not so clearly disclosed are the evil conditions suffered by "coolie" workers.

► In their conquest of Southeast Asia the Japanese advanced over a tenth of the world's surface at an amazing speed. The story of this campaign, however, masks the divisions and dilemmas of the Japanese command. Should India be invaded? Australia? Should resources be held back for a strike against the USSR? As it happened, the Japanese tide flowed thinly over precisely the area known as "Southeast Asia" before meeting its limit in New Guinea and Melanesia. With the Battle of Midway, June 1942, its ebb in effect began.

But a glance at the datafile shows at once how hopeless the Japanese cause was if the eagle of America moved. The British and Dutch had not believed that the Japanese would dare to take on the USA. Nor did the Americans themselves. Americans professed to be anti-imperialist. The Philippines since 1916 had been governed by reliable Filipinos. In 1935, as if to upstage developments in British India, the USA had promised full independence in 14 years' time.

Yet Churchill, who had proclaimed in 1940 that the British Empire should survive another thousand years, signed in August 1941 the Atlantic Charter which declared the "right of all peoples" to choose the form of government under which they lived. An official information handout for British troops in 1943 announced self-righteously, in line with government professions: "To demands from other Powers that we should 'hand over the colonies' we reply that we are already in the process of handing them over to their only true owners, the Colonial peoples themselves.... Self-government is better than good government."

The British had long before developed the "Dominion" model, by which colonies could become self-governing under the British throne. This made it easier to contemplate decolonization than in France or Holland, where the communists were at one with the political right in believing that their empires must be regained.

▼ Japanese troops hail yet another victory in early 1942. It was a triumph to transform into victorious warriors such bespectacled persons as seen here. An Australian soldier who became their prisoner said later: "Whatever their other qualities might be, to me they are – with envy – the brave Japanese."

The defeat of the Europeans

On 8 December 1941, the day after Japan bombed the US naval base at Pearl Harbor, Japanese forces attacked Hong Kong, Thailand, Malaya and Burma. A series of devastating and rapid advances followed. Four months later the Japanese were masters of Southeast Asia. Organized US resistance in the Philippines continued until May 18 but proved fruitless. Economic ties between European possessions in Southeast Asia and their respective "motherlands" were completely cut, even in those which remained nominally under European rule, such as the possessions of Portugal and Vichy France, which were not at war with Japan. But the Portuguese enclave at Macao on the Chinese coast was dominated by Japanese "advisers", and the Portuguese part of the island of Timor shared the same fate as the Dutch section. In French Indochina the Japanese did what they liked.

They stationed troops in Tongking (North Vietnam) in September 1940 after a Franco-Japanese Treaty acknowledged their "preeminent position in the Far East". The first of a series of commercial treaties the following May provided that all Vietnamese rice, corn, coal, rubber and minerals available for export should go to Japan, and in November all Vietnamese enterprises were ordered to work for the Japanese. Meanwhile Japanese troops had spread through the country. The French lost their last leverage in Vietnam when their troops and administrators were attacked by the Japanese in April 1945. Covertly the Japanese had encouraged local nationalism. Communists led by Ho Chi Minh dominated the Vietminh popular front formed in 1941, built up under Vo Nguyen Giap (a guerrilla leader in the north), and were poised to dominate the whole country when war ended.

The twilight of the British Raj

Part of the mystique surrounding British rule in India was dissipated when thousands of US servicemen arrived in the subcontinent during the war and astonished whites and natives alike by surviving without sun helmets. In sterner ways the war also hastened recognition by Britons that they must go. Denouncing Japanese cruelty implied support for the softer values of liberalism and humanitarianism against the credo that might was right. If Japanese expansionism, like that of Germany and Italy, was to be seen as immoral, where did that leave British imperialism? The UK's allies the USA and the USSR were both professedly anti-imperialist. Under any leader other than Churchill, the British might have conceded Indian independence during the war.

Churchill detested the nationalist Congress Party and its venerated leader Mahatma Gandhi. He would have agreed with Lord Linlithgow, the Viceroy who wrote early in 1939 that British "evacuation in any measurable period of time" was "unthinkable."

Linlithgow personally declared war on behalf of all India in the September of that year. The response of Indians was very varied. Millions volunteered for the Indian army (troops from which could still be reliably used after the war to

The Japanese Conquest of Southeast Asia 1941–42

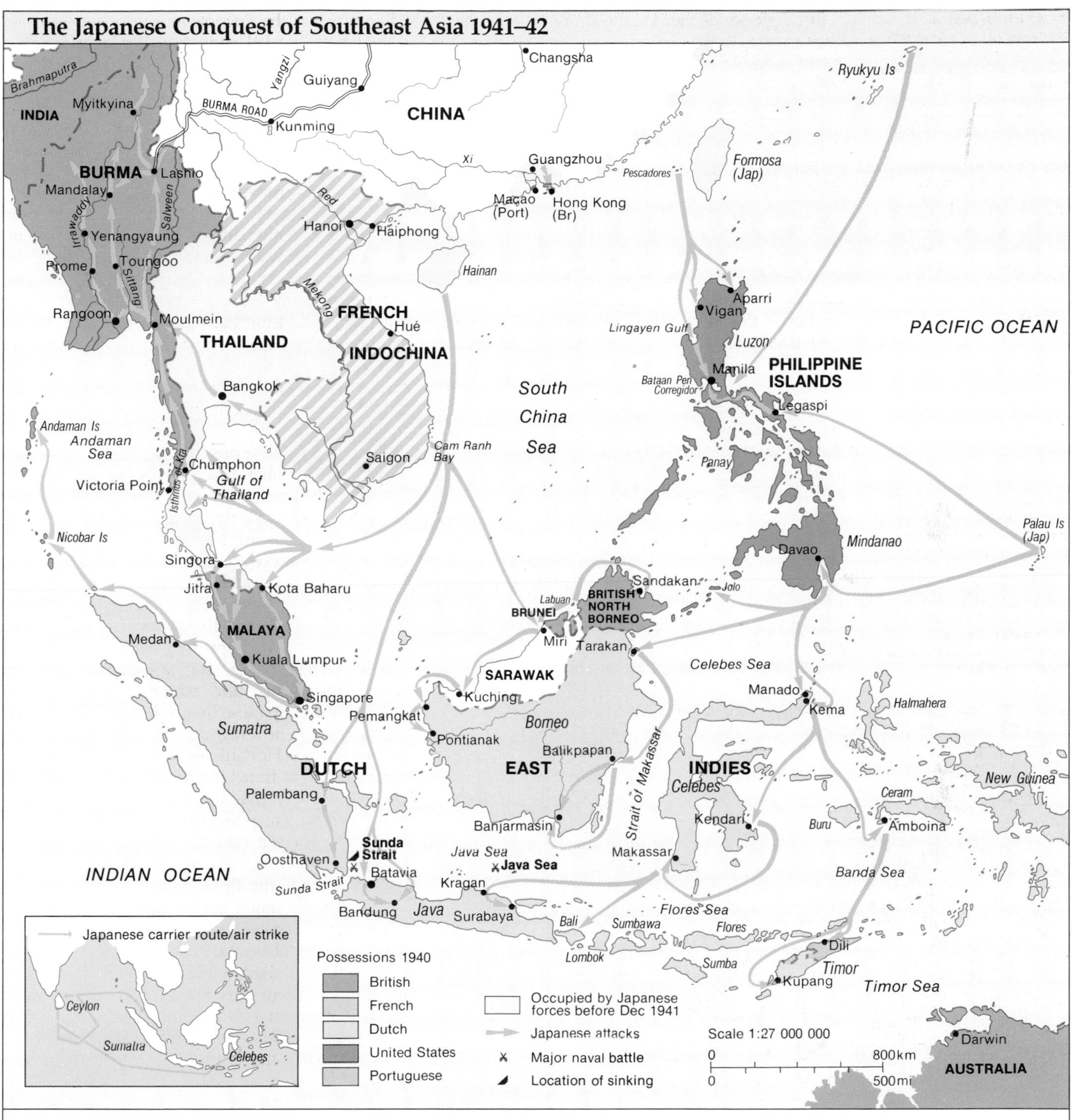

1941

HONG KONG
December 8
Japanese bombard Victoria and land on Dec. 19. The British surrender on Dec. 25.

MALAYA
December 8
Japanese capture Victoria Point airfield in southern Burma (Dec. 16). The main army breaks the British line at Jitra (Dec. 13). The British retreat to Singapore island.

PHILIPPINES
December 10
Conquest of Luzon. Japanese land at Aparri, Vigan (Dec. 10) and Legaspi (Dec. 12). The main force lands at Lingayen Gulf on Dec. 22.

BRITISH NORTH BORNEO, SARAWAK, BORNEO, CELEBES AND AMBOINA
December 16
Units land at Miri and expand along the coast. Others seize Davao in Mindanao (Dec. 20) and Jolo island (Dec. 25). Two forces capture main towns of Borneo and the Celebes. Amboina falls on Jan. 30, 1942.

1942

SINGAPORE
February 1
After an intensive bombardment Japanese land on Feb. 7. The British surrender on Feb. 15.

SOUTHERN SUMATRA, TIMOR AND BALI
February 14
The Japanese isolate Java. On southeast Sumatra they seize airfields and refineries near Palembang (Feb. 16) and occupy Oosthaven (Feb. 20). They bomb Darwin (Feb. 19), and land on Bali (Feb. 18) and Timor (Feb. 19).

JAVA
February 26
Two task forces converge on Java. One defeats an Allied fleet in the Battle of the Java Sea (Feb. 27) and lands troops at Kragan (March 1). The other lands troops near Batavia and sinks two Allied cruisers (Battle of the Sunda Straits March 1.) Bandung surrenders on March 9.

SOUTHERN BURMA
January 20
Japanese cross Thailand and take Moulmein (Jan. 31). Rangoon is occupied on March 8. Meanwhile British troops have withdrawn.

NORTHERN SUMATRA, ANDAMAN AND CEYLON
March 23
Japanese forces occupy northern Sumatra and the Nicobar and Andaman Islands (March 23). They bomb Colombo (April 5) and Trincomalee (April 9).

BURMA
March 30
A Japanese force advances along the Sittang and defeats Chinese at Toungoo (March 30). A second takes Lashio (April 29), then Mandalay. The third follows the Irrawaddy takes Prome (April 2) and the Yenangyaung oilfields (April 19). Chinese and British abandon Burma (April 30).

THE PHILIPPINES
January 1
US and Filipino forces on the Bataan Peninsula surrender on April 9. Corregidor falls on May 6.

help restore the French and Dutch briefly to rule in Vietnam and the East Indies). India, thanks to "native" enterprise, was the world's sixth largest steel producer. Its industry came to supply three-quarters of the needs of British troops in the Middle Eastern theater as well as all its own army's requirements, save tanks and big guns. More than half of the 950 men in the elite Indian civil service were now "natives". Constitutional advance before the war had put 11 huge provinces under the control of elected cabinets of Indian ministers.

But in protest at Linlithgow's unilateral declaration of war, Congress ministers, who controlled 6 of the 11 provinces, resigned. Nor was Congress impressed by Linlithgow's "August Offer" of 1940 when Britain's position in the war against Germany was dire. Indians would be allowed to devise their own constitution when the war ended, and meanwhile Indian politicians would be invited to join the Viceroy's Executive and Defence Councils. The campaign of civil disobedience which Gandhi launched in October led to 14,000 going to jail.

American pressure on Churchill to decolonize merely made him privately more stubborn. The Japanese advance in 1942 made conciliation of Congress seem urgent, but when Churchill sent Sir Stafford Cripps, a member of his war cabinet, to India in March, there is every reason to believe that he wanted the mission to fail. Cripps' offer was to be of Dominion status (equal to Canada's

Australia and New Zealand at War

On the face of it Australia was the most loyal of Dominions. Prime minister Robert Menzies declared war without consulting his Parliament, announcing on radio, "Britain is at war, therefore Australia is at war." But even he found Churchill's hauteur toward Australia hard to stomach, and after he fell in 1941 a Labour prime minister of working-class Irish Catholic origins succeeded him ("bad stock", Churchill muttered privately). Australians had long feared Japan. Yet after Pearl Harbor, Churchill treated John Curtin and his colleagues as if they were panicky children. He was furious when Curtin signed a newspaper article declaring, "Australia looks to America, free of any pangs as to our traditional links with the United Kingdom." But Labour was reelected in 1943.

Australia faced real peril. Towns in the North were bombed. Papua, governed from Australia, was invaded. Nineteen Australians died when Japanese submarines appeared in Sydney Harbor in May 1942.

Meanwhile, Australia was the only possible base for US operations in the Southwest Pacific. Nearly a million US servicemen sojourned in the country. The USA in its turn seemed prone to treat Australian views too lightly. Even Curtin, close to MacArthur, worried about the danger of US domination. On their side, US diplomats continually complained about "Digger" addiction to class struggle and strikes.

New Zealand, on the other hand, remained besottedly loyal to King and Empire. More than a year before Pearl Harbor a Home Guard was created on the British model. New Zealanders deprived themselves, even of potatoes, to send the UK all the food they could. Despite calling up a quarter of all males for military service (who incurred the Commonwealth's highest casualty rate), New Zealand achieved the highest agricultural productivity of any combatant nation. Women were heavily mobilized. For instance, "Women Herd Testers" replaced men, with equal pay.

▲ Australian troops: from a population of 7 million, 550,000 served in the forces overseas. Over 23,000 were killed in action; 8,000 died in Japanese prisoner of war camps. Volunteer soldiers fought bravely and widely. Australian tradition forbade conscription for overseas fighting, but Curtin managed to get a compromise through parliament – conscripts could be sent to a restricted "Southwest Pacific Zone" which did at least include all of New Guinea.

◀ India contributed at peak over 2 million servicemen – roughly as many as the rest of the British Empire and Commonwealth combined. Of these 24,338 were killed, 11,754 declared "missing", and 64,354 were wounded. Among nearly 80,000 captured, perhaps 17,000 joined the Indian National Army formed by Subhas Chandra Bose to fight alongside Japan. Of 6,000 INA men sent to the front, nearly one third died and a quarter surrendered or deserted.

or Australia's), but only after the Japanese were defeated. Gandhi reportedly called this a "post-dated cheque on a failing bank". Cripps, a peacetime friend of Nehru, exceeded his remit in suggesting that the viceroy's Executive Council with its Indian preponderance might act as a genuine cabinet. But Churchill would not permit this offer to stand. Congress leaders felt cheated. Gandhi called on the British to "Quit India". The Congress leaders were imprisoned. A wave of popular protest swept the country, in which the railway line between Delhi and Calcutta was blown up so that supplies could not reach the Burma front; police stations were sacked; policemen killed; telephone wires cut. About 1,000 people, almost all Indians, died. The British did not "quit", but they were shaken.

A new viceroy, Sir Archibald Wavell, arrived in September 1943, personally determined to press ahead with rapid constitutional advance. But though he released Gandhi the following May, Churchill would not allow him to meet the Mahatma. Not until May 1945 would Wavell get permission to bring into his Executive Council members of Congress and of the Muslim League, which had won less than 5 percent of the Muslim vote in 1937.

The League, led by Mohammed Ali Jinnah, had made much capital out of the situation created by British and Congress intransigence. Committed to establishing a separate Pakistan ("land of the pure") for the 100 million Muslims who formed a quarter of India's population, it had been gratified by Cripps' pledge that any province could opt out of the proposed Indian Union. Churchill had allowed this because he wanted to upset Congress and cement the loyalty of the Muslims who were disproportionately numerous in the Indian army. His refusal to move decisively towards the Dominion status for India which had been clearly foreshadowed in 1935 helped to ensure that when independence finally came, the subcontinent would be divided.

Meanwhile the war brought horror to Bengal. Matters had been so organized that this province had been spared famine for 60 years. But the Japanese occupation of Burma cut off supplies of rice. The pressure of military demands on the rail road system made Punjab rice hard to bring in. Nehru claimed that distribution problems had been exacerbated by the panicky destruction of thousands of river craft lest these should fall into Japanese hands. The famine killed 3.4 million people. Such a disaster struck at the claim that the British deserved credit for governing wisely in the natives' interests.

Japan renews the conquest of China

The oldest empire would prove unconquerable. The Chinese people would eventually sort out their own destiny. Meanwhile the Japanese invasion led to a three-way division of the country. The Japanese controlled the seacoast and the great cities on or near it; but Jiang could not be dislodged from his southwestern capital, Chongqing. Mao's communists made steady gains from their northern base in Yan'an.

American liberals liked to imagine that Mao's Red Army were not real communists, merely "agrarian democrats". In fact they were pursuing a brilliant, ruthless strategy on behalf of a version of Marxism and they also recognized the authority of Moscow – though the USSR was never able to enforce it.

Mao believed that the people were to revolutionaries what water was to fish – the element in which they moved. The Japanese raped and murdered: Jiang's conscripts looted and bullied: Mao's men paid their way, helped with the harvest, respected women and property. This created safe bases. Whoever attacked the Reds attacked the people. The Japanese, pushing forward, left large areas to their rear thinly garrisoned. The Reds moved in their troops and political organizers, set up self-governing authorities and organized militias. Grateful peasants were given the holdings of landlords who had fled or collaborated.

The second phase was "strategic stalemate". The Reds wore down Japanese strength by guerrilla action. In 1937 Mao's forces had been confined to some 90,000sq km (35,000sq mi) of arid and poor land in northern Shaanxi, with a population of 1.5 million. By the end of 1943 they effectively controlled 570,000sq km (220,000sq mi) with a population of 94 million. Since the Red Army was waging revolutionary war, clashes

▼ A Japanese poster for Indian consumption. The text, in Hindu and Bengali, reads: "Don't follow the English blindly. Forget the differences in religion and be united to bring about the independence of India." Addressing white opponents, Japanese propagandists were hampered by idiomatic difficulties. (One line read: "The remaining British planes took to their heels.") Their attempts to play on soldiers' sexual neuroses produced unintentionally hilarious results, as in a leaflet which showed a "Slick Yank" fornicating with a Melbourne girl and exclaiming, "Take your sweet time at the front Aussie – I got my hands full right now with your sweet tootsy at home."

with Jiang's troops were not ruled out and did occur. Mao and Jiang watched each other as closely as either did the common enemy. By February 1944 Jiang was deploying up to half a million of his best troops against communist-held areas in the northwest.

The Americans tried to convince themselves that Jiang was a liberal democrat. US General "Vinegar Joe" Stilwell was installed as his chief of staff. The Allies wished to make China a base for offensive operations against Japan, preferably with active help from Jiang's forces. But the Japanese advance into Burma blocked the route *via* Mandalay, the "Burma Road", by which supplies had reached Jiang. The only alternative was an airlift over the Himalayas. It was attempted but not till December 1943 did the monthly tonnage of supplies reach five figures.

Neither the British, preoccupied in Europe, nor the USA, fighting back in the Pacific, would release the forces required to clear the Japanese off the "Burma Road". A three-pronged attack on Burma agreed by combined chiefs of staff in 1943 was abandoned. Instead a more limited operation was devised to drive a new road south from Assam to join the northern section of the old one. In August 1944 Stilwell, with two American-trained Chinese divisions based in Assam, succeeded in establishing the new "Ledo Road". Overland traffic to China by this route began in January 1945.

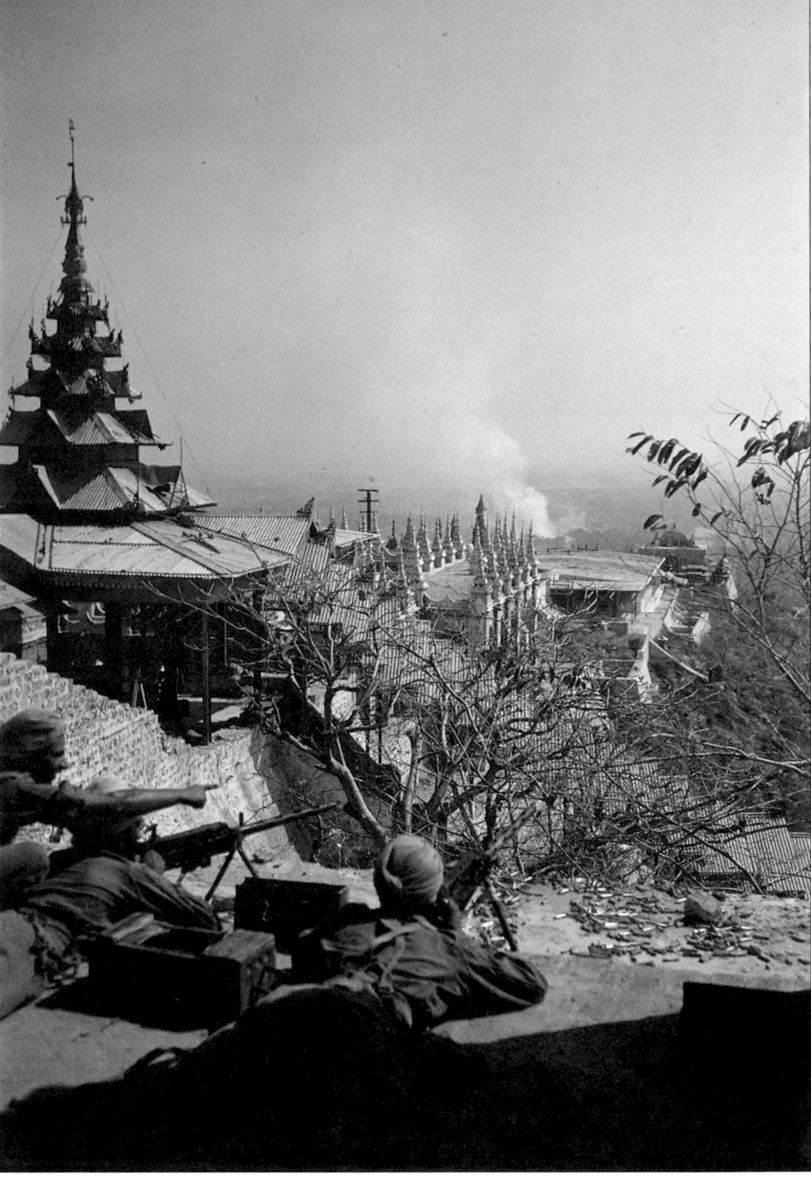

Japan's last campaign in China

In April 1944 the Japanese had begun a huge campaign in China. By now their aim was essentially defensive, to remove China as a base from the Americans. "Operation Ichi-Go" committed 1.8 million Japanese troops against Jiang's 6 million. It showed up in a devastating light the corruption and ineptitude of the Nationalists. Henan was a province ravaged by famine due to a long drought after Jiangs's government had commandeered most of the 1942 grain crop for taxes. Millions had died of hunger and disease. When the Japanese struck, Chinese units in Henan were under strength, so that officers could pocket pay for nonexistent soldiers. The commander and his men had been heavily involved in black market trade with the Japanese. The Chinese 12th and 13th armies fought each other as they fled. Local peasants with pitchforks disarmed 50,000 Chinese troops. In three weeks 60,000 Japanese smashed a Chinese force of 300,000 and reached their objective, the northern section of the railroad. By late November the US 14th Air Force had been deprived of six bases and the Japanese were within 320km (200mi) of Chongqing. Jiang, however, was still no pushover. He gathered all the troops he could, including some flown from the Burma front, halted the Japanese at Kuizhou and managed to drive them back.

Chongqing, its population swollen from 200,000 to over a million, was a theater of intrigue and graft. US Lend-Lease became a basis for rampant corruption. Inflation soared to astronomic levels. The peasantry had been taxed to the limit, so the government simply printed more notes. The peasants, salaried classes and intellectuals

◀ Indian machine gunners on Pagoda Hill during the battle for Fort Dufferin in Mandalay, March 1945. The city was taken by the British after a fierce 12-day siege. The British campaign in Burma aimed to restore British prestige in Asia.

◀▼ At a base in China (bottom left) one of Jiang's soldiers guards US fighter planes. The "Flying Tigers", commanded by General C. L. Chennault, were flamboyantly successful against the Japanese.

▶ A confident Guomindang poster: "The more we fight, the stronger we get. The more the enemy fights, the weaker he gets".

▼ Chongqing in 1941, damaged by Japanese air raids. In this formerly quiet trading town at the confluence of the Yangzi and Jialin rivers, "Vinegar Joe" Stillwell would seethe over the behavior of Jiang Jieshi, whom he privately dubbed "the Peanut Dictator".

were all alike alienated from Jiang's regime. However, 27 Japanese divisions – a third of the whole army – were tied up in China in 1943 and 1944. In this way Jiang served the Allied cause.

US policy toward China was based on misapprehensions. Traditionally the USA favored what was called the "Open Door" policy – all nations should have equal economic opportunity within China and the territorial and administrative integrity of the country should be respected. In 1943 the USA relinquished "extraterritorial" rights which had given its citizens special privileges in China. The desire to see Jiang and Mao as Jeffersonian democrats was a reflection of this idealism. Roosevelt strove to bestow "Great Power" status on China. Jiang was one of the "Big Four" at the Cairo conference in November 1943 where Roosevelt committed himself to returning Formosa (Taiwan) to China and preventing any Soviet seizure of Chinese territory in the north. In return he got Jiang's promise that he would reach a settlement with Mao. Roosevelt did not keep America's side of the bargain, nor Jiang his. Nevertheless in 1945 China became one of the five permanent members of the Security Council of the new United Nations.

THE FALL OF SINGAPORE

Winston Churchill called it "the worst disaster and biggest capitulation in British History". One of his fiercer critics, the military theorist Basil Liddell Hart, blamed the debacle on Churchill himself and accused him of not really expecting danger, of underestimating the Japanese.

Most British officials, soldiers, planters and businessmen in Asia had underestimated Asiatic peoples in general. Their racialist contempt for "little yellow men", against whom their clubs maintained a color bar, was unlikely to foster loyalty to the British Empire among Chinese and Malay subjects. Their power depended not on their subjects' love, but ultimately on the Royal Navy. Whites had faith in the great Singapore naval base, completed in 1938 at a cost of over 60 million pounds, the largest single construction project carried out in the British Empire between the wars. It symbolized the UK's will to remain in great power in the East and to guarantee the safety of kith and kin in Australia and New Zealand. But its guns faced seaward; no use against land attack.

British behavior in Penang (in northwest Malaya) was symptomatic. Even after the Japanese, on Pearl Harbor day, invaded the Malayan Peninsula, British inhabitants complacently flocked to the bar of the Eastern and Oriental Hotel. But soon air raids brought terror and chaos. Whites – and whites only – were evacuated in haste. Almost all British officials, doctors and nurses withdrew, leaving Malay and Indian subordinates to make terms with the conquerors and serve the sick.

Refugees poured into Singapore, whence key skilled personnel, women and children were evacuated by ship. By the New Year of 1942 the last British troops from the mainland had crossed to the island. The Japanese followed them. By 14 February, Singapore City was closely besieged, with perhaps a million people crammed within a radius of 5 km (3mi). Water supplies, damaged, would soon run out. Next day the commanding officer, General Arthur F. Percival, surrendered. The largest army ever assembled by Britain in the Far East, 130,000 British, Empire and Commonwealth troops, became prisoners of war of 50,000 "little yellow men".

Singapore's population was mainly Chinese in origin, and these "yellow" men could expect no joy from Japanese rule. Nor did capturing Singapore win the war for Japan. But the facade of Western imperialism had been blown away like balsa wood. A high US official had recently warned that if Singapore fell it would "lower immeasurably the prestige among Eastern peoples of 'the white race', and particularly of the British Empire and the United States". So it proved. In India the nationalistic *Bombay Chronicle* suggested in December 1941 that the Western powers would punish Japan. By March it was sneering at "blunders and inefficiency" in the defense of Singapore and hailing "heroic" China as Asia's hope for the future.

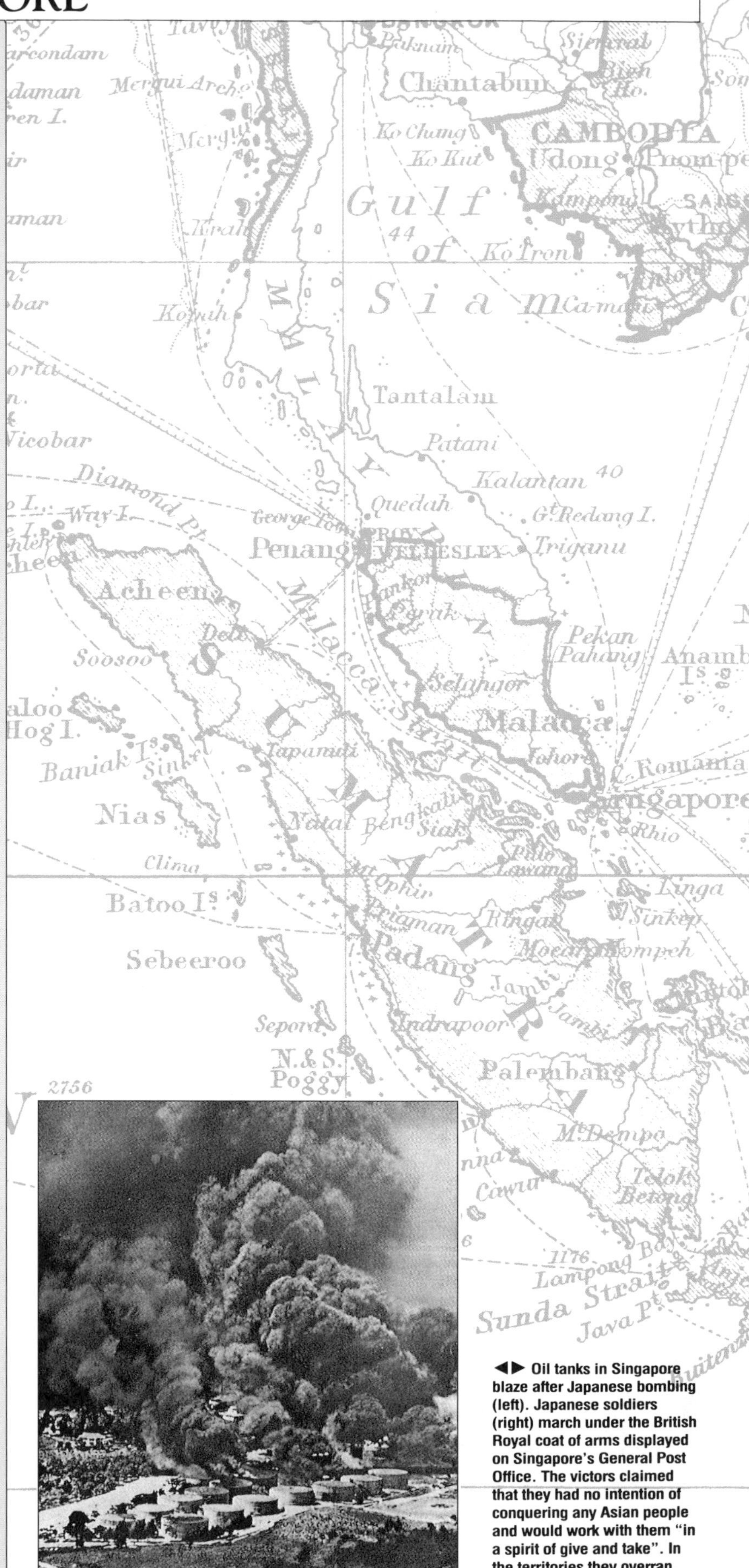

◄► Oil tanks in Singapore blaze after Japanese bombing (left). Japanese soldiers (right) march under the British Royal coat of arms displayed on Singapore's General Post Office. The victors claimed that they had no intention of conquering any Asian people and would work with them "in a spirit of give and take". In the territories they overran, Chinese in particular learnt otherwise.

▲▶ Singapore provided ample pleasures for whites such as the soldiers who sit (right) on the Esplanade by the harbor with its Chinese fishing vessels. In the celebrated Raffles Hotel (above), as the Japanese advanced, whites danced to Dan Hopkins' Band. Right up to surrender, there was plenty of food – even fresh meat, from Chinese-kept pigs. To thwart Japanese thirst, a million and a half bottles of liquor were destroyed.

◀ An early 20th-century map clearly shows the nodal position of Singapore in Asian sea-routes.

Datafile

The German invasion of the Soviet Union, launched the greatest land war in world history, involving the Red Army in 1,418 days of war on battlefronts which varied from 3,000 to 6,200km (1,860–3,900mi). Red Army campaigns destroyed, captured, or disabled 607 enemy divisions, more than three times the total of German and satellite divisions destroyed in North Africa, Italy and western Europe.

Soviet losses

Millions

5 4 3 2 1 0

1942 1943 1944

▲ The Red Army suffered horrendous casualties in the course of the Soviet-German War. During the very first three weeks of war the Red Army lost 98 divisions, destroyed or disabled. The German army took over 3 million prisoners, many destined to die from starvation or execution, by the end of 1941.

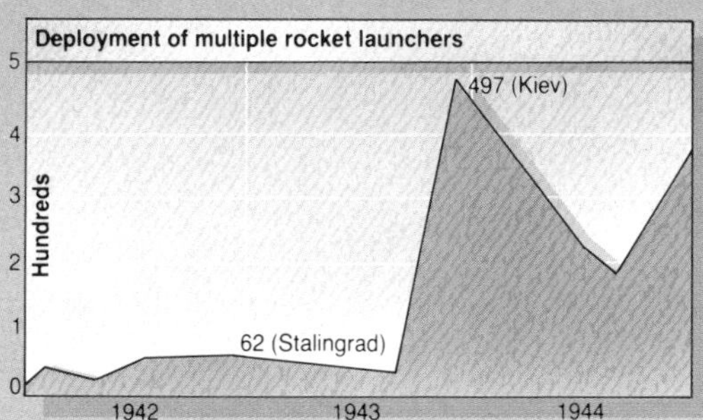

▲ The Red Army's "wonder weapon" was first used in July 1941 with devastating effect upon unsuspecting German troops. The BM-13 MRL (Multiple Rocket Launcher, popularly known as "Katyusha") fired 16 132mm caliber missiles, each weighing 42.5kg with a range of 8,470m, in 7–10 seconds.

▶ The initial Soviet numerical superiority in tanks was restored slowly after the devastating losses of the 1941 battles. At Stalingrad in November 1942 the odds shifted in Soviet favor as new tanks rolled out of the Urals factories, while at Kursk in July 1943 growing Soviet strength brought new "tank armies" into the field.

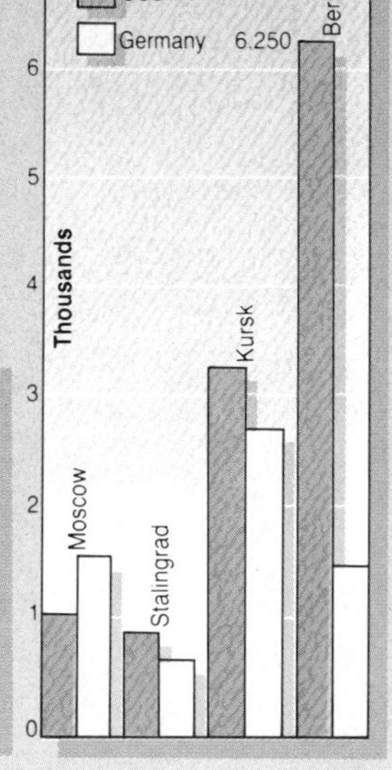

▶ German forces in the East reached their peak in winter 1942 at the giant Battle of Stalingrad. There were constant Soviet complaints that in the absence of a "Second Front" in the west the German command was free at critical moments to transfer divisions from western Europe to bolster the Eastern Front.

Axis deployments

Divisions

300 250 200 150 100 50 0

266

Eastern front

Other fronts

1942 1943 1944

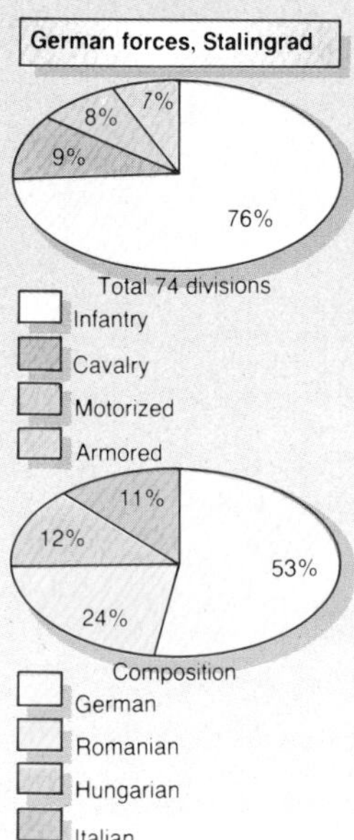

◀▶ In the critical battle for Moscow at the end of 1941, neither side enjoyed any great superiority in manpower, the German forces near to exhaustion and the Red Army drawing on very limited reserves. Both sides were fighting near to the limits. The same condition prevailed at Stalingrad on 19 November 1942 when the Red Army launched its counter-offensive, 1,000,500 Soviet troops opposed to 1,011,500 German and Axis troops, though the 28 Axis divisions (Italian, Romanian and Hungarian) were militarily inferior to Soviet and German forces. These supporting divisions could not protect the German flanks and it was here that the Red Army struck first.

Battle forces

USSR

Germany

Millions

2.5 2.25 2.0 1.75 1.5 1.25 1.0 0.75 0.50 0.25 0

Moscow Stalingrad Kursk Berlin

At 0300 hours on Sunday 22 June 1941 Adolf Hitler unleashed 3.5 million German soldiers, supported by 3,680 tanks and 2,770 combat aircraft, to execute Operation "Barbarossa" – the code name for the massive surprise attack on the Soviet Union. He initiated nothing less than the greatest land campaign in the history of the world.

Planning for the invasion of the USSR had begun shortly after the collapse of France in June 1940, the Soviet-German Nonaggression Pact signed in August 1939 presenting no obstacle whatsoever. On 5 December 1940, in the wake of foreign minister Vyacheslav Molotov's ill-starred November visit to Berlin, Hitler issued "Directive 21", "Case Barbarossa", deliberately invoking a medieval memory (the nickname of the 12th-century emperor Frederick I) for a war to be waged with deliberate medieval cruelty. Early in 1941, to those who even hinted at the magnitude and the risks involved, Hitler implacably declared his intention to fight and "to flatten them [the Russians] like a hailstorm".

Hitler's attitude to Stalin and the USSR

Hitler had long regarded Bolshevism not only as an ideological enemy but as a deadly physical menace from which Germany and indeed the whole of Europe must be saved, by ruthless eradication of the whole system and its Stalinist satraps. Barbarossa embodied not only a military campaign but an ideological crusade given added ferocity by the infusion of Nazi racist creeds, seeking the extermination of the Slav *Untermensch*, the Slav "subhuman", whose fate was determined by the "special guidelines" developed side by side with the military plans. The ideological struggle must proceed at the same time as "the clash of arms". Heinrich Himmler's SS would undertake "certain special duties" including the annihilation of Bolshevik commissars and the Communist intelligentsia". Hitler personally commanded that his senior officers put any ideas of "soldierly comradeship" out of their minds in this war.

For Hitler Stalin represented only a "cold-blooded blackmailer" with whom he intended to finish. Soviet policy, ie Stalin's attitude, bespoke menace, as he invested eastern Poland as his "share" of the Soviet-German Pact, moved into the Baltic states (Estonia, Latvia, Lithuania) with more attendant cruelties in June 1940 and in the spring of 1941 resisted German policies and actions in the Balkans. Worse, in late March and early April 1941 Stalin appeared to be carrying through a covert mobilization of his Red Army on the western frontiers of the USSR.

In the three years before the invasion Stalin had attempted to prevent isolation. To make an alliance, however, was a difficult proposition. In

THE EASTERN FRONT

1938 British and French appeasement of German ambitions had aroused Stalin's suspicion. In August 1939 Soviet negotiations with the British and French for military cooperation against Germany collapsed. Stalin feared that the USSR might be left to fight Germany alone. So when Nazi Germany offered Stalin a nonaggression pact he could scarcely refuse, especially as the agreement also gave the USSR a free hand in eastern Europe and Finland. In 1941, while German troops pushed into the Balkans, Stalin carefully insured himself against a two-front war by concluding the Soviet-Japanese Neutrality Pact on 13 April 1941. Both dictators played games of the utmost perfidy, Hitler out of manic hostility, Stalin from arrogance, conceit and misjudgment. Though he received multiple warnings of a German attack, Stalin persuaded himself that Hitler would never embark on a two-front war. In the west the UK remained undefeated. Hitler suspected an Anglo-Soviet "plot". Stalin feared British intrigues either to entangle him in war with Germany or else to agree on an Anglo-German separate peace, leaving Hitler a free hand in the east. The flight of Rudolf Hess to Scotland (10 May 1941) only intensified this particular paranoia. The warnings went unheeded.

▼ Motorized German troops in Russia, 1941. The 148 invading German divisions inflicted massive defeats on the Red Army. The German Panzer divisions hopelessly outclassed the Soviet tank forces. German infantry accomplished prodigies of fighting and marching.

The Soviet-Finnish War

The Soviet-Finnish War, also known as the "Winter War", lasted from 30 November 1939 to 13 March 1940. Four Soviet armies attacked Finland. Soviet troops advanced to the defensive positions of the Finnish "Mannerheim Line" on the Karelian isthmus but were unable to reduce the Finns. A second phase, from the end of December 1939 to 12 March 1940, culminated in a massive but costly Soviet effort to smash through the "Mannerheim Line". The Finns had to sue for peace, but Soviet aims could not be realized in full.

The war cost the Soviet Red Army 200,000 dead, the loss of 1,600 tanks and 634 aircraft. No less than 1.2 million Soviet troops were employed in what General Zhukov later called the "acid test", a test which both the Soviet leadership and German intelligence believed the USSR had failed. In mid-April 1940, on the orders of the Politburo, the Main Military Soviet met to discuss the lessons of the Winter War and the serious shortcomings in Soviet performance. An immediate consequence was the removal of Voroshilov as defense commissar and his replacement by Timoshenko whose Order No. 120 of 16 May 1940 listed Soviet failures. Officer training, especially that of junior officers, needed drastic improvement, as did that of general troop training for hard fighting. Timoshenko wanted an army realistically trained and better equipped, able to fight like the Finns. Iron discipline must also go hand in hand with intensive training.

Based on Soviet performance against the Finns, German intelligence concluded that the Red Army was simply not capable of fighting a modern war. In particular, the damage inflicted in 1937–38, when Stalin purged the Soviet command, could now be seen and (in one view) could take 20 years to repair. This gigantic military machine simply operated in the mass: organization, leadership and equipment were unsatisfactory. The principles of leadership were sound, but the commanders inexperienced. The communications system and troop transportation were bad, the fighting quality of Soviet troops in heavy combat doubtful. Russian tanks were badly armored, the Red Army "leaderless".

To some degree the views of both sides, Soviet and German, coincided. Both, however, drew different disastrous conclusions: the Soviet side that the damage could be quickly repaired, the German that it would never be. German perceptions of Soviet weakness added to the case for invading the USSR. Soviet actions to improve the Red Army subsequently contributed to holding and defeating the German army.

The Eastern Front 1941–42

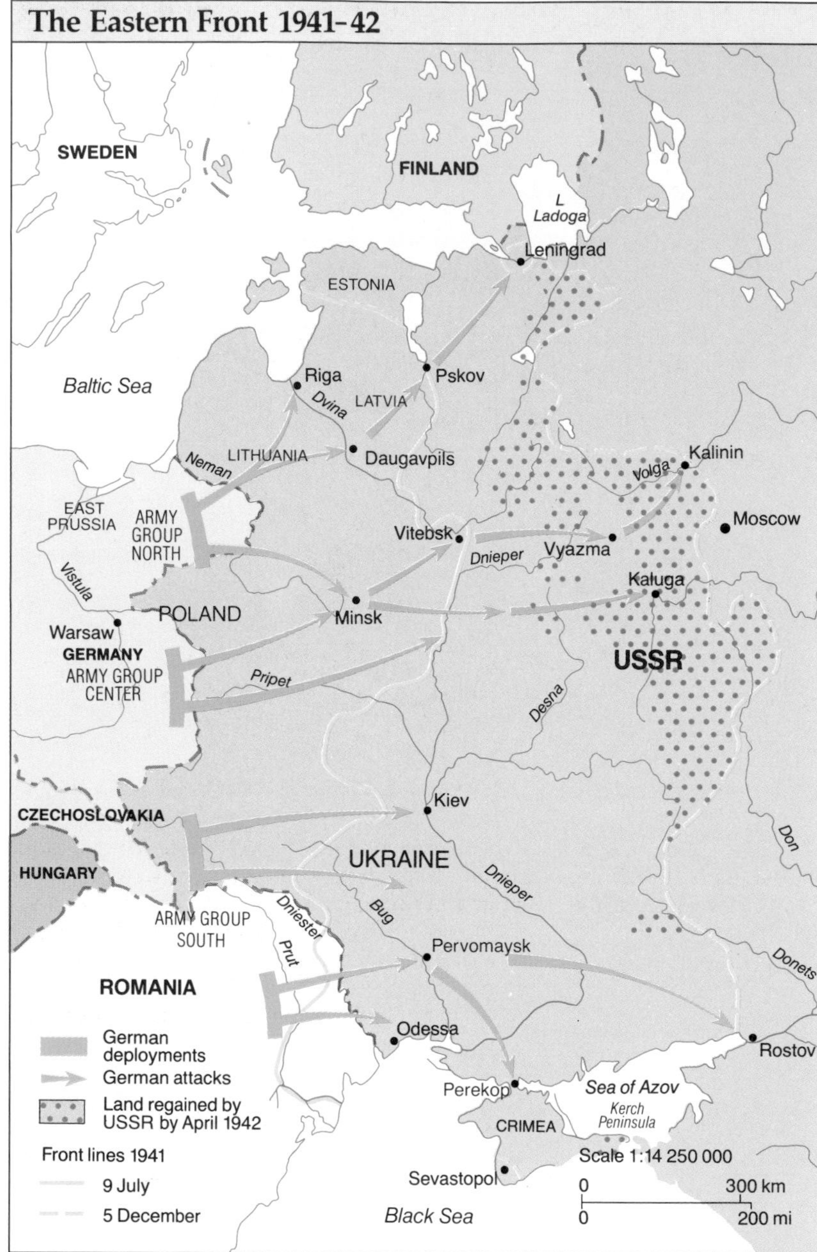

1941

June 22
(to July 9) The first phase of Barbarossa, the German invasion of the USSR. Hitler launches three army groups on a Blitzkrieg offensive: Army Group North, toward Leningrad; Army Group Center, toward Moscow: Army Group South toward Kiev and the Ukraine. Group North captures Daugavpils on June 26, Riga on June 30 and Pskov on July 9. Group Center encircles Soviet troops west of Minsk and captures Vitebsk on July 9. Less progress is made in the south.

July 9
(to Dec. 5) The German attack on the cities and advance into the Ukraine. In the north, German armies advance quickly toward Leningrad which they reach on Sept. 8, but they cannot breach the city's defenses and are forced to begin a siege. In the center, armies approach Moscow from the west and southwest. The Germans take Vyazma on Oct. 7, Kaluga on Oct. 12 and Kalinin on Oct. 14. But they have not progressed fast enough: Moscow's defenses are being strengthened and the Russian winter has begun. In the south German forces capture Kiev (Aug. 14) and sweep through the Ukraine and into the Crimea (though Sevastapol holds out). They capture Rostov at the mouth of the Don on Nov. 21, but are forced out eight days later.

December 5
(to Jan. 5 1942) The Soviet counteroffensive. To relieve pressure on Moscow, Soviet forces attack on the Kalinin Front (north of Moscow) and on the West and Southwest Fronts (south of Moscow). Army Group Centre is forced back. The Red Army retakes Kalinin on Dec. 16 and Kaluga on Dec. 30. In the Crimea Soviet troops begin to land on the Kerch Peninsula (Dec. 26) to relieve Sevastapol.

1942

January 5
(to March 31) The Soviet Winter offensive. Following the success of the initial counterattacks Stalin decides to launch an offensive on all Fronts. The Red Army makes gains in the north and central areas but is stalled in the south. Eventually the offensive peters out. (Kerch is held until May 1942, after the loss of which Sevastapol falls to the Germans, in June.)

Josef Stalin 1879–1953

Secretary-general of the Soviet communist party from 1922, Stalin ran the Soviet war effort as an extension of his personal dictatorship through a centralized machine. The state defense committee, established on 30 June 1941, under Stalin's direction forced through the total mobilization of the country, while the *Stavka* (high command) and the general staff with the Front commands under his control ran military operations. After the resounding Soviet victory at Stalingrad Stalin assumed the title of marshal of the Soviet Union but took greater account of military advice. As a military commander he was mediocre, even less than competent; as a war leader, associated with Churchill and Roosevelt, he showed qualities of determination, clear-sightedness and a stubborn defense of Soviet interests.

Germany's threefold drive into the USSR

For the attack in the east the German command mustered 153 divisions (including 19 Panzer divisions) divided into three main Army Groups, North and Center deployed north of the Pripet Marches, and South, aimed at Kiev. Additional support came from Romanians, Italians, Slovaks, Hungarians and latterly the Spanish. Lined up along the huge front, half mobilized, imprisoned within Stalin's orders to avoid "provocations", the Red Army deployed 170 divisions supported by a huge tank park of 17,000 machines, many of them aging or unworkable, and 8,000 antiquated aircraft. The battle-tested German divisions tore through these badly deployed and poorly commanded Soviet forces. In the north Leningrad was closely invested, to suffer 900 days of ghastly siege conditions. At the center the Soviet Western Front caved in: only in the south did Army Group South have to fight its way through Soviet armor, which was trying to hold the Germans from Kiev.

The German offensive inflicted vast losses on the Red Army – 98 of the original force of 170 Soviet divisions disappeared from the order of battle, destroyed or savagely mauled. Corps vanished, divisions shrank to brigades. With millions made prisoner by the Germans, Stalin's own son among them, Red Army strength plummeted to its lowest wartime strength in November 1941. Poorly trained, barely armed "worker-militias" took up positions to face crack German units. Leningrad and Murmansk were under constant threat and fire. Army Group Center had smashed through Soviet reserves in October and was driving for Moscow, which was in the grip of evacuation. Stalin, now self-appointed supreme commander, stayed, charging General Georgii Zhukov with the defense of the capital. In the south Kiev had fallen and with it much of the Ukraine, the huge Soviet "breadbasket". Industrial production dipped disastrously with plants either lost to the Germans or dismantled in a gigantic improvised industrial evacuation to the deep interior, which put 80 percent of Soviet equipment "*na kolesakh*", "on wheels".

By the beginning of December 1941 the Red

▶ Soviet troops on the Volkhov Front, about 400km (250mi) northwest of Moscow. On 5 December 1941 the Red Army attacked the flanks of the German armies pressing on Moscow.

◀ On 22 June 1941 three huge German Army Groups with four tank armies in the lead invaded the Soviet Union, advancing several hundred kilometers within weeks, blockading Leningrad, occupying the Ukraine and closing in on Moscow itself. The Soviet counterblow at Moscow in early December 1941 developed into a major counteroffensive which continued until the spring of 1942.

Army was again filling out with men and even machines. Red Army manpower had climbed back to a figure of 4,196,000. The recovery from the frightful losses had been made possible largely by the mobilization of fresh reserves from the interior and the reorganization of existing armies, from which specialist troops and particularly artillery assets were withdrawn to build yet more reserves. Already 124 divisions had been disbanded due to heavy losses and reformed, in many instances as rifle brigades. The formation of 286 new rifle divisions proceeded apace, while in deepest secrecy special reserve armies were carefully husbanded in and around Moscow. At the beginning of November 1941 Soviet reserves consisted of three armies; at the beginning of December that figure rose to eight armies with 44 rifle divisions and 14 cavalry divisions, plus a further 13 rifle brigades. And the first tanks were coming from the relocated factories. Already two limited counterstrokes, at Tikhvin in the north and Rostov in the south, had registered some success. Stalin literally hoarded reserves in the Moscow area, 30 brigades with 9 reserve armies forming. The "Battle for Moscow" now took a new turn, even as some German scouts could see the Kremlin towers. During December 5–6 a limited counterstroke from the Kalinin and Western Fronts designed to fend off an immediate threat to Moscow suddenly rolled against weakened and extended German forces and continued to roll. The Russian recovery had begun, assuming the form of a counteroffensive and resulting by 13 December 1941 in a Soviet communiqué announcing the repulse of the German attempt to encircle the capital.

Now Soviet ambitions, in particular those of Stalin, overreached themselves. On 17–18 December Stalin ordered the Volkhov and Northwestern Fronts to strike against Army Group North; and amphibious landings on the Kerch peninsula aimed to relieve Sevastopol. Against the advice of General Zhukov and Marshal Shaposhnikov, Stalin contemplated a vast multifront counteroffensive against all three German Army Groups. Rather than concentrating on Army Group Center, Stalin planned a great "war-winning" attack to recover all he had lost. War production, not yet fully recovered, could not sustain such a scale of operations, but Stalin brooked no argument. Late in January 1942 Soviet armies made deep penetrations but lacked the resources to strike for distant objectives. German resistance increased. In the south the Soviet offensive stalled. In March 1942 Soviet offensive operations ceased, Soviet forces having in the previous three months attacked

Heinz Guderian 1888–1954

A German pioneer of armored warfare, Guderian commanded the 2nd Panzer Group and the 2nd Panzer Army in the invasion of the USSR. For an unauthorized withdrawal in December 1941 Hitler retired him. He was recalled as inspector general of Panzer troops in 1943. Guderian opposed the German offensive against the Kursk salient in order to conserve German tanks but was overruled. Hitler dismissed him in March 1945.

Aleksandr Vasilevskii 1895–1977

The "brains" behind most of the successful Soviet operations, Vasilevskii began the war as deputy chief of the general staff and head of operations. He became chief of the general staff in June 1942. However, much of his time was spent at the Front as a "*Stavka* representative", planning and coordinating operations. In February 1945 he took over a direct front command and was made a member of the *Stavka*.

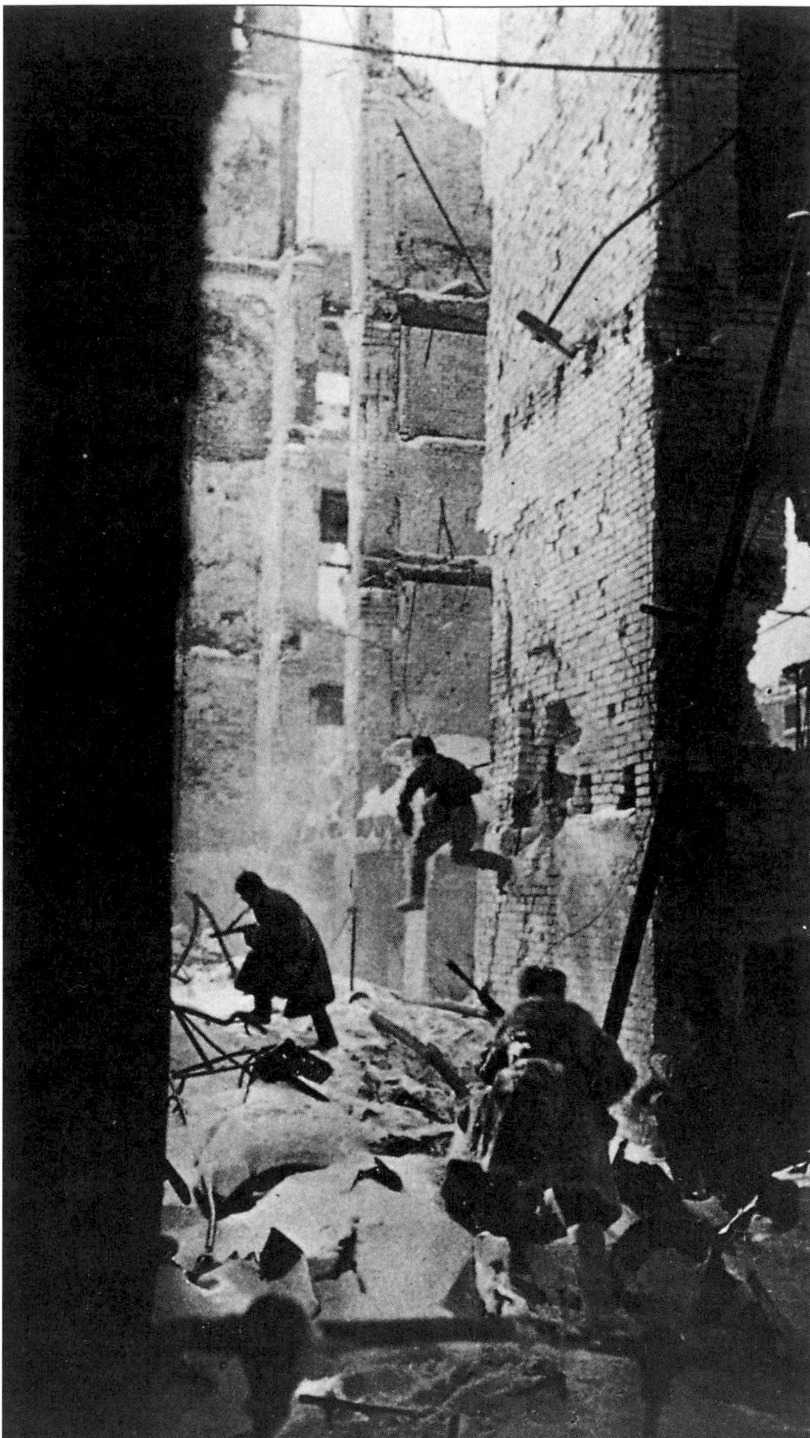

▲ **The epic battle of Stalingrad engulfed more than 2 million men, trapping the German 6th Army in ferocious fighting inside the ruined city. Day and night battles raged for single houses, cellars and factory floors. Chuikov's 62nd Army clung on desperately to tiny bridgeheads on the western bank of the Volga, denying the Germans full control of the city. On 19 December 1942 the Soviet counteroffensive opened, finally trapping the 6th Army.**

across 1,600km (1,000 mi) of front, destroyed or damaged 50 German divisions, cleared the Moscow and Tula regions and in places penetrated to a depth of 320km (200mi).

After the German invasion of the USSR the British prime minister, Winston Churchill, had offered British assistance for the USSR, which was confirmed in an agreement signed on 12 July 1941. Additional support was provided by the USA, especially after Hitler declared war on the USA on 11 December 1941. From the outset Stalin adopted a peremptory tone with his new-found allies. His stated war aims included the recovery of the USSR's frontiers of 1941, which included eastern Poland. This demand caused difficulties for the UK, which had gone to war against Hitler because of his violation of Poland. Stalin also demanded a "Second Front", a cross-Channel attack against German forces in western Europe, but Allied strength was insufficient. Stalin repeated his demand in 1942 and 1943, but his pressure was resisted and rebuffed. He took this as almost a breach of faith and his innate suspiciousness was heightened.

Hitler changes direction

By November 1941 it was clear that the German Blitzkrieg had failed. German planners now worked on the details for an offensive in summer 1942. Hitler now decided to go for the economic jugular rather than the political heart. A southern campaign would trap Soviet forces and be followed by a drive into the Caucasus and an assault on Leningrad. The main objective was the Volga river, to cut a Soviet lifeline, but it was the oil of the Caucasus which drew Hitler on and he intended the "Stalingrad operations" merely to protect the deep drive to the oil. German armies would make the main attack, the Axis "satellite" armies would hold the flanks. In spring 1942, as the armies of both sides released themselves from the clinging spring mud, the *rasputitsa*, Stalin also contemplated offensive operations, with the object of recapturing Kharkov. Marshal Timoshenko would strike from the Izyum salient won in January, but it would be a risky move and was criticized by Marshal Shaposhnikov.

Here was a prelude to further disaster. German infantry held the front while a Panzer group attacked the flank and encircled Timoshenko's forces, losing a quarter of a million men taken prisoner, plus more than 1,000 tanks. On 18 May 1942 the German Spring offensive opened and carved its way into the eastern Ukraine and on to the upper Don. Soviet defenses, weakened by such loss, fell apart. The crisis which produced Stalin's draconian order No. 227 of 28 July 1942: "Not a step back". Persuaded now that the German attack in the south was not a feint to draw Soviet forces away from Moscow, Stalin ordered his best commanders – Zhukov, Vasilevskii and Voronov – southward. For all of Stalin's alarm and resort to "the discipline of the revolver", the Red Army was neither finished nor in hopeless flight, a view entertained quite mistakenly by Hitler. On the contrary, it was the German 6th Army under von Paulus which was marching to its doom, toward Stalingrad. At this juncture, in mid-July 1942, Hitler split his forces, dividing Army Group South in Group "A" and Group "B", the former to drive into the Caucasus, the latter with the 6th Army to invest Stalingrad. After an agonizing siege Sevastopol fell to von Manstein's 11th Army on 3 July whereupon Hitler transferred the 11th Army to Leningrad. The scene was almost set for the terrible drama of Stalingrad.

German attempts to capture the city began on 19 August, 1942. Four days later German troops reached the Volga north of the city. Luftwaffe bombing set the city ablaze only to have the defenders set up defensive positions in the ruins. Hand-to-hand fighting ensued for the major buildings and installations – the Tractor Factory, Red

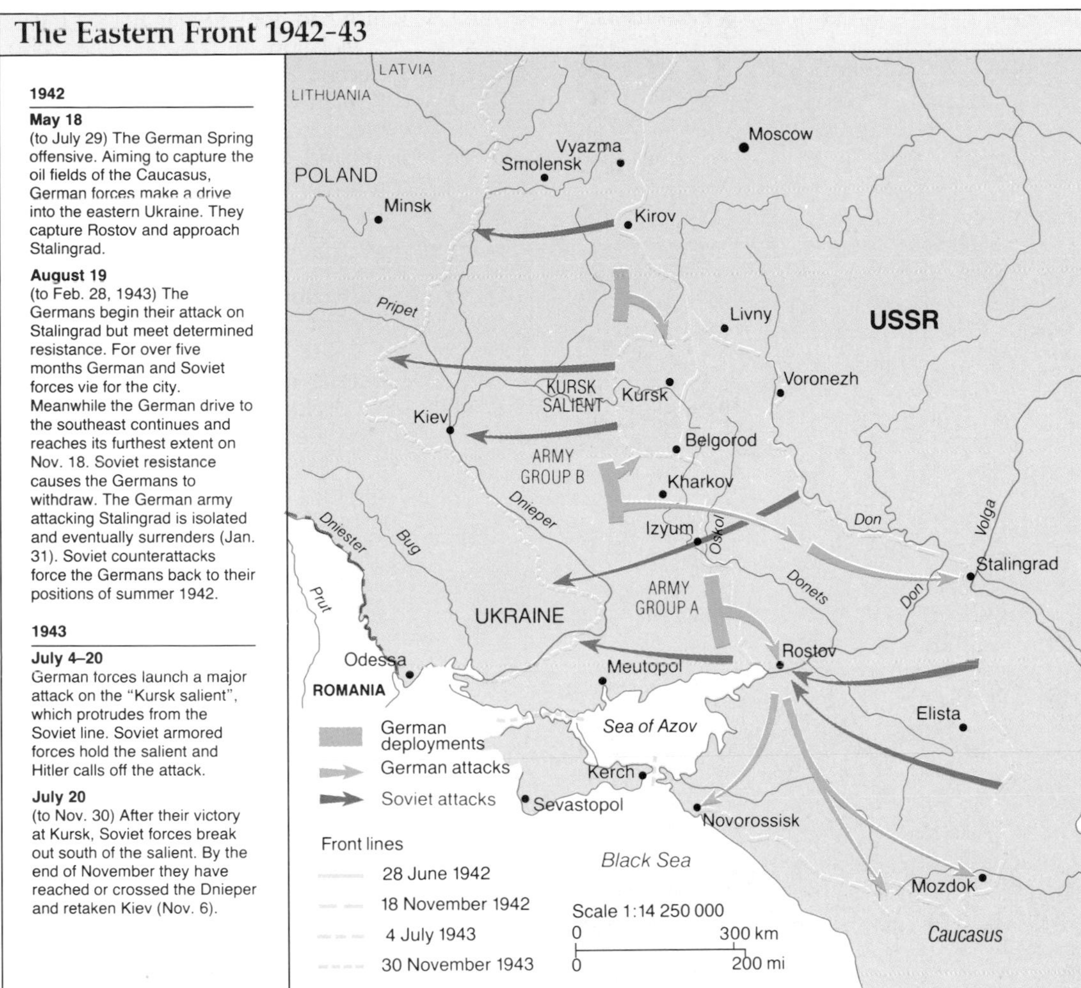

The Eastern Front 1942-43
1942
May 18
(to July 29) The German Spring offensive. Aiming to capture the oil fields of the Caucasus, German forces make a drive into the eastern Ukraine. They capture Rostov and approach Stalingrad.
August 19
(to Feb. 28, 1943) The Germans begin their attack on Stalingrad but meet determined resistance. For over five months German and Soviet forces vie for the city. Meanwhile the German drive to the southeast continues and reaches its furthest extent on Nov. 18. Soviet resistance causes the Germans to withdraw. The German army attacking Stalingrad is isolated and eventually surrenders (Jan. 31). Soviet counterattacks force the Germans back to their positions of summer 1942.
1943
July 4–20
German forces launch a major attack on the "Kursk salient", which protrudes from the Soviet line. Soviet armored forces hold the salient and Hitler calls off the attack.
July 20
(to Nov. 30) After their victory at Kursk, Soviet forces break out south of the salient. By the end of November they have reached or crossed the Dnieper and retaken Kiev (Nov. 6).
LATVIA
LITHUANIA
POLAND
Minsk
Smolensk
Vyazma
Moscow
Kirov
Pripet
Livny
USSR
KURSK SALIENT
Kursk
Voronezh
Kiev
Belgorod
ARMY GROUP B
Kharkov
Dnieper
Dniester
Bug
Izyum
Oskol
Don
Volga
Stalingrad
Prut
UKRAINE
ARMY GROUP A
Donets
Don
Odessa
ROMANIA
Rostov
Meutopol
Sea of Azov
Elista
Kerch
Sevastopol
Novorossisk
Black Sea
Mozdok
Caucasus
German deployments
German attacks
Soviet attacks
Front lines
28 June 1942
18 November 1942
4 July 1943
30 November 1943
Scale 1:14 250 000
0 300 km
0 200 mi

◀ **The Soviet victory at Stalingrad was the turning-point of the Soviet–German war. The quantity of Soviet arms and men now gained the advantage over the Germans' superiority in skill and equipment.**

▼ **In the early morning of 1 February 1943 Radio Moscow announced the German surrender at Stalingrad. The newly promoted Field Marshal von Paulus and his staff were taken prisoner. Twenty German and two Romanian divisions had been destroyed; the Soviet command claimed 91,000 prisoners.**

Barricades factory, Red October Foundry, the "commanding height" of Mamayev Kurgan – followed by close-quarter combat for rooms, cellars, even sections of wall. The vital ferries chugged across the Volga, bombed and shelled but never stopped. Gradually the Soviet positions shrank. Vasili Chuikov took command of 62nd Army and stayed put himself, virtually clinging to the river bank. During the first half of November 1942 ice clogged the Volga, the ferries stopped, all that was left to the Soviet defenders was to "stick it out". Both sides felt the crisis. Von Paulus, for his part, understood the significance of Soviet forces massing on his flank and sought permission from Hitler to pull back to a winter line. Hitler refused: von Paulus was categorically forbidden to abandon Stalingrad. A final German attack was planned for November 18. The next day, in morning fog, Soviet tanks crashed into the two Romanian armies and then the Italian army holding 6th Army's northern and southern flanks. Within four days, on 23 November, 6th Army was encircled.

The 6th Army suffered death by slow starvation. Almost 300,000 soldiers were trapped and 30,000 wounded were entombed, cut off from care, in ruined cellars. Malnutrition increased the ravages of disease. But the fate of Stalingrad was linked to the possible doom facing German armies deep in the Caucasus. There had been a Soviet breakthrough against the Italian 8th Army south of Stalingrad. Coupled with a German surrender at Stalingrad, the way would be open to cut off German Army Group A and close its escape route at Rostov. At the end of December 1942 Hitler authorized withdrawal from the Caucasus and the "race for Rostov", full of fateful implications, was on. The German 6th Army at Stalingrad, palsied and dying, fought on, tying down Soviet forces needed to seal the German escape route from the Caucasus. In the event, amidst fierce battles, German 1st Panzer Army moved the 650km (400mi) to Rostov and safety, leaving 17th Army in the western Caucasus to hole up in the Kuban. Von Paulus, newly promoted field marshal, played out the final act with his surrender on 31 January 1943.

Erich von Manstein 1887–1973

A soldier since 1906, Manstein, was given command of 11th Army in the Crimea. In July 1942 it captured Sevastopol, and was moved northward to assault Leningrad. But soon Manstein's "fire brigade" was rushed south again and Manstein took over Army Group Don to check the Soviet advance westward from Stalingrad. Manstein failed to relieve the encircled Germans at Stalingrad but stabilized the Front. In March 1943 he retook Kharkov, and attacked the Kursk salient but on 30 March 1944, with Army Group South under threat, Hitler dismissed his "master tactician".

I will cite only one example to illustrate the intensity of this battle (during Kursk). Nearly thirty tanks were advancing on a battery commanded by Captain G.I. Igishev. The gunners engaged them in unequal combat. Their four guns let the enemy approach to within 600–700 metres before opening fire. They destroyed 17 tanks, but by this time only one gun was left of the battery, with three men to serve it. They continued to fire, putting two more heavy tanks out of action. The enemy was forced to retire. Thanks to the heroic combined operations by all arms the attack was repulsed.

K.K. ROKOSSOVSKII

Soviet forces gain the upper hand

The full fury of the Soviet counteroffensive broke over the German southern wing, though Stalin's attempt to break into the Kuban defensive perimeter and trap 400,000 German troops failed. Manstein held Rostov until 14 February 1943 and redeployed 1st Panzer, while Soviet plans called for a drive on Kharkov to thrust deep into the German rear and complete the Soviet encirclement east of the river Dnieper. In a series of parrying blows and counterstrikes, Manstein chopped off the Soviet spearheads and planned late in February to retake Kharkov, abandoned temporarily on 15 February. One month later the Germans again held Kharkov and could even think of driving north on Kursk. Manstein proposed an attack to eliminate the salient just formed and holding six unprepared Soviet armies. But the mud came and indecision overcame the German command. The Kursk salient jutted out intact, the next German target.

In the late spring and early summer of 1943 a strange, unaccustomed lull settled on the Eastern Front. At Stalingrad the German army had suffered a grave psychological blow but it had not yet been defeated decisively in the field. Both sides now fixed their eyes on the Kursk salient. Hitler intended to attack it as the first step in his summer offensive. Zhukov at the beginning of 1943 advised Stalin that there would be no further drive to the Volga, rather an attack toward Moscow taking the most immediate route – striking first through the Kursk salient. For once Stalin listened and agreed in spite of misgivings to delay his own offensive and build a formidable defense in the salient. The Soviet defense in depth ran for more than 160km (100mi), built around six zones and massively mined.

For his planned attack Hitler demanded more mighty Tiger tanks, and so constantly delayed his offensive, thus allowing the Red Army more time to prepare. When the blow came on 4 July 1943 the German army pitted 42 divisions and the cream of the Panzer arm against 54 Soviet divisions with 12 tank corps. Striking from the northern and southern faces German armor chewed its way through Soviet defenses, bringing about the greatest tank battles in history and culminating in the enormous armored joust at Prokhorovka on 12 July 1943. The next day Hitler called off the attack, transferring a Panzer corps to Italy on July 17 and ordering two Panzer divisions north to stave off a Soviet threat to Army Group Center. As Hitler "temporarily" canceled his attack on the salient, the Red Army loosed its own considerable forces in the south, making for the river Dnieper on a front constantly broadening, dropping a large airborne force on the Bukhrin bridgehead on 24 September 1943. But it was left to General N.F. Vatutin to strike out for Kiev, which fell early in November 1943. The Soviet offensive slowed but German forces had been pushed back to the Dnieper and the German Panzer arm gravely mangled as a result of Kursk. The stalemate had been broken. From now on the German army was condemned only to retreat.

Stalin now demanded the liberation of the Ukraine and the smashing in of the entire German southern wing. In January 1944 Koniev carried through the "Korsun encirclement", a "touch of the Stalingrad whip" and the continuation of his "mud offensive". After the reduction of the Korsun "pocket" in late February, Zhukov at the beginning of March took over 1st Ukrainian Front and speedily sliced 4th Panzer Army in two; on 17 March Koniev was over the river Dniester and moving to encircle 1st Panzer Army. In the

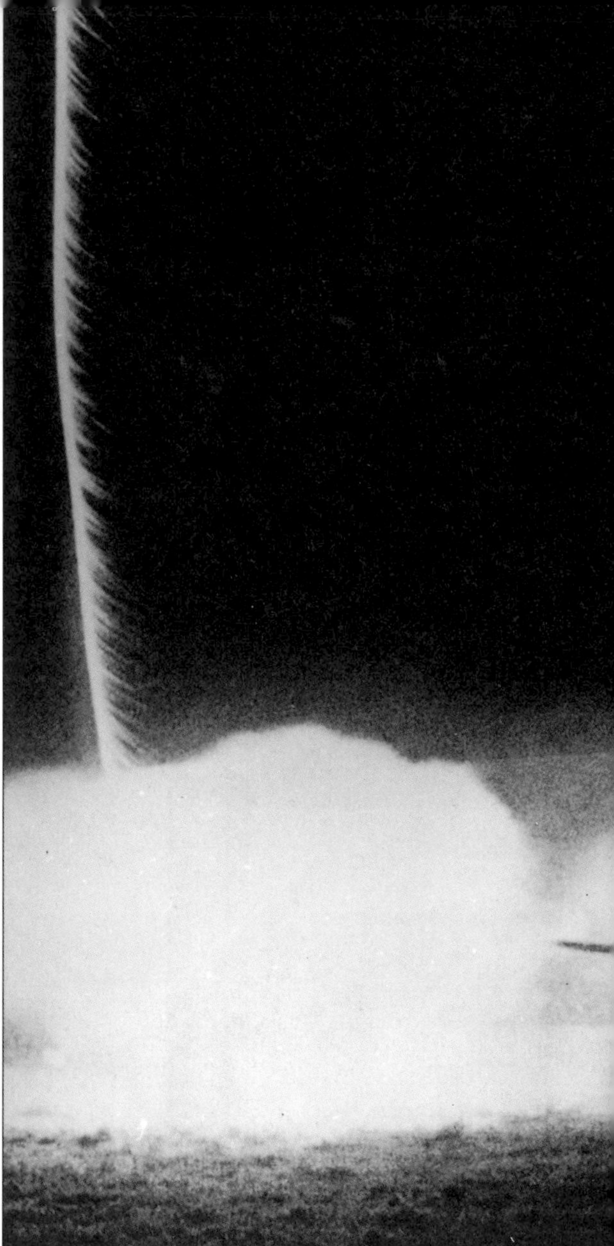

▼ **Soviet mounted troops pull a gun through the spring mud. Both the Red Army and the German army had to battle the mud brought annually in the spring thaw, though Soviet tanks, with their broad tracks, found it easier to sail these glutinous seas. Usually the *rasputitsa*, the thaw, brought an enforced lull on the main battlefronts but in the late winter of 1977 General Koniev launched his "mud offensive" in the Ukraine. Exploiting his wide-tracked tanks and American-supplied 4-wheel and 6-wheel trucks, as well as sturdy peasant carts, he carried out a major encirclement, while German tanks and motor vehicles, road-bound as they were, struggled against the mud.**

Crimea the 17th Army, withdrawn from the Kuban bridgehead, was another Soviet target, in an attack begun on 7 April 1944. At the northern extremity of the Eastern Front at long last, after months of innumerable horrors, in January 1944, a major Soviet offensive rammed the Germans away from Leningrad.

The Soviet defeat of Nazi Germany

With Army Group North pushed back and Army Group South in tatters only Army Group Center remained relatively inviolate, the Red Army's oldest and most formidable enemy. Stalin now planned its complete destruction. His Operation Bagration was timed to open on a memorable date: 22 June. By now the Second Front had materialized, but Stalin's plans were his own. Four Soviet Fronts with more than 2.5 million men attacked German positions headon; one by one, the German armies splintered and broke, 3rd Panzer at Vitebsk, 4th Army on the Dnieper, 9th Army at Bobruisk. By mid-July German forces had been swept from Belorussia and the Red Army had moved into northeast Poland.

In late August 1944 a strategic earthquake boiled up in the Balkans completely destroying the

▲ **On 5 July 1943, after a hesitation which proved well-nigh fatal, Hitler launched the German armies in the attack on the massively fortified Kursk salient. In what became known as the "death ride" of the 4th Panzer Army, German and Soviet armor clashed in the greatest tank battle of World War II. The nimble but outgunned Soviet T-34 medium tanks, as seen here, literally charged into the ranks of the mighty German Tiger tanks. Such ferocious engagements, made at the cost of great Soviet losses, finally broke the power of the German Panzer troops.**

◀ ***Za Stalina, Za Rodinu!*** **– "For Stalin, For the Motherland!" Red Army troops, like the civilian population, fought under this slogan: Red Armymen used it as a battle-cry, soldiers scrawled it on tank turrets. Here Stalin rides high over Soviet tank-borne infantry symbolically pressing westward.**

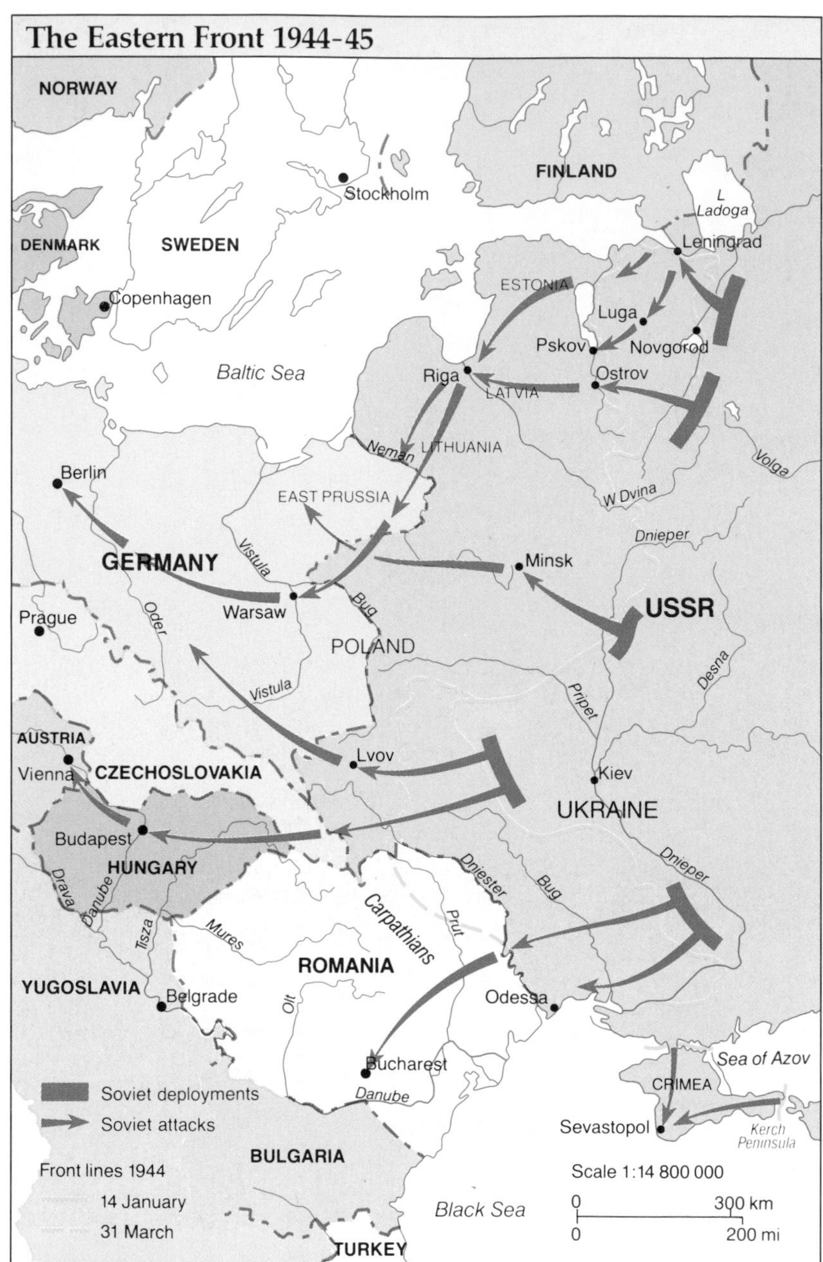

1943–44

December 24
(to mid May 1944) The Soviet winter campaign, consisting of three major offensives: in the Ukraine, the Crimea and in the north. The main offensive is the Soviet bid to regain the Ukraine. Despite fierce German resistance, Soviet forces push forward and on March 17 General Koniev and Soviet troops cross the Dniester. By mid April they are approaching the Carpathians. The northern offensive begins on Jan. 14. Soviet forces attack near Leningrad and then near Novgorod. Leningrad is freed from siege on Jan. 26. Soviet forces advance south to Luga, but are unable to capture Pskov and Ostrov. In April Soviet forces attack Germans sealed in the Crimean Peninsula. Sevastopol is recaptured on May 12.

June 22
(to Aug. 23) Operation Bagration, the Soviet offensive in the central part of the Front. The offensive breaks up the German Army Group Center. On July 26 Soviet troops reach the Vistula in Poland and then move toward Germany. Further south Soviet forces take Lvov and push toward the Vistula.

August 20
(to Sept. 8) A Soviet attack on Romania causes the country to change sides. Soviet troops enter Bucharest on Aug. 31. On Sept. 8, Bulgaria declares war on Germany and places its troops under the command of the Red Army.

October
Soviet forces move into Slovakia, Hungary, and Yugoslavia. Soviet forces and Tito's partisans clear Belgrade on Oct. 19. Soviet forces attack Budapest on Oct. 30 but the city resists until February 1945.

1945

January 12
(to Feb. 24) Soviet troops advance in Poland and move into East Prussia. Warsaw falls on Jan. 17. By the end of Feb. Soviet troops are on the Oder, only 60km (37mi) from Berlin.

February 24
(to May 8) The fall of the major cities. A Soviet invasion of Austria leads to the fall of Vienna on April 7. An offensive launched against Germany on April 16 takes the Red Army to Berlin: the city is encircled on April 25 (the same day as Soviet and US troops meet at Torgau on the Elbe) and surrenders on May 1. On May 8 Prague capitulates.

German southeastern theater. A Soviet offensive into Romania, beginning on 20 August, trapped 16 German divisions in the Jassy-Kishinev operation. Romania capitulated without delay and promptly changed sides, perfidy which doomed more Germans. Bulgaria followed suit; Soviet units moved into Yugoslavia to operate with Tito's partisans; and plans were afoot to drive into Hungary. Malinovskii reached Budapest on 30 October 1944 with no less than 64 divisions, completing the encirclement by the end of December but initiating yet another ghastly siege. In the north Finland had also surrendered.

The Red Army was now poised to advance into the Reich itself, the "lair of the Fascist beast". Already in November 1944 Stalin had nominated the commander to capture Berlin, Marshal Zhukov. Halted since July 1944 on the Vistula, staring at a Warsaw in the flames of revolt but rendering no assistance, the Red Army now mustered five Fronts for the grand assault on the Reich. The "Vistula-Oder operations", launched on 12 January 1945, brought Zhukov to the River Oder at the end of January, to Küstrin, a mere 60km (37mi) from Berlin. Koniev had invested most of Silesia. Despite this immediate threat Hitler insisted on sending 6th Panzer Army to Hungary for an offensive designed to win his way back to the Danube.

In the south the German offensive in Hungary failed. The way lay open to Vienna which fell to Tolbukhin on 7 April. Now it wanted little more than a week for the gigantic Soviet offensive aimed at Berlin, which Stalin knew would not be an Anglo-American objective. Three Soviet Fronts – 2nd Belorussian, 1st Belorussian and 1st Ukrainian – stood ready on the Oder-Neisse river line, a mass of 2.5 million men, more than 6,000 tanks, 41,000 guns and 7,500 aircraft. Zhukov opened his attack on 16 April with a gigantic cannonade which could be heard in Berlin itself. Koniev crossed the Neisse and raced from the south for Berlin. Zhukov, his mass of men and armor stumbling over themselves, fought to break through from the east while Rokossovskii to the north blocked 3rd Panzer Army. The encirclement of Berlin was complete on 25 April 1945, while on the River Elbe Soviet and American troops linked up, cutting the Reich in half. After days of bitter and often bloody street fighting, on 30 April the

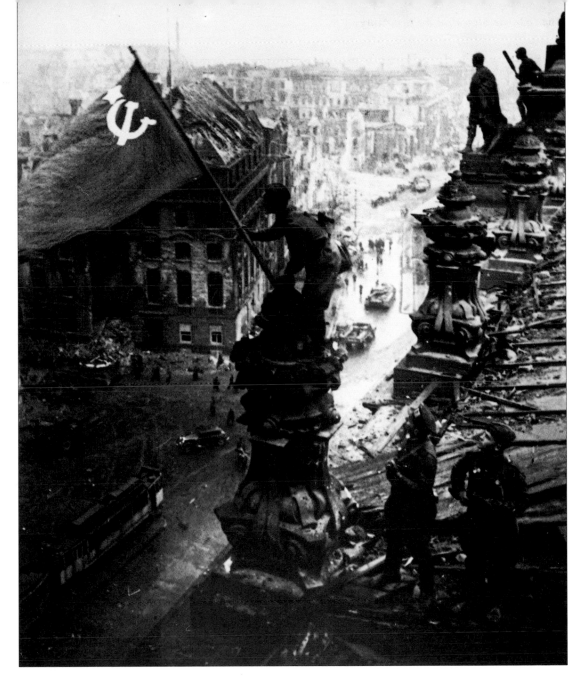

◀ Bulgarians greet units of the Soviet 3rd Ukrainian Army Group. At the end of August 1944, with Soviet armies flooding through Romania, Bulgaria faced the prospect of a Soviet invasion. Never an ardent ally of Hitler and his campaigns in Russia, Bulgaria proceeded to disarm Germans withdrawing from Romania, and, on 8 September 1944, declared war on Germany. Bulgarian troops were placed under Soviet command. Marshal Fedor Tolbukhin's 3rd Ukrainian Group crossed the frontier between the lower Danube and the Black Sea, and occupied the whole country without resistance.

◀◀ 1944 was the year of the Red Army's "ten decisive blows", launched initially on the flanks. In the north Leningrad was freed from blockade; in the south, large-scale operations recovered much of the Ukraine. In the center, in June 1944, the Red Army smashed Army Group Centre, and struck westward toward Poland.

▶ Toward noon on 30 April 1945 Soviet regiments took up positions for the final assault on the Reichstag in the heart of Berlin. After a day of heavy fighting and bombardment, two sergeants, Yegorov and Kantariya, succeeded in planting the red victory banner high over the building at 22.50 hours. It was the symbol of the Soviet triumph over Nazi Germany.

Walter Model 1891–1945

One of Hitler's "ruthless generals", Field Marshal Model took over the 9th Army on the Eastern Front in January 1942 and led it in the massive 1943 German operation against the Kursk salient, unsuccessfully as it proved. After holding commands in the north and the Ukraine, Model joined Army Group Center in April 1944 but he was unable to halt the massive Soviet offensive of summer 1944. He committed suicide in 1945.

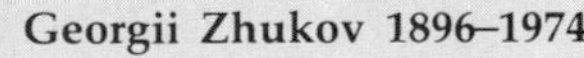

Georgii Zhukov 1896–1974

The "general who never lost a battle", Zhukov joined the *Stavka* (GHQ) on 23 June 1941 and remained a member throughout the war. In spring 1944 he controlled the operations that smashed German Army Group South. At the beginning of 1945 he planned and directed the mighty Vistula-Oder operations, driving into Germany, and then, in the final assault on Berlin, commanded the 1st Belorussian Front.

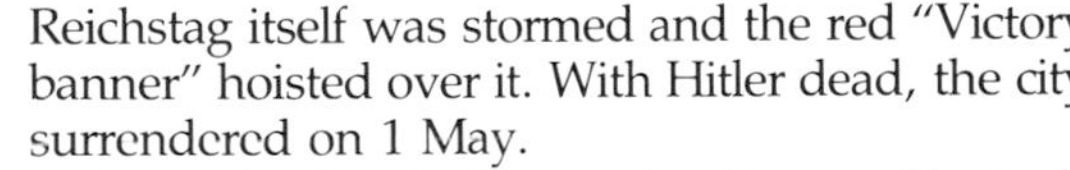

Reichstag itself was stormed and the red "Victory banner" hoisted over it. With Hitler dead, the city surrendered on 1 May.

It remained only to race for Prague. Koniev's tank armies wheeled away from Berlin and sped into Czechoslovakia from the east. On 8 May 1945 Marshal Koniev transmitted a capitulation order to all German units in western Czechoslovakia, accompanying it with a massive artillery barrage. On 9 May Soviet units linked up in Prague.

Stalin frequently reminded his wartime allies that the Red Army and the Soviet people were carrying the heaviest (and the bloodiest) burden of the fighting against Nazi Germany. In 47 months of fighting the Red Army inflicted more than 70 percent of the total losses in men and weapons suffered by the German army, destroying in the process 214 divisions, capturing 56, disabling 143 divisions and accepting the surrender of 93.5, making a grand total of 607 divisions. The corresponding tally for Anglo-American forces amounted to 176 enemy divisions.

Datafile

Compared to the fighting in France in 1940 or 1944 or on the Eastern Front (1941–45) the statistics for the Mediterranean theater seem unimpressive. Germany and the USA regarded the area as a sideshow and only Italy and the UK committed substantial forces. But the campaign did serve as a testing ground where British forces could develop new techniques and regain confidence after the debacle of 1940.

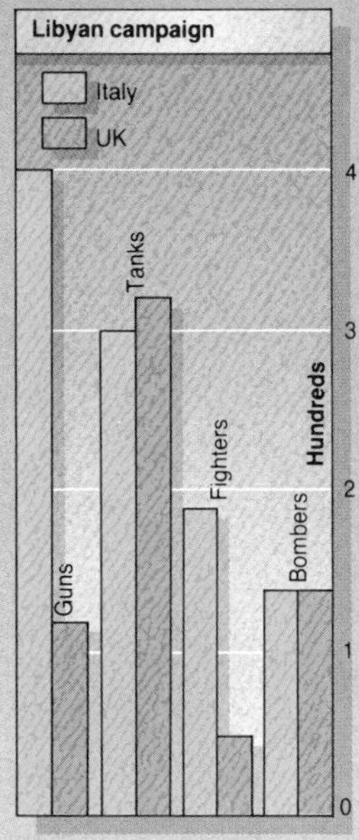

◀ In December 1940 the numerical superiority of Italian over British forces in North Africa was crushing. However, organization and technique were more important than numbers. Most of the 320 British tanks were concentrated in the highly trained 7th Armoured Division; Italian tanks were distributed among infantry units.

▶ Montgomery did not attack at El Alamein (23 October 1942) until he had achieved nearly a 2:1 numerical superiority. Unlike the highly mobile actions which had thus far characterized the desert war, El Alamein was an attritional battle: over 50 percent of Axis forces became casualties, most as prisoners.

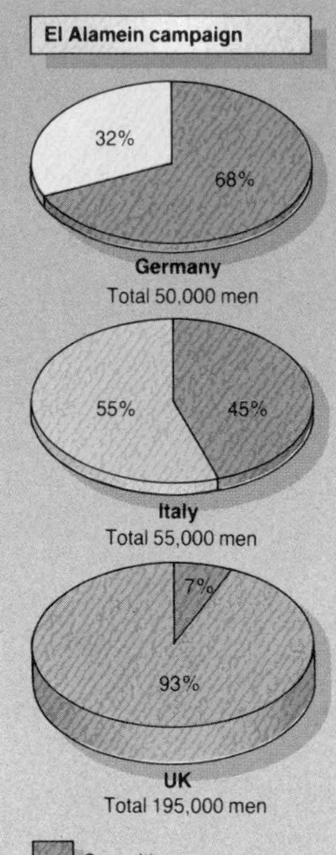

▼ The North African campaign was very important in restoring British confidence but its effect on the outcome of the war was marginal. Germany lost nearly as many men (800,000) in Russia in the first six months of Operation Barbarossa as all the combatants did in nearly three years in North Africa.

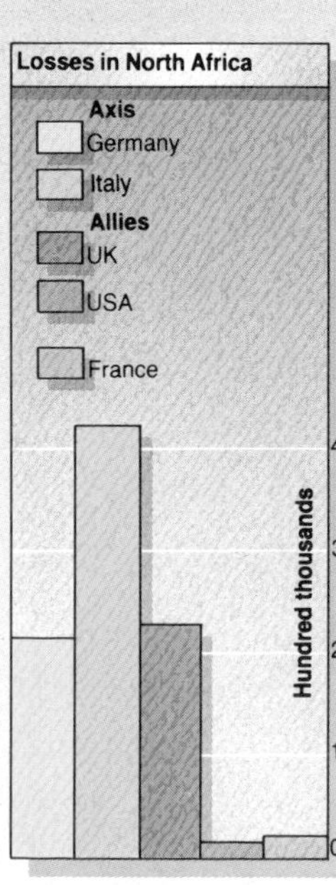

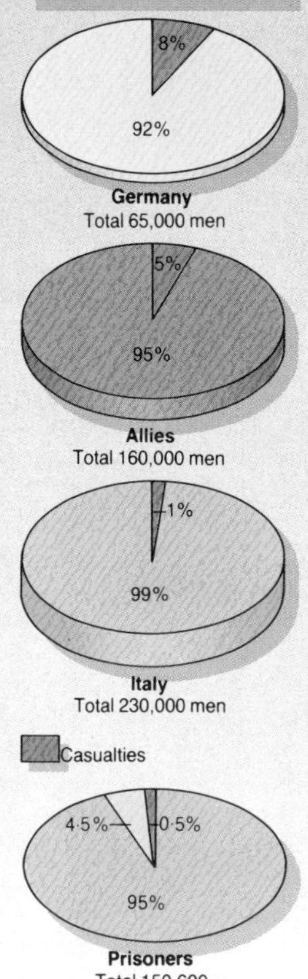

El Alamein forces

UK

Axis

Guns Tanks Aircraft Thousands 2 1 0

▲ Material deployed at El Alamein. As with manpower, British preponderance in materiel was crushing: 400 of the 600 Axis tanks were Italian and incapable of meeting the more heavily armored and gunned British tanks with any hope of success. Similarly the majority of the 675 Axis aircraft were Italian and inferior.

◀ In the invasion of Sicily (July–August 1942) Axis defenders outnumbered the Allies by nearly 2:1, but the reluctance of the Italian forces to fight was reflected in the small number of battle casualties as opposed to the vast number of prisoners taken by the Allies (143,500 of the 230,000 strong Italian garrison).

War came to the Mediterranean basin on 10 June 1940, when Italy declared war on the Allies. Italy's dictator, Benito Mussolini, was jealous of Hitler's spectacular successes and hoped that Italy, too, could achieve massive and glorious victories that would mark a revival of the "New Roman Empire". For Mussolini his decision was disastrous. Only crushing defeat, humiliation and political nemesis followed. For Hitler the decision was distracting and he too paid a heavy price for Mussolini's impulsive act. Italy was wholly unprepared for war.

The first British advance and retreat

The war in the Mediterranean followed the same pattern as that in western Europe. A period of inactivity was followed by rapid and crushing movement. The Italian army in Libya appeared very imposing – on paper: 250,000 men. The British commander-in-chief Middle East, General Sir Archibald Wavell, had only 36,000 men in Egypt. In September 1940 the Italians began a timid invasion of Egypt, halting after only 80km (50mi) at Sidi Barrani.

Wavell believed that the Italian 10th Army should be attacked at Sidi Barrani and driven

THE WAR IN THE MEDITERRANEAN

back to Libya. The results of this audacious decision were momentous. Within two days five Italian divisions had been destroyed. He reached the coastal hamlet of Beda Fomm and cut off the Italians. In 62 days the Western Desert Force had annihilated 10 divisions, capturing 130 prisoners, 380 tanks and 845 guns. It was a great feat of arms, but incomplete. O'Connor urged that he be allowed to press the advance and seize Tripoli. Wavell refused: the Germans were menacing Greece.

Wavell's forces were spread very thinly. Not only was he advancing in Libya, but since January he had also been attempting to reconquer Italian Abyssinia. Now he was required to send troops to aid the Greeks. In 1939 the British government had guaranteed Greek independence along with that of the ill-fated Poles. The Greeks were reluctant allies. The Greek dictator, General Ioannis Metaxas, refused to do anything that might antagonize the Germans. British help had to be covert. On 29 January 1941 Metaxas died and within a month the Greek government at last acknowledged that it needed British troops to counterbalance the growing German threat. These could only come from one source: O'Connor's Western Desert Force.

▼ Heirs to the Roman Legions? The abysmal performance of Italian troops in the opening campaigns of the desert war forced German intervention. These members of the Afrika Korps were captured at Tobruk in November 1941. In the later campaigns the Italians fought bravely, though their equipment was poor. But relations with the Germans were always tense.

The British-supported Greek campaign (March–April 1941) was a disaster, mitigated only by another naval victory over the Italians, at Cape Matapan on 28 March. The Germans invaded Greece and Yugoslavia on 6 April. The Greek army of 14 divisions refused to give up its gains in Albania. The British and three Greek divisions took up position on the incomplete Aliakmon Line. A gap emerged between this position and the main Greek army through which the Germans surged and within two days Salonika had fallen. The British then withdrew to Thermopylae and thence to Piraeus. Evacuation – a repeat of Dunkirk – was the only option. The Royal Navy

Benito Mussolini 1883–1945

Dictator, or *Il Duce*, of Italy. He had seized power in 1922 during the "March on Rome." Always inclined to swagger and boast, he degenerated from being a dominating figure on the international scene to a buffoon and helpless puppet. The Italian declaration of war (10 June 1940) was followed by a string of humiliating defeats. But Mussolini's own position was not threatened until 1943. In July, after the Allied invasion of Sicily, at a meeting of the fascist grand council, a motion was passed (19–7) depriving him of power. Imprisoned, Mussolini, was rescued only to preside over a German puppet government in the north. With the collapse of German authority, he was captured and executed unceremoniously.

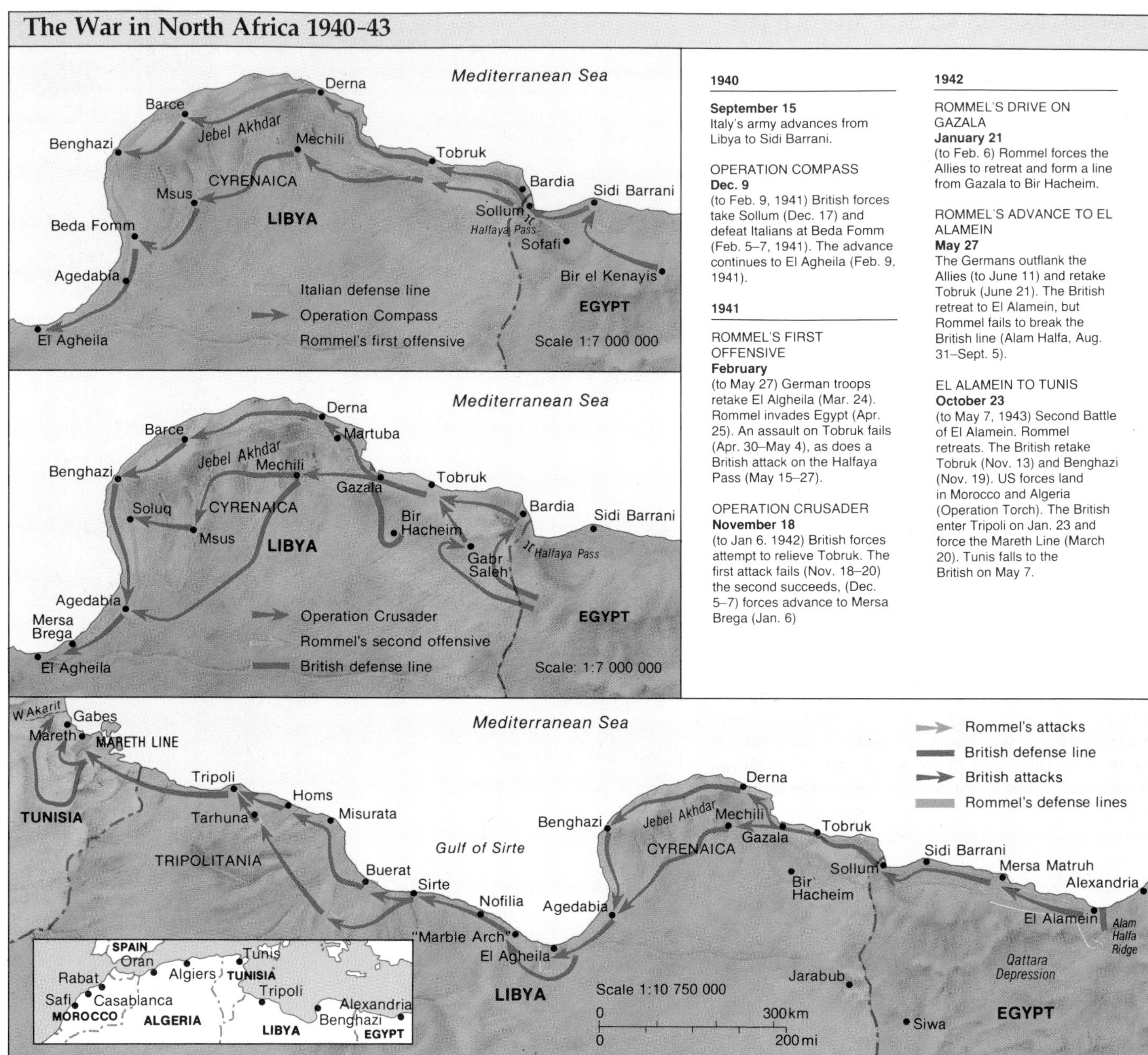

▲ The "pendulum war" in North Africa. Both the Allied and Axis armies were small by the standards of the Eastern Front. For the first two years of the campaign neither side could secure a decisive victory as their offensives lost momentum when they advanced further from their bases and drove the enemy back on his. It was only after summer 1942, when American equipment began to arrive in abundance, that the margin of quantitative superiority swung decisively in the Allies' favor.

▶ Rommel's precarious supply lines – stretching from Tripoli to the Egyptian frontier – forced him to rely heavily on airpower. The aircraft shown here are mainly JU52s – slow but reliable "work horse" transport aircraft.

achieved a second "miracle" in getting the troops away in face of German air superiority. The same exercise was to be repeated once Crete had been lost the following month.

The easy victories of O'Connor had been transformed into bitter defeat once the overwhelming might of the Germans had been brought in to redress the balance. The British had been driven from the northern shores of the Mediterranean. Yet worse was to follow. In February 1942 German troops arrived in Tripoli.

There can be little doubt that if the full might of the Wehrmacht had been deployed in the Mediterranean then British power in this region would have been destroyed in the autumn of 1941. British forces were too weak and poorly equipped to withstand German professionalism. The following years were to show that Britain's commanders had quite enough difficulty coping with the forces that were reluctantly dispatched by the German army high command (OKH). It considered North Africa a "side show", a tiresome distraction from the main objective – the invasion of Soviet Russia. Before December 1942 German forces in North Africa never amounted to more than three divisions. But they were commanded by a redoubtable officer and a master of mobile warfare: Erwin Rommel.

Rommel seizes the initiative

Within six weeks "this obscure general", as British Intelligence referred to him, had launched a counteroffensive. The British were caught flat-footed. Their best troops had been sent to Greece, and so had their commanders. Lieutenant-General Philip Neame was brought in to take command. His troops were inexperienced at armored warfare and so was he. He was not only intellectually ill-equipped but psychologically unprepared for the challenge he would face. He established an unimaginative HQ kilometers behind the frontline. Rommel undertook O'Connor's advance in reverse. Benghazi fell on 4 April and Derna three days later. Wavell was unhappy with Neame's performance and sent O'Connor to "advise" him, a most unhappy compromise as it turned out. Going up to Derna both Neame and O'Connor were captured. The loss of O'Connor was a disaster. He had shown himself to be one of the very few British commanders who were equal to the demands of mobile warfare: perhaps the only senior British general who would have been a match for Rommel.

The only glimmer of hope for Wavell was that Tobruk continued to hold out. All that he had achieved in four months had been cancelled in one. The rout in North Africa could have been greater had Rommel not been obsessed with taking Tobruk. He refused to consider anything else. On 23 April 1941 General Franz Halder, chief of staff to OKH, wrote in his diary that "things are in a mess... Rommel is in no way equal to the task. He rushes about the whole day between his widely scattered units... and fritters away his resources". General Friedrich von Paulus was dispatched to report on the desirability of continuing the siege. Von Paulus advised that no more troops should be sent to Rommel until the situation improved.

This message was decoded by Ultra and sent to Wavell by Churchill. The "Tiger" convoy carrying 300 new tanks had also arrived in Alexandria. Churchill, conscious of growing British strength and Rommel's weakness, pushed Wavell into a premature offensive. The result was Operation Battleaxe. Wavell launched his armor up the Halfaya Pass in a vain effort to destroy the German armor and break through to Tobruk. Rommel preferred to hold his armor in reserve and repel the British attack with his antitank guns – especially the deadly 88mm antiaircraft gun which he used in an antitank role. The British armor was devastated – over half of it was lost in the battle. Rommel then launched his own armor in a counterattack at Sidi Omar but the Western Desert Force withdrew before the jaws of Rommel's trap slammed shut.

The first British victory

After this defeat Wavell was replaced by the commander-in-chief India, General Sir Claude Auchinleck, a commander of fine character and high reputation, but a complete stranger to GHQ Cairo. The British forces were reorganized as the 8th Army, and Lieutenant General Sir Alan Cunningham was brought fresh from his triumphs in Abyssinia to take command. Cunningham was a tactician of some ability, but like so many British generals, quite inexperienced in the techniques of armored warfare. He was ordered by Auchinleck to plan Operation Crusader to occupy Cyrenaica and relieve Tobruk.

This operation, more than any other, illustrates the interdependent character of the North African campaign. By the autumn of 1941 the plight of the British island of Malta, lying astride Rommel's lines of communication, and the only British outpost remaining in the central Mediterranean, was becoming desperate. Its relief also had to be considered. This could be aided by seizing the

Claude Auchinleck 1884–1980

A product of the Indian (not British) army, after service in Norway and India, Auchinleck was appointed commander-in-chief Middle East, on 22 June 1941. He was a general of first-rate tactical flair, but was a poor judge of character. He led the 8th Army to the first British victory over a German-commanded force in Operation "Crusader". But he was held responsible for the defeats of 1942, and his victory at First Alamein was not acknowledged. He was sacked in August 1942 and returned to India.

Archibald Wavell 1883–1950

A cultivated man and well-read , Wavell was forced to make war pulled in opposite directions, and yet he succeeded. Apart from Operation "Compass," he put down a pro-fascist coup in Iraq. Just as O'Connor was poised to make a move on Tripoli, Wavell participated rather more readily than was later acknowledged in the ill-fated intervention in Greece. "Things fell apart", Rommel counter-attacked, and Wavell's response "Battleaxe", was unsuccessful. He was moved to India and rose to the viceroyalty.

It was fairly clear to me that there had been consternation in Whitehall... when I was getting ready for the final blow. Casey had been sent up to find out what was going on; Whitehall thought I was giving up, when in point of fact I was just about to win. I told him all about my plans and that I was certain of success; and de Guingand spoke to him very bluntly and told him to tell Whitehall not to bellyache.

BERNARD MONTGOMERY, EL ALAMEIN, 1942

Libyan ports and assisting the Royal Navy in their capture. But the British army could only move on these ports under the cover of air power, and to guarantee that cover more airfields in Cyrenaica had to be captured. Yet the desert air force could not use these until the 8th Army had driven Rommel from Cyrenaica. Consequently, the operations in this campaign revolved around the northern communication centers.

The burden of responsibility lying on Cunningham's shoulders was so great that he decided to give up smoking, and this put his nerves on edge. The plan he evolved for Crusader was highly unsatisfactory. He intended to destroy Rommel's armor in a great tank battle. The lessons of Operation Battleaxe had not been learnt. The 8th Army was to advance rapidly, taking Rommel by surprise, and then halt and wait for the German armor to deploy. To surrender the initiative on the first day of an offensive was foolish in the extreme, and so it was to prove.

Although Cunningham had 455 tanks to Rommel's 412, he only had 72 antitank guns to Rommel's 192. This equation had great importance, and Cunningham should have paid more attention to it. He was successful in achieving surprise but not in gaining a decisive victory over the enemy's armor. A chaotic battle of attrition developed around Sidi Rezegh. Rommel made a "dash for the wire" deep into Cunningham's rear. "It was the contagion of bewilderment and fear and ignorance", wrote the war correspondent, Alan Moorehead. "Rumors spread at every halt, no man had orders". Cunningham seemed to lose his nerve and counseled retreat. Auchinleck relieved him, replacing him with his chief of staff, Neil Ritchie.

The crisis of the battle passed. Rommel's thrust was too weak and he was forced to retreat. Tobruk was relieved and the 8th Army arrived triumphantly at El Agheila. The scale of the victory was substantial. Rommel had suffered 24,500 casualties and lost almost 400 tanks. Operation Crusader was the first major British victory over a German-commanded force in World War II. It was sorely needed but of short duration. The opening of the war against Japan deprived Auchinleck of badly needed reinforcements and

▲ **Although Rommel operated in oil-rich areas, throughout his campaigns in North Africa his fuel supply operated on a shoestring. Operating from Italy, Axis oil tankers put in to Tripoli, unloaded their precious cargoes and the oil was then moved to the front by road. The Allies tightened their stranglehold during 1943.**

Rommel launched a counterattack which drove 8th Army back to the line Bir Hacheim-Gazala.

Rommel enters Egypt

The desert war now entered its period of crisis. The British and their allies had built a large fortified line, based on minefields and several strong points or "boxes". These appeared to enjoy no relationship with the armored, mobile forces. Rommel determined to break through this line before it grew too strong. On 26 May he attempted to outflank the Gazala Line from the south. The Free French forces resisted gallantly at Bir Hacheim but were overwhelmed. Rommel advanced into the British rear and "dug in" in the "cauldron". This amounted to a "bridgehead" out of which Rommel intended to advance once he had refuelled and worn down the British armor in a series of defensive actions. On 4–5 June British armor was indeed dissipated in a number of uncoordinated piecemeal attacks; by 12 June only 70 British tanks survived. With massive air support Rommel struck towards Tobruk. Its capture on 21 June was a severe moral blow for Churchill's government. Rommel was promoted to field marshal and his troops celebrated "with captured tinned fruit, Irish potatoes, cigarettes and canned beer". On 23 June Rommel crossed the Egyptian frontier. Two days later Auchinleck relieved Ritchie

Erwin Rommel 1891–1944

A man who had no other interest but soldiering, Rommel was a brilliant cavalry general, tactician, and exponent of armored warfare. He took over the Deutsche Afrika Korps (DAK) in 1941. The skill he exhibited in commanding this outnumbered force earned him the admiration of his adversaries, and Auchinleck was forced to issue a general order deprecating this tendency. His success was owed to his eye for ground, his personal style of command, leading "from the front", and the important realization that armored warfare revolved around the antitank gun just as much as around the tank. Temperamentally inclined to neglect matters such as logistics, his formula was much less successful against a prudent general like Montgomery. He committed suicide in 1944.

It was a battle of attrition. Its main aim was to contain the enemy's armor and wear down Rommel's infantry by a process of "crumbling". Once these had been destroyed, Rommel's defensive line would collapse and the armor could break out. The plan ran into difficulties and Montgomery was forced to modify it. The Australians launched a diversionary attack to distract Rommel's reserves. Once these had been committed the armor broke through and the pursuit began on 1–2 November. In Britain El Alamein was hailed as a great victory. But its results were disappointing. Rommel escaped and the long battle exhausted the victors almost as much as the vanquished.

On 8 November 1942 the Anglo-American 1st Army landed in North Africa – Operation "Torch". Rommel was forced to withdraw back to Tunisia. Montgomery advanced cautiously and on 23 January 1943 occupied Tripoli. The 1st Army became bogged down in the Tunisian mountains. Rommel succeeded in routing the Americans, but Montgomery, forewarned by Ultra intelligence, was well prepared. For no loss the 8th Army destroyed 50 of Rommel's precious tanks and then closed up to the Mareth Line. After an imprudent frontal assault, Montgomery traversed this with a classic "left hook" which saw the 8th Army safely into the Tunisian mountains above Gabes.

The noose was tightening around the Axis forces in North Africa. Rommel left Africa on 9 March 1943 and was succeeded by General Hans-Jürgen von Arnim. Alexander now commanded the 1st and 8th Armies of the 15th Army Group. The Cape Bon peninsula was a great natural fortress. Alexander gave the Germans no respite to build up their defenses, and following a great

▲ US infantry advancing through the Tunisian hills, 20 January 1943. American standards of training and tactics in North Africa were initially very poor and shocked visiting British officers. Some claimed that the Americans "were our Italians". After their defeat at Kasserine Pass, the US 2nd Corps was transformed by Patton and Bradley into a first-rate fighting force.

and took personal command of 8th Army. He withdrew to the El Alamein position. There in July he fought a confused action called First Alamein. Rommel was halted but not defeated. "It is unlikely", Auchinleck telegraphed London, "that an opportunity will arise for resumption of offensive operations before mid-September". This was not good enough for Churchill and he determined to go to Cairo and sort out affairs there personally, accompanied by the chief of the imperial general staff, General Sir Alan Brooke. Auchinleck was dismissed and returned to India as commander in chief. He was replaced by General Sir Harold Alexander; a few days later – after Churchill's first choice had been killed in an air accident – Lieutenant-General Bernard Montgomery was appointed to command the 8th Army.

Montgomery and El Alamein

Montgomery's first aim was the restoration of the 8th Army's morale. He succeeded brilliantly. A man of pronounced individuality, he exuded confidence, and at El Alam Halfa he fought a model defensive battle in which, for once, German armor attacked British antitank guns and was repulsed. Still, Montgomery would not be hurried into launching his own offensive. A thorough and prudent general, he began his attack on 23 October 1942: the Second Battle of El Alamein.

► Crowds celebrate the liberation of Tunis in May 1943. A French colony, Tunisia had been spared the horrors of war until the German occupation in response to the "Torch" landings in November 1942. By the standards of Western Europe, their ordeal had been mercifully short.

▼ **After the invasion of Sicily, the German forces escaped intact. Montgomery's crossing of the Straits of Messina, Operation "Baytown", accompanied by intricate preparations and a massive artillery bombardment, had something of the air of a "comic opera". The only casualty was an escaped leopard from Reggio Zoo which took a shine to brigadier and had to be shot. In the north, it was quite another matter. Mark Clark's 5th Army was almost driven into the sea before the capture of Naples.**

offensive down the Merjerda Valley, Tunis fell on 7 May. Alexander signalled to Churchill: "Sir, it is my duty to report that the Tunisian Campaign is over. All enemy resistance has ceased. We are masters of the North African shores."

From Africa to Italy

From North Africa the war now spread once more to the northern shores of the Mediterranean with the invasion of Sicily, which was agreed to by Churchill and Roosevelt at the Casablanca conference in January 1943. The amphibious landing operation went well, but was followed by bitter fighting in the mountains north of Catania. Patton advanced westward to Palermo and arrived in Messina a few hours before Montgomery. The fall of Sicily led also to the fall of Mussolini. Churchill and Brooke argued that this victory should be exploited by an invasion of the Italian mainland to knock Italy completely out of the war. The idea of Italy as a "soft underbelly" did not appeal to the Americans, but they agreed reluctantly to the continuance of the war in the Mediterranean so long as it remained subordinate to Operation "Overlord", the invasion of France. General Mark Clark would land with the US 5th Army at Salerno in September; Montgomery would cross the straits of Messina and advance up the Italian "toe".

Everything that could go wrong at Salerno did go wrong and the Allies were almost driven back into the sea. Montgomery, who was sulking because 8th Army had been denied the Salerno

The Allied Invasion of Italy 1943-45

1943

July 10
(to Aug. 17) Operation Husky: the British and US invasion of Sicily. The British objective is Messina, but they are checked south of Catania (July 18). They attempt to go west of Mt. Etna but move slowly: Catania is captured on Aug. 5. The Americans meet fierce resistance but as German units withdraw they take Palermo (July 22) and then reach Messina first (Aug. 17).

September 3–20
Operation Baytown. As a diversionary strike, British forces advance up the "toe". Auletta is reached on Sept. 19, Potenza on Sept. 20.

September 9–19
Operation Avalanche. A British-American force lands in the gulf of Salerno. It meets fierce German resistance. Heavy bombardment forces the Germans to withdraw to the Gustav line.

September 9
(to Nov. 30) Operation Slapstick. British forces land at Taranto and move quickly to Foggia (Sept. 27). Thereafter progress is slow. They cross the Sangro river at the end of Nov. (28–30) but have insufficient strength to attack the Gustav Line.

September 12
(to Dec. 31) Allied forces advance from the Salerno area. British forces enter Naples (Oct. 1), US troops Benevento (Oct. 2). The Germans withdraw in stages and holed the Bernhard Line (Nov. 5–15). The Allies cross the Garigliano (Dec. 2–6) but exhaustion keeps them south of the Rapido.

1944

January 17
(to June 4) The Allies attempt to break the Gustav Line at Monte Cassino. Three attacks fail (Jan. 17, Feb. 15–18, March). Meanwhile an attempt is made to outflank it with landings at Anzio (Jan. 22), but the Allies are besieged there until late May. The Gustav Line is finally broken 30km (20mi) east of Cassino on May 11 and Poles storm Cassino on May 18. This success enables troops from Anzio and Cassino to break out: they enter Rome on June 4.

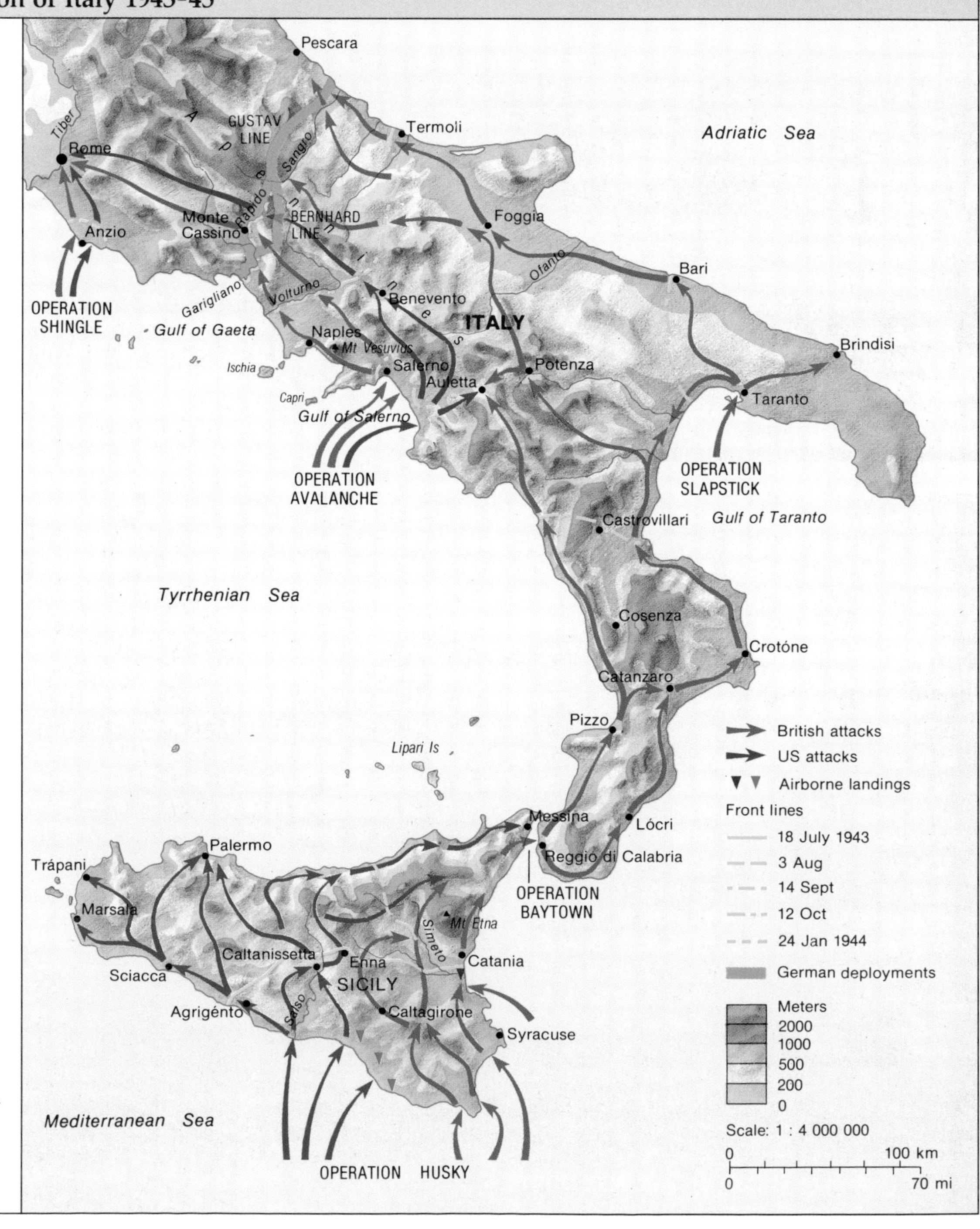

mission, did not hurry to Clark's aid. The US 5th Army survived; Naples fell in October 1943. Thereafter the pace of operations slowed down.

Stalemate and Allied success

The Italian peninsula was ideal country for defensive operations: narrow, mountainous, with rivers running from west to east. The Germans, commanded by Field Marshal Albert Kesselring (a former airman), constructed defenses to protect Rome, the Gustav Line. The result was stalemate. Alexander fought four great battles below Monte Cassino. He used massive artillery bombardments and strategic bombers in a ground-support role, but could not break through. The Italian mountains were among the worst terrains the Allies fought over. The slogging match at Cassino resembled the Somme, though the casualty bill was somewhat shorter. Another amphibious effort, at Anzio, almost ended in catastrophe.

Alexander's greatest victory, Operation Diadem, opened on 11 May 1944. A greatly strengthened 8th Army smashed through the center of the Gustav Line. The left-hand thrust, Clark's 5th Army, should have wheeled behind the rear of the Germans and enveloped them, but Clark let the Germans escape, preferring the glory of the capture of Rome, which fell on 4 June. Kesselring withdrew with his army intact to a new defensive line south of Bologna, the Gothic Line. Alexander was determined to continue the offensive and advance on Vienna. The Americans were adamantly opposed to this scheme.

After the Yalta conference in February 1945, Alexander was instructed to seize any opportunities that presented themselves. He acted decisively and secured the complete military victory that had eluded him the year before. A fine deception scheme persuaded the Germans to expect an amphibious scheme on the Adriatic coast. The 5th Army in fact struck west of Bologna and the 8th Army through the Argenta Gap. The Germans were encircled south of the River Po, the armies linking up appropriately at a town called Finale. The war in the Mediterranean was at an end. It had been a major contribution to the Allied victory, providing that badly needed front behind which to build up strength and experience for the decisive battle in the West. Thereafter it distracted some 25 German divisions. The campaigns in Italy and North Africa were the decisive proving grounds for the Allied armies.

I studied the rough terrain ... in the hope of finding some of our forces still hanging on ... There were no such hopeful signs, but I did see eighteen tanks... For a moment I hoped that they were ours, but... I soon discovered that they were German ... we were again in the utmost danger of being split apart and crushed.

GENERAL MARK CLARK
SALERNO, 1943

▲ Partisans in Bologna in May 1945. Most were leftwing and so were suppressed by the British, who remembered their experience in Greece in 1944 when a communist coup had followed liberation.

◀▲ In terms of ships and troops employed the invasion of Sicily was the largest amphibious operation in history – even larger than Overlord. Other amphibious operations – Salerno and Anzio – were not to go so smoothly.

◀ The great abbey of Cassino in ruins after its capture in May 1944. Its majestic presence looming over the junction of 5th and 8th Armies exerted an eerie psychological hold over Allied troops who felt that they could not move unobserved. Whether the aerial bombing was justified is controversial. What was certainly not justified was that the bombing should be uncoordinated with the ground attacks – allowing time for the Germans to recover.

ALLIANCES AND VICTORY IN WORLD WAR II

The USA, USSR and UK defeated the Axis powers for three main reasons: first, because their strategic and military cooperation was closer and more effective than the enemy's; secondly, because the Soviet armies and people fought with such persistent courage while the West was mobilizing its forces; and thirdly, because the Axis powers made avoidable political and military blunders.

The broad success of Anglo-American cooperation owed much to General George Marshall, the US army chief of staff. From the bombing of Pearl Harbor onward, Marshall insisted on two principles: that a combined Anglo-American chiefs of staff organization should plan strategy, and that in each major theater all Allied forces and all three services should operate under a supreme Allied commander. In addition Roosevelt and Churchill, the heads of government, met frequently for personal conferences.

In Europe and the Mediterranean both the USA and UK contributed substantial forces and cooperation was close if not always harmonious. In the war against Japan, however, the US Central Pacific and Southwest Pacific Commands conducted largely American campaigns (with help from Australia and New Zealand), separate from the British Southeast Asia Command.

The positions in regard to Anglo-American cooperation with the USSR and China were different, partly for political, partly for geographical reasons. Stalin in the USSR and Jiang Jieshi in China were political and military dictators who would not brook the loss of sovereignty which effective military integration requires. Moreover, the Eastern and Chinese fronts were separated from contact with the Western Allies by enemy-controlled territories and Chinese forces could only hold defensive lines.

So far as Stalin was concerned, however, Allied cooperation meant simply that the Western Allies should supply as much material and arms as they could, and invade Western Europe as soon as possible. For the West it meant that the Red Army should hold the bulk of Germany's armies until the Allies were ready to strike in the West.

The position on the Axis side was totally different. Neither of the two dictators, Hitler and Mussolini, were natural cooperators. They frequently disagreed on political and military objectives and several times launched important strategic or political initiatives without consulting each other. Liaison officers were exchanged and there were occasional meetings between heads of state or senior ministers or officers, but little coordination of strategy or real cooperation.

With Japan, too, there was a lack of genuine consultation; which makes it all the more surprising that Hitler allowed himself to be dragged into open war with the USA at a moment chosen by the Japanese. There were no joint war plans with Japan, and a total absence of strategic cooperation. The Japanese and the European Axis fought their widely-separated campaigns with little regard for each other.

CONFERENCES OF ALLIED LEADERS

1941

OFF NEWFOUNDLAND August 9–12
Churchill (UK), Roosevelt (USA)

WASHINGTON Dec. 22–Jan. 14, 1942
Arcadia conference. Churchill (UK), Roosevelt (USA)

1942

WASHINGTON June 17–21
Churchill (UK), Roosevelt (USA)

LONDON July
Churchill (UK), H.L. Hopkins (USA, Roosevelt's envoy), G.C. Marshall (US chief of staff), E.J. King (US chief of naval staff)

1943

CASABLANCA Jan. 12–23
Symbol conference. Churchill (UK), Roosevelt (USA)

WASHINGTON May 11–25
Trident conference. Churchill (UK), Roosevelt (USA)

QUEBEC Aug. 11–24
Quadrant conference. Churchill (UK), Roosevelt (USA)

MOSCOW Oct. 18–30
Foreign ministers: V. Molotov (USSR), C. Hull (USA), A. Eden (UK)

CAIRO Nov. 23–27; Dec. 2–7
Churchill (UK), Roosevelt (USA), Jiang Jieshi (China)

TEHRAN Nov. 28–Dec. 1
Churchill (UK), Roosevelt (USA), Stalin (USSR)

1944

QUEBEC Sept. 10–17
Octagon conference. Churchill (UK), Roosevelt (USA)

1945

YALTA Feb. 4–11
Argonaut conference. Churchill (UK), Roosevelt (USA), Stalin (USSR)

POTSDAM July 14–Aug. 2
Terminal conference. Stalin (USSR), H. Truman (USA), Churchill replaced by C. Attlee (UK)

▲ The two dictators, Hitler and Mussolini, together. Hitler was grateful to Mussolini for his support of German seizure of Austria in 1938 and respected him as a colleague and senior statesman. Mussolini, however, was jealous of his more powerful partner. Consequently, Axis consultation and cooperation tended to be irregular.

◄▲▼ Cooperation between the Allies occurred at all levels. Below, Stalin, Roosevelt and Churchill meet at Tehran; left, Eisenhower, Montgomery and their staff plan for D Day; above, Soviet and US troops meet in defeated Germany.

Datafile

At dawn on 6 June 1944 British and American forces began to land on five beaches in Normandy, in a colossal amphibious operation. It was the culmination of a year's planning and the start of Operation Overlord.

The success of Overlord depended above all on three key factors. First, Operation Fortitude, the elaborate deception plan which convinced the Germans that Normandy was not the objective. Secondly, the achievement of aerial supremacy over northern France, which meant that the invasion fleet could put the troops ashore without interference from the Luftwaffe. And thirdly, the ability of the Allied navies to move and keep supplied a force which was the equivalent of the British city of Birmingham. If in the first six weeks any one of these operations had gone wrong the Germans would have pushed the Allies back into the Channel, with consequences which are incalculable.

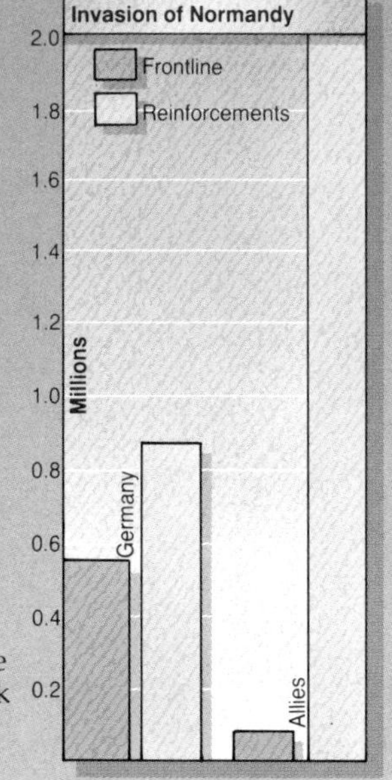

▲ A graphic illustration of Overlord's dependence on Operation Fortitude. The comparatively small number of men the Allies were able to land in the first few days would have been pushed into sea had the Germans concentrated their forces against them. Fortunately Hitler was convinced that Normandy was a feint.

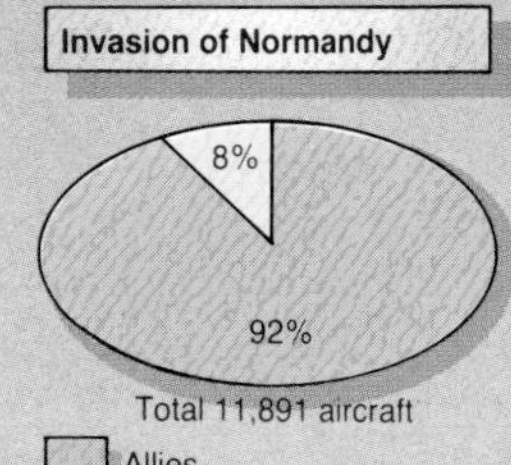

◀ The elimination of the Luftwaffe. In the spring of 1944 the Allies shifted their aerial offensive from Germany to northern France and after a three-month battle of attrition achieved a crushing superiority. Allied air forces could then provide aerial defense and hamper German efforts at resupply and redeployment.

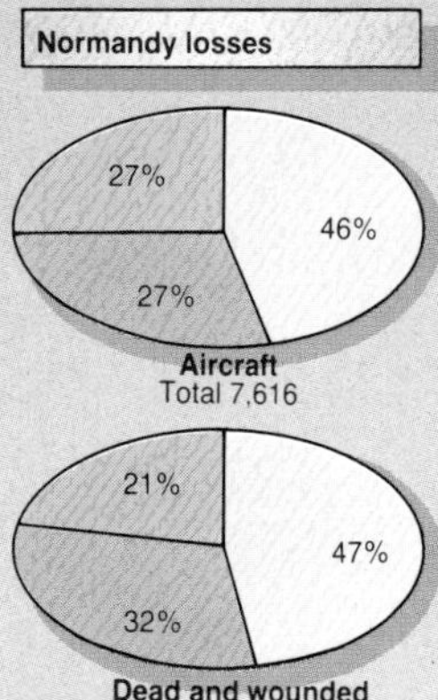

◀ The Normandy fighting was a bloody attritional struggle which in intensity resembled the fighting of World War I. The Allies expected an early breakthrough but became bogged down in the bocage country which was excellent for defense. Allied and German casualties were approximately the same.

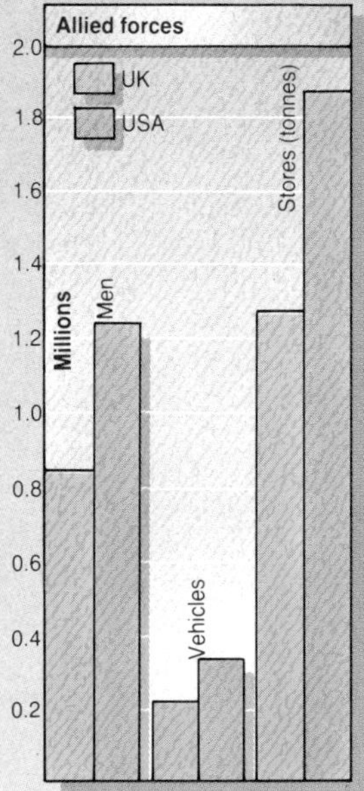

▶ The Allied buildup of 6 June to 31 August 1944 was the most immense supply operation in history. This flood of manpower and materiel eventually enabled the Allies to crack the German defenses. Even though the Mulberry harbors proved a disappointment, Allied supplies passed unhindered over the beaches.

The overriding problem facing the British army after its expulsion from France in 1940 was not when it would return but *how*. Within days of Dunkirk, Churchill issued orders for the planning of an invasion to begin. But the prodigious difficulties that would face any force crossing the English Channel – perhaps the most tempestuous passage of water anywhere in the world in proportion to its size – were a serious deterrent to planning. The fact also had to be faced that, on its own, the British army was not strong enough to undertake the task.

Instead, the Royal Marine commandos were ordered to begin raiding in an attempt to weaken the enemy. By 1941 their raids were sizeable, and included attacks on the Lofoten Islands, Spitsbergen, South Vaagso and Maaloy, and in 1942, St Nazaire. But such pinpricks on the periphery of the Nazi empire were no substitute for a full-scale onslaught on "Fortress Europe".

The origins of "Overlord"

No matter how forcefully Lord Beaverbrook and others might press for an invasion, however, the wish alone offered no solution to overcoming the massive problems such an operation faced. This was brutally confirmed by the failure of the Dieppe raid in August 1942, when a force of 6,000, mostly Canadians, left behind 3,670 casualties.

The entry of the USA into the war changed drastically the material balance. But an amphibious operation, if launched prematurely against an enemy as skilful as the Germans, would end only in disaster. If an invasion of France was repulsed, a Dieppe on a grander scale, the consequences for the Allies would have been incalculable. Nonetheless, in 1942 President Roosevelt, in his casual way, gave Stalin a promise that the Allies would open a second front in Europe that year. Roosevelt's friend and adviser, Harry L. Hopkins, was dispatched to London on a mission to begin preparations, accompanied by an unknown major-general, Dwight D. Eisenhower. As the bulk of the force for this enterprise would be British, the chief of the imperial general staff, Alan Brooke, had no difficulty in brushing aside the somewhat hazy arguments of the Americans, and Roosevelt was persuaded that the main Allied effort should be directed to the North African landings in November 1942. Here American troops could be "blooded" without risking heavy casualties.

Nevertheless, the US army's doctrine was dedicated to the principle of massive concentration; the overwhelming use of firepower and material resources to defeat the enemy at the earliest opportunity in the decisive theater of war. George Marshall, the US chief of staff, was suspicious that further Mediterranean adventures

THE SECOND FRONT IN EUROPE

would sap Allied strength to little profit. No serious planning for a cross-Channel invasion began, however, until after the Casablanca conference (12–23 January, 1943). Here a joint Allied staff was agreed upon and a chief of staff to the supreme Allied commander, COSSAC, was appointed – Lieutenant-General Sir Frederick Morgan – but no supreme commander. At the Washington conference in May 1943 it was finally agreed that Operation "Overlord" should be given the highest priority. Morgan had inherited Eisenhower's preliminary planning which emphasized a landing on the tip of the Cotentin Peninsula, near Cherbourg. The only alternative was the Pas de Calais, which enjoyed the shortest sea route. But this had the disadvantage of an inadequate springboard for launching the assault, as it would have to depend exclusively on Dover and Newhaven. Normandy was much more central, all the southern English ports could be used and the direction of attack would be less obvious to the Germans. Morgan planned for an assault on a frontage of three divisions, with a buildup to 18. It was reported that the Germans had 60 divisions deployed in the West.

At last, in November 1943 at the Cairo Conference, the command structure was approved. As the operation was scheduled for May 1944 this was not a moment too soon. Eisenhower was chosen as the supreme commander and Air Chief Marshal Tedder as his deputy. Montgomery was appointed to command the ground forces (21st Army Group). For the invasion there would be two armies, US 1st Army (under Lieutenant-General Omar Bradley) and Br 2nd Army (under Lieutenant-General Sir Miles Dempsey). Montgomery's first act was to scrap Morgan's plan, asserting that a three-divisional frontage was much too weak. Montgomery's criticisms were undoubtedly correct.

Montgomery's concept envisaged a five-divisional attack on a front demarcated by the Rivers Vire and Orne. Three further airborne divisions covered the flanks, two on the right flank to ensure the early capture of Cherbourg. The total host comprised 50,000 men, with a

▼ Wasted strength: the German Atlantic Wall. So convinced were the Germans – and this included Rommel – that the Allies would land in the Pas de Calais region, they concentrated their defenses in this region.

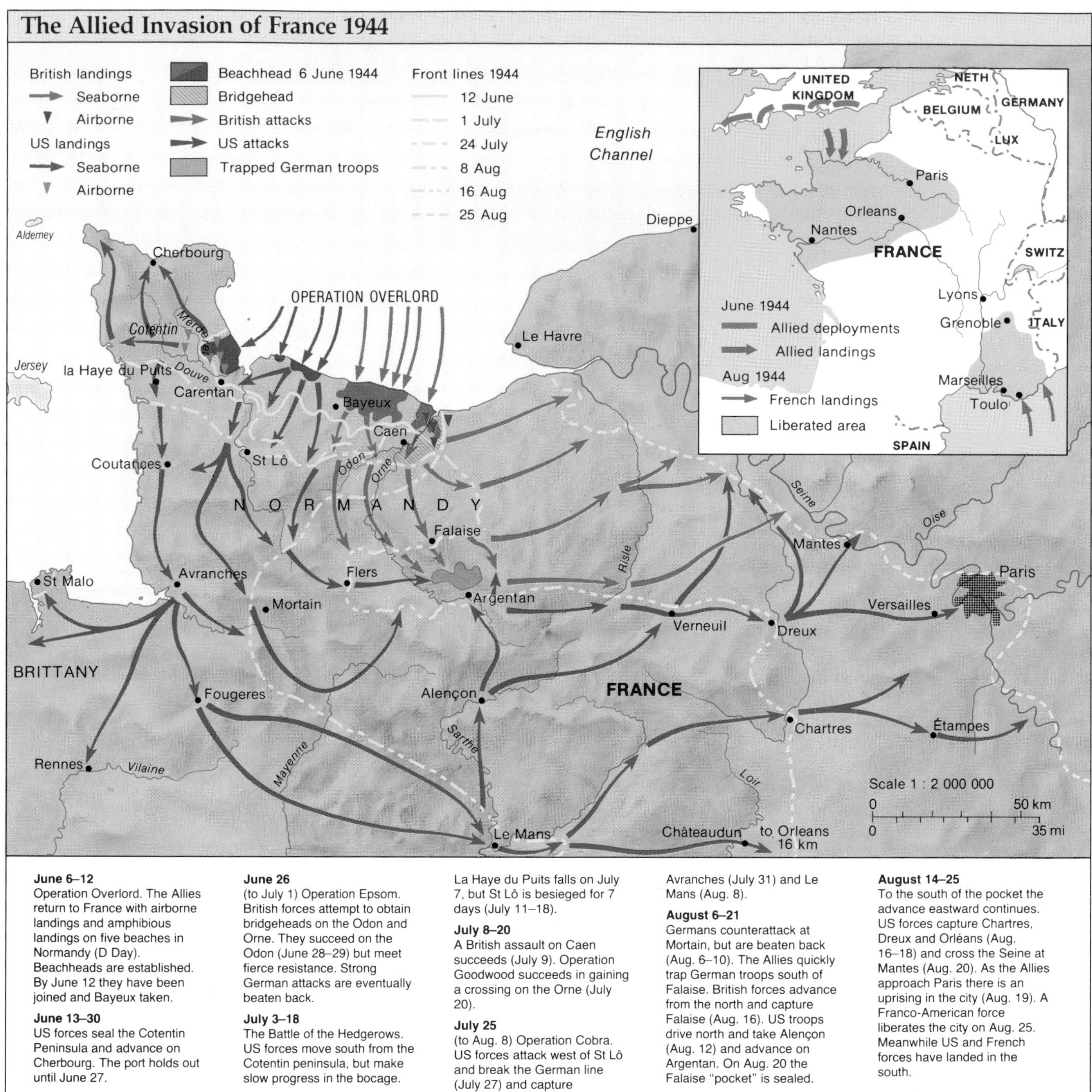

June 6–12
Operation Overlord. The Allies return to France with airborne landings and amphibious landings on five beaches in Normandy (D Day). Beachheads are established. By June 12 they have been joined and Bayeux taken.

June 13–30
US forces seal the Cotentin Peninsula and advance on Cherbourg. The port holds out until June 27.

June 26
(to July 1) Operation Epsom. British forces attempt to obtain bridgeheads on the Odon and Orne. They succeed on the Odon (June 28–29) but meet fierce resistance. Strong German attacks are eventually beaten back.

July 3–18
The Battle of the Hedgerows. US forces move south from the Cotentin peninsula, but make slow progress in the bocage. La Haye du Puits falls on July 7, but St Lô is besieged for 7 days (July 11–18).

July 8–20
A British assault on Caen succeeds (July 9). Operation Goodwood succeeds in gaining a crossing on the Orne (July 20).

July 25
(to Aug. 8) Operation Cobra. US forces attack west of St Lô and break the German line (July 27) and capture Avranches (July 31) and Le Mans (Aug. 8).

August 6–21
Germans counterattack at Mortain, but are beaten back (Aug. 6–10). The Allies quickly trap German troops south of Falaise. British forces advance from the north and capture Falaise (Aug. 16). US troops drive north and take Alençon (Aug. 12) and advance on Argentan. On Aug. 20 the Falaise "pocket" is sealed.

August 14–25
To the south of the pocket the advance eastward continues. US forces capture Chartres, Dreux and Orléans (Aug. 16–18) and cross the Seine at Mantes (Aug. 20). As the Allies approach Paris there is an uprising in the city (Aug. 19). A Franco-American force liberates the city on Aug. 25. Meanwhile US and French forces have landed in the south.

further 2 million to follow in 39 divisions, 138 major warships, 1,000 minesweepers and auxiliary vessels, and 4,000 landing craft or ships. It was to be the greatest seaborne armada in history, the most hazardous operation of its kind.

The Allied conquest of northern France

The most careful planning may sometimes be overturned by the whims of fate. The weather on 4–5 June was atrocious. Eisenhower shouldered the awesome responsibility of ordering the invasion on 6 June with the mundane instruction: "Let's go". His gamble succeeded. All five divisions got ashore safely, though the casualties on Omaha beach were severe. The major problem by the end of June was whether the Allied build-up could outmatch the German. Here the weather turned against the Allies and one of the "Mulberry" harbors (artificial jetties) was destroyed – though this was perhaps less serious than the dislocation of shipping.

The German CinC West, Field Marshal von Runstedt, planned to hold his reserves back until the Allies had committed themselves. Rommel, commander of Army Group B, was conscious of the strength of Allied air power, and urged that the beachhead be destroyed before it grew too strong. A compromise was reached. The infantry was posted near the coast and the armor was held back. In addition, both field marshals had been fooled as to the real direction of the Allied assault. Both expected it to come in the Pas de Calais area. This was an intelligence triumph of immense significance. A brilliant deception plan combined

▲ Montgomery's plan was to draw the German armor on to the British flank, so enabling the Americans to break out on the left. In operations Epsom and Goodwood he fought two great battles of attrition. He did not break out himself, though he often gave the impression that he could have. But these battles made the American advance to Avranches possible.

with the "turning" of German agents convinced the German high command. Thanks to Ultra the Allies could read German signals intelligence and knew that the plan was working. Over half the German strength remained in the north waiting for a phantom army.

Montgomery's "master plan" envisaged that the British army on the left should draw the German armor on to it, permitting the American forces on the right to break out. He had also hoped to take the road center at Caen on the first day. He failed, but its seizure was vital to the security of the expanding bridgehead. Caen, like Tobruk, assumed symbolic importance. Montgomery organized two massive offensives, "Epsom" and "Goodwood" to seize Caen and draw Rommel's armour to the left. He employed massive artillery bombardments and strategic bombers to provide a "colossal crack" at the enemy's defenses. Such a weight of firepower only increased the devastation and rubble blocked the roads, impairing the advance. Mobility was sacrificed for hitting power, but this was unimportant so long as the German armor continued to move eastward on to the British Front.

In this aim Montgomery was successful. Seven of the nine Panzer divisions faced the British 2nd Army by the third week of July. But as usual he claimed too much. He said that the plan had gone "without a hitch", which was clearly untrue. He also claimed that he had never intended to break out on the left flank. This may have been true as far as the letter of the plan was concerned, but not its spirit. Fundamentally, Montgomery was contemptuous of American military ability and believed that he could break out toward the Seine *and* draw the German armor simultaneously on to his front. In this he was wrong, and his failure to break out permitted his enemies, Morgan and Tedder, to gain the ear of the supreme commander, and urge that he be sacked.

Meanwhile, the American forces moved slowly toward St Lô, another road junction. The fighting here was laborious. The terrain – the *bocage* – was ideal defensive country: high hedgerows, winding lanes and numerous hamlets. St Lô eventually fell on 18 July. On 26 July the US 1st and 3rd Armies broke out of the Normandy beachhead in Operation "Cobra". Hitler now took a dramatic hand in the campaign. He had earlier replaced von Runstedt with von Kluge and ordered a counterstroke toward Avranches. By advancing in the opposite direction to the Americans, the Germans guaranteed their envelopment. The Allies monitored the anxious signal traffic between Hitler and von Kluge eagerly. The Germans were entering a trap of their own making. Then on 20 July occurred the "July Plot" against Hitler. Von Kluge disappeared from view and then committed suicide. But the real significance of these events was that the German command system was paralysed.

The Germans hastened to escape in terrible confusion through the Falaise Gap. By the end of August the salient was finally eliminated. The Battle of Normandy was a tremendous victory. The Germans had suffered half a million casualties, including 210,000 prisoners, and lost 2,000 tanks and self-propelled guns. The Allied pursuit began. Patton advanced into Lorraine and Montgomery took Brussels. When on 1 September Eisenhower took over command of the Allied ground forces from a reluctant Montgomery, it seemed that the war would be over in 1944.

▲ American soldiers come ashore on Utah beach. The American experience during the Overlord landings was mixed. At Utah the GIs walked ashore, meeting only light resistance. At Omaha beach they met strong resistance, and gained a toehold very slowly and with heavy casualties. The sea was covered with bodies – like gruesome seaweed.

▼ Vehicles of US 1st Corps are landed at Utah beach. The great array of amphibious landing vehicles added considerably to the flexibility of amphibious operations. Amphibious vehicles permitted the landing of troops and mobile fire power much more quickly than during World War I.

▲ US 82nd Airborne Division dropping over Nijmegen. The concept of operations for the Battle of Arnhem, Operation Market Garden, belied Montgomery's reputation as a cautious commander. It was a bold stroke to secure the bridges over the Rhine and thus secure lines of communication to outflank the Rühr from the north. Most bridges were captured, but the vital bridge at Arnhem was "a bridge too far". The British troops who landed there were isolated and virtually annihilated.

Dwight D. Eisenhower 1890–1969

Eisenhower was the first of a new generation of "political generals", whose strength was politics rather than operational command. Before taking command in the Mediterranean after Operation "Torch" in November 1942, he had not seen a shot fired in anger. He overcame this inexperience with a skill for making difficult personalities work together. He was particularly adept at developing a "team" which dropped national prejudices. Not a great strategist, Eisenhower was the model coalition leader, tactful and cheerful.

Bernard Montgomery 1887–1976

"Monty" had dedicated his life to the profession of arms. He neither smoked nor drank and had no other interests outside the British Army. He was the most proficient operational commander produced by the British army in World War II. Although prudent and enjoying vast material superiority over the enemy, he could put together an ambitious plan on a great scale and see it through to a successful conclusion.

From Belgium to Germany: debates and disaster

Though great in magnitude, however, the Battle of Normandy was another incomplete triumph. The Germans had escaped, and it can be argued that they should not have done. The advance into Brittany was an unnecessary diversion from the main front. Too many Allied commanders showed undue caution in exploiting their success. There is justice in both of these criticisms. But the Allies were, to some degree, prisoners of geography. Because the Americans had to cross the Atlantic and use western British ports, the maneuver wing perforce had to be on Montgomery's right, and not his left. To guarantee the annihilation of Army Group B, the maneuver wing should have been on the left flank so that it could engage the Germans west of the Seine and cut their lines of communication. But to sustain *their* lines of communication across the Atlantic, the Americans needed the Brittany ports.

The extent of Allied success ushered in the most acrimonious debate yet over the course of Allied strategy. Montgomery, perhaps the only Allied commander who was the conceptual equal of his German counterparts at the operational level, urged that a narrow but concentrated thrust be mounted in the north to seize the Ruhr and advance on Berlin. However justified this concept was at the operational level, Eisenhower, highly attuned to political factors, realized that such an operation, carried out by predominantly Commonwealth forces commanded by a British general, was unacceptable to American public opinion. In any case, Eisenhower was temperamentally disposed to military operations in which "the whole team" kept "kicking the ball forward" all along the line. This "broad front", in so far as it can be described as a concept, was his preferred style. It minimized political difficulties.

Montgomery nevertheless gained Eisenhower's permission to mount Operation "Market Garden" to seize the bridges over the Maas and Waal. These would serve as a springboard for an offensive toward the Ruhr that autumn. The American airborne landings at Eindhoven and Nijmegen secured their objectives. The landings of the 1st British Airborne Division at Arnhem did not. The troops were landed too far from the bridges and their fighting power was diffused over too great an area. Of the 10,000 men who landed only 2,163 survivors broke out.

This setback dealt Montgomery's case for a narrow front a severe blow. Indeed, his attention had been so focused on the Rhine that he had overlooked the paramount importance of the port of Antwerp, which he needed to supply his columns. An intricate and time-consuming operation was required to break the German blockade of the port. But Montgomery's greatest enemy was himself. He took no pains to conceal his disdain for the supreme commander. He behaved in such a truculent manner that Eisenhower, urged on by Tedder, moved to have him relieved. Montgomery's chief of staff, de Guingand, realized the danger and after a dramatic flight to Montgomery's tactical HQ persuaded him to write a diplomatic note agreeing to abide by Eisenhower's instructions. The supreme commander was placated and Montgomery survived – this time by the skin of his teeth.

The issue over which Montgomery and Eisenhower clashed was the need, as Montgomery saw it, for a ground forces commander interposed between the supreme commander and his army group commanders. Montgomery only had one candidate in mind for the job: himself. If introduced, this measure would have denied Eisenhower any effective direct control over his armies. Montgomery wanted to see more "grip" over the course of operations. At the operational level, Montgomery was probably right. The Allied forces were overstretched and barely coordinated. The Germans were now fighting on their own soil and contesting every centimeter.

It did not occur to the Allies that this was a prelude to a major counteroffensive. A new commander, Field Marshal Walter Model, had rebuilt Army Group B. It now comprised two Panzer and one infantry army. It was poised for an onslaught toward Antwerp via Namur. This was to be a reprise of the victorious thrust through the Ardennes in 1940. But then the Germans had enjoyed air superiority.

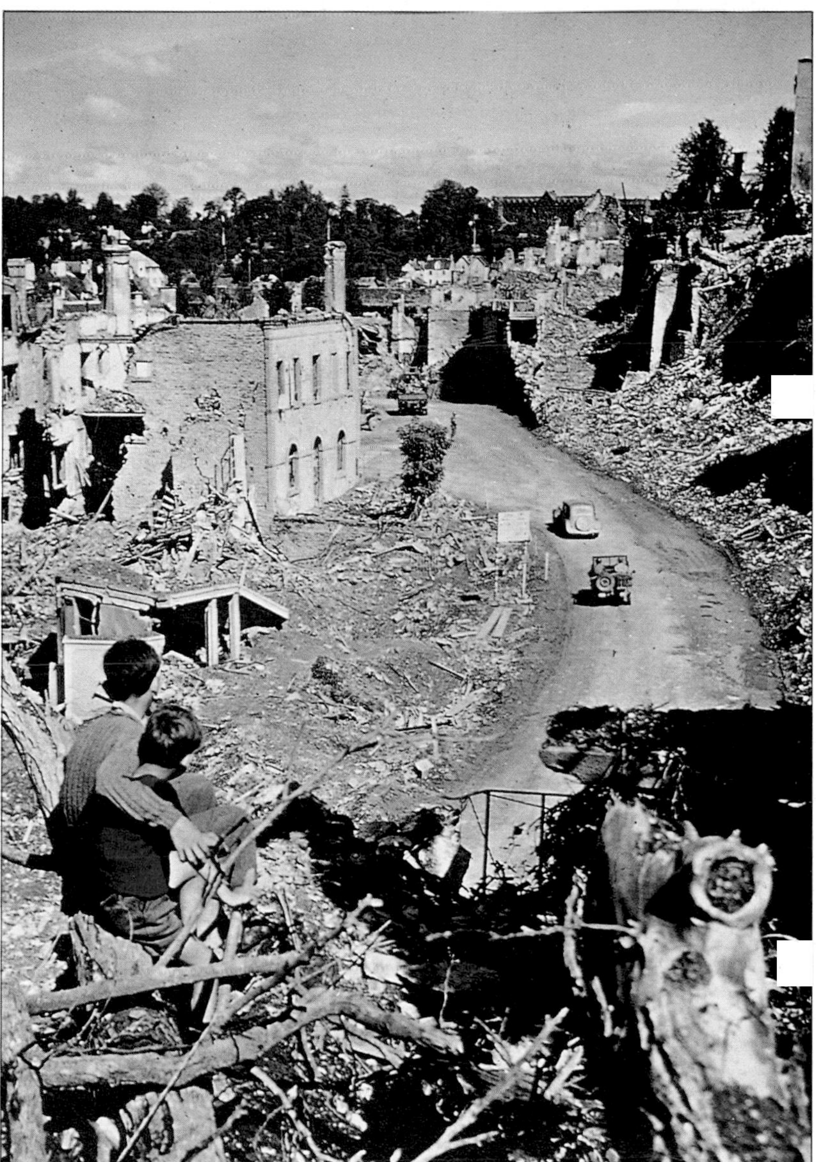

▲ The fruits of liberation, St Lô in July 1944. The only antidote the Allies could find to fanatical German resistance during the Battle of Normandy was overwhelming firepower and systematic destruction. Consequently, Allied offensives were accompanied by massive artillery bombardments which destroyed towns, villages and roads on a greater scale than in 1914–18. During Operation Cobra, the breakout, strategic bombers were used to supplement artillery fire.

Omar N. Bradley 1893–1982

Bradley had risen to prominence on taking over US II Corps in Tunisia. Unlike the flamboyant Patton, Bradley was a reserved and diffident man. A skilled tactician, Bradley was initially Patton's subordinate, and his rise to command 12th Army Group left their relations, at best, as ambivalent. Bradley performed soundly during the Battle of Normandy. During the Ardennes counter-offensive he appeared to lose his grip, but he survived and in 1945, during the advance to the Elbe, commanded the largest force ever raised by the United States.

George S. Patton 1885–1945

Patton was a ruthless cavalry general, at his best in the pursuit. His skill was less evident in the bloody battles in Lorraine in 1944. Although from a wealthy family and himself well-read and cultured, Patton developed the almost obsessive element in his personality. He became the embodiment of the "fire-eater": vulgar, insensitive and foul-mouthed. Hardly the natural diplomat, his snide asides at the British and his temper – notably the notorious "slapping" of a shell-shocked patient in Sicily in 1943 – almost destroyed his career.

I found Miley, Commander of the 17th Airborne, in Bastogne. While there we had considerable shelling, including airbursts. The flashes of our own guns and those of the enemy in the gathering darkness against the white snowfields were very beautiful, but not too reassuring. In my diary I made this statement on the afternoon of January 4, and it is significant, as it is the only time I ever made such a statement: "We can still lose this war."

GEORGE S. PATTON

The Germans' last offensive

As Montgomery wrote, the Allied armies "sprawled all over the place". There was no reserve. Even Ultra intelligence gave no hint of the attack. When the Germans struck on 16 December the surprise was complete. But the German offensive soon ran into difficulties, as progress in the center was less than hoped for. The Americans held Bastogne, an important road center – a thorn in the German rear. Montgomery had to be given command of all forces north of the breakthrough. This amounted virtually to the ground forces command he had advocated. Montgomery provided a steady hand and slowly but surely took control of the battle and provided himself with a reserve from 21st Army Group.

Eisenhower now planned a counteroffensive in which both Bradley and Montgomery were to advance simultaneously. This Montgomery considered a waste of strength. He could not understand that this move was required for political rather than military reasons. The USA was by now the dominant partner in the Alliance. American generals were calling the tune.

Patton relieved Bastogne. The Germans withdrew, and on 16 January 1945 American forces linked up at Houffalize, effectively pinching out the German "Bulge". The significance of the battle was that it shattered Hitler's last strategic reserve. By throwing this against the western Allies rather than the Russians, Hitler guaranteed that a great chunk of Germany, including Berlin, would fall to the Red Army.

Across the Rhine – to victory

West of the Rhine, the Germans now faced seven armies. In a great pincer movement reminiscent of the Normandy campaign, Montgomery advanced to the Rhine (Operation "Veritable"), drawing what remained of German strength northward to permit Bradley to smash his way to the Rhine (Operation "Grenade"). This strategy exceeded expectations when Bradley seized an intact bridge at Remagen. But the material inferiority of the German army did not diminish its fighting power, and the fighting during "Veritable" was some of the bitterest of the war.

On 22 March Patton crossed the Rhine and Montgomery the following day. On regrouping Eisenhower decided that once the Ruhr was encircled, US 9th Army, previously under Montgomery, would revert to Bradley. The US 12th Army Group would now comprise 48 divisions, the largest single command in American history, and the source of patriotic pride. The political impossibility of a British officer commanding this force was now apparent.

▶ It is a much quoted axiom that in war an attacker must have a 3:1 superiority. The Allied crossing of the Rhine illustrates the massive superiority the Allies now enjoyed – amounting on some fronts to 10:1. It was not so much numbers but equipment which overwhelmed the last legions of the Third Reich

The Battle of the Bulge 1944-45

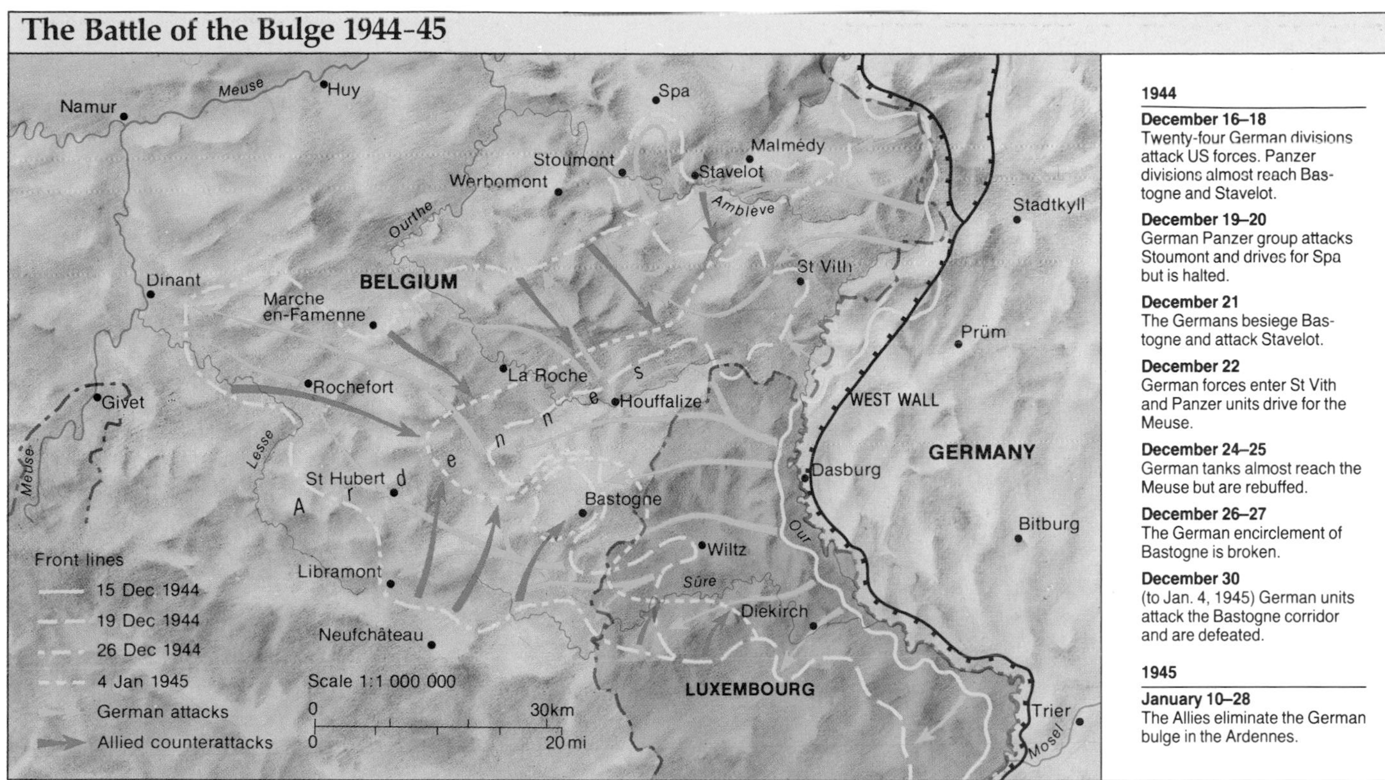

1944

December 16–18
Twenty-four German divisions attack US forces. Panzer divisions almost reach Bastogne and Stavelot.

December 19–20
German Panzer group attacks Stoumont and drives for Spa but is halted.

December 21
The Germans besiege Bastogne and attack Stavelot.

December 22
German forces enter St Vith and Panzer units drive for the Meuse.

December 24–25
German tanks almost reach the Meuse but are rebuffed.

December 26–27
The German encirclement of Bastogne is broken.

December 30
(to Jan. 4, 1945) German units attack the Bastogne corridor and are defeated.

1945

January 10–28
The Allies eliminate the German bulge in the Ardennes.

After the great victories west of the Rhine, the movement of Allied armies was barely contested. In a great double envelopment the US 9th and 1st Armies completed the encirclement of the Ruhr. On 18 April this offered up 400,000 prisoners – the greatest single capitulation of the war. On 21 April Model committed suicide. Hitler now lived in a world of daydreams.

As the end neared, Eisenhower's generalship became increasingly cautious. The main objective of the US 12th Army Group, he ordered, was to be the Elbe. The 21st Army Group was to protect Bradley's flank and move on Lübeck. Churchill protested and claimed that Eisenhower's objective should be Berlin. Roosevelt supported Eisenhower in overruling Churchill's protests. They claimed that Berlin was no longer of strategic importance. This was untrue. Here lurked Hitler, still exerting his will over the German people.

What was more in doubt was not the importance of Berlin but the practicality of its capture. By 31 March the nearest Allied troops were still 440 km (275 mi) away, while the Russians were only 65 km (40 mi) from the city. Any drive to Berlin also had to take into account the possibility that American and Russian troops might accidentally clash. As ever, Eisenhower erred on the side of caution. As his armies closed to the Elbe, Eisenhower knew that the Allies had already agreed at Yalta on the division of Germany into zones. Any further advance eastward would have had to be withdrawn after the war had ended.

The war in Europe moved to a close. On 1 May 1945 Hitler's death was announced. Admiral Dönitz succeeded him as *Führer*. On 4 May Montgomery received the unconditional surrender of all German forces in Holland, Northwest Germany and Denmark. In dealing with Eisenhower, Dönitz attempted to extend the negotiations so that the greatest number of German soldiers and refugees "will find salvation in the West". Jodl, his plenipotentiary, tried to play for time. Eisenhower threatened to seal the Western Front, and Jodl had no choice but to sign the instrument of unconditional surrender on 8 May.

▲ Hitler's last throw. The immediate object was the Meuse crossings at Namur. The plan was over-ambitious, and could not be sustained even with surprise and poor visibility which grounded the Allied air forces.

◄ American troops greeted with joy by the citizens of Aachen. Elsewhere their encounters with the former citizens of the Third Reich were less enthusiastic.

▼ Montgomery in solemn mood signing the 21st Army Group instrument of surrender. Nothing was done to spare the sensibilites of the emissaries. On being introduced to the field marshal, each member of the German delegation received the withering reply, "Never heard of you."

Datafile

In summer 1945 Japan had under arms more than twice the number of men with which it had started the Pacific conflict, and it was clear that most of them would fight like the garrisons of the Pacific islands and sacrifice their lives for relatively light (in proportional terms) American casualties. These casualties, of course, were viewed very differently by the Americans, who feared that an invasion of the Japanese home islands would cost hundreds of thousands of lives. But Japan's power to resist was being undermined, by a submarine blockade and by a remorseless aerial offensive which the Japanese airforce could no longer oppose effectively.

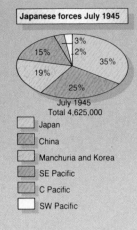

Japanese forces July 1945

3%
2%
35%
25%
19%
15%

July 1945
Total 4,625,000

Japan
China
Manchuria and Korea
SE Pacific
C Pacific
SW Pacific

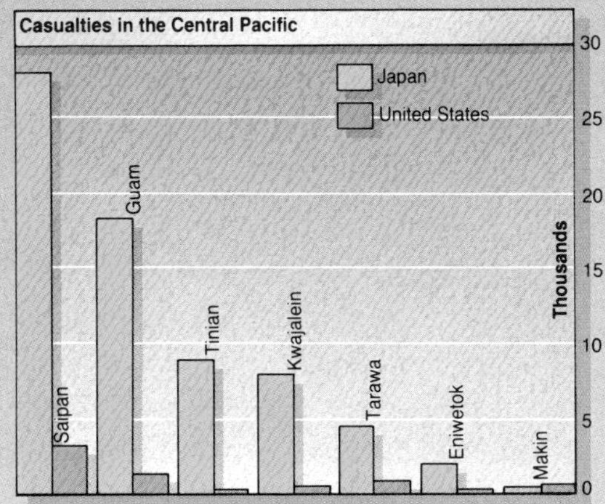

▲ Despite the fact that Allied invasion forces were assembling in the Philippines and Okinawa, in July 1945 more than 2 million of Japan's 4.6 million soldiers were still deployed on the Asian mainland. The relatively large forces still holding out in the Central and Southwest Pacific and in Southeast Asia were cut off and could be of no assistance in the event of an invasion of the home islands, so the Japanese government attempted to mobilize civilians to face the Allies.

Bombing raids on Japan

B29s per raid
600
400
200
0
Tokyo 9 March
Osaka 13 March
Kobe 15 March
Nagoya 16 May
Tokyo 23 May
Tokyo 25 May
Yokohama 29 May
Kobe 5 June
Osaka 7 June
1945

▲ The casualty statistics for the Central Pacific "island hopping" campaign graphically illustrate the Japanese method of sacrificing every man of a garrison in order to maximize the loss to an invader. The knowledge that their enemy would fight to the end had a demoralizing effect on even elite US units.

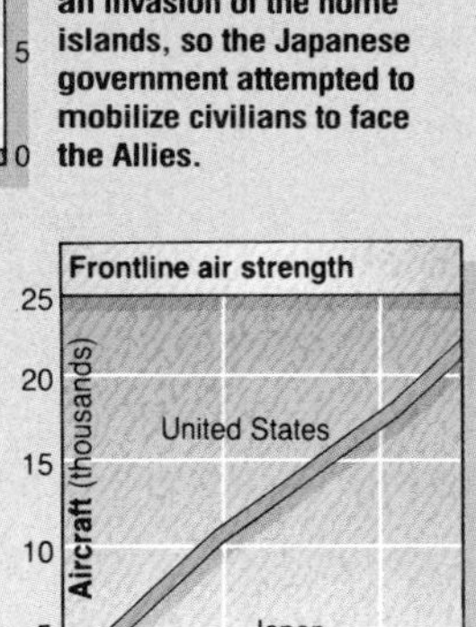

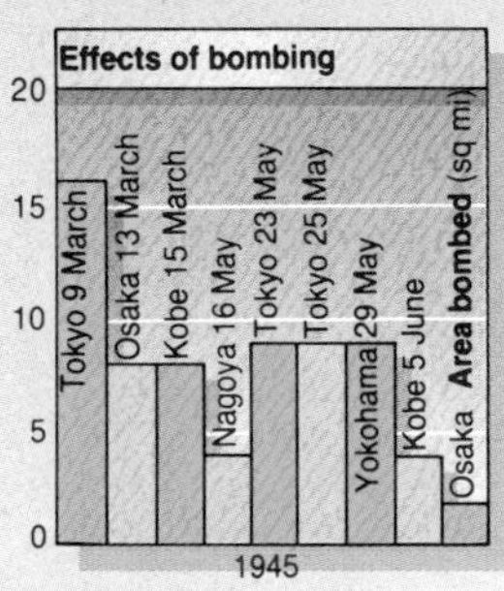

▼ Initial US high-level bombing raids from June 1944 did little damage to Japanese cities. On 9 March 1945 the first low-level incendiary raid was carried out and within three months nearly 40 percent of Japan's five major cities, Tokyo-Yokohama, Kawasaki, Osaka, Kobe and Nagoya, had been destroyed.

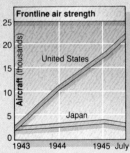

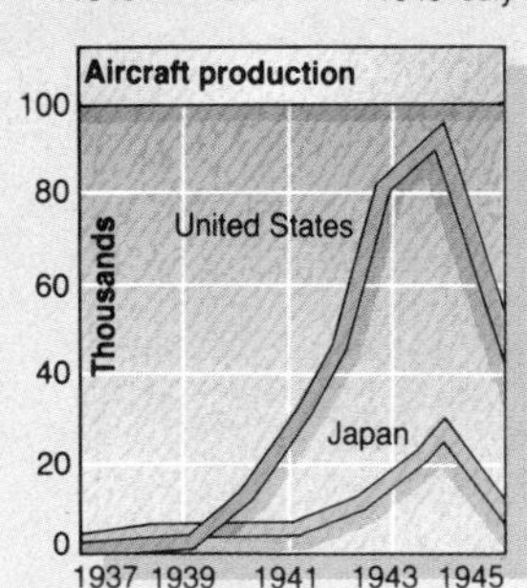

▲ A comparison of the nine heaviest raids showing the number of bombers employed and the area destroyed. The most devastating raid was the first, though the largest came on 23 May 1945 when 562 B-29s hit Tokyo. By this time much of the city had been burnt out so the raid was relatively less effective.

Major Japanese cities

38%
62%
Total 275 sq mi

Area remaining
Area bombed

▲ A graphic presentation of aerial supremacy. By 1945 America deployed five times as many war planes as Japan did. It was not just the B-29s which wreaked havoc. By the summer of 1945 fighters from Iwo Jima also and fighter bombers from American and British carriers roamed at will in the skies over Japan.

Between March and June 1943 America prepared for the next round. The Japanese, determined to hold an 8,000km (5,000mi) perimeter running from Attu via the Marshalls and Gilberts down to the Central Solomons and Eastern New Guinea, concentrated Yamamoto's Combined Fleet at Truk, a central position from which any penetration could be countered. For their part the Americans planned a two-pronged breakthrough. MacArthur's forces, supported by Halsey, were to advance on Rabaul (Operation Cartwheel) in June while Nimitz's forces would drive into the Gilberts in the fall. From March onward Japanese airstrikes sought to disrupt Allied preparations but incurred crippling losses while Allied squadrons penetrated ever deeper into Japanese territory. On 18 April 1943 one such foray sent Admiral Yamamoto, who was flying from Rabaul to Bougainville, to a fiery death.

From island hopping to leapfrogging

Halsey's forces – 1 New Zealand and 6 American divisions, 1,800 aircraft, and the 3rd Fleet which included 2 carriers and 6 battleships –

THE DEFEAT OF JAPAN

The US plan for a two-pronged offensive
The attack on New Georgia and the consequent change of strategy
June 1944: the capture of the Marianas
The return to the Philippines
The bloody capture of Iwo Jima and Okinawa
Japanese preparations for the Allied invasion of Japan
The dropping of the atomic bombs and Japan's surrender

struck Japanese-held New Georgia on 2 July. Despite a superiority of 4:1, it took five weeks to capture the enemy base at Munda, only 16km (10 mi) from the beachhead. At sea the Japanese displayed equal tenacity, though their loss of 17 ships for 6 American showed that the balance of skill had shifted.

The increasing effectiveness of their navy notwithstanding, the Americans knew that if the reduction of all Japan's outposts proved as costly, the war could drag on into the 1950s. The joint chiefs of staff (JCS) now decided on a fundamental change in strategy, from "island hopping" (taking each base in succession) to "leapfrogging" (bypassing large Japanese concentrations and using naval and air forces to seal them up). It fell to Halsey's forces to put this into effect. On 14 August his assault forces sailed past heavily defended Kolombangara, the next island in the Solomon chain, and the following morning landed on Vella Lavelle to the northwest, surprising the small garrison. A brilliant success, this operation set the pattern. Bougainville, the next target, was defended by 60,000 Japanese. On 27 October

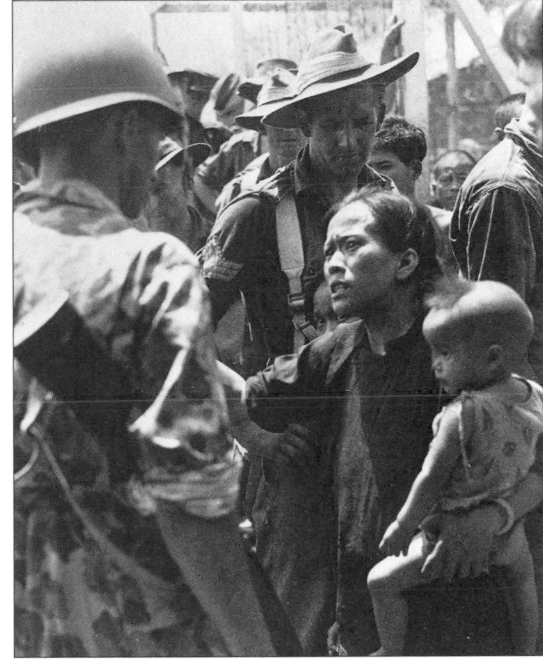

► ▼ Australia played a key role in the Pacific War. Right, "Diggers" and "Yanks" care for bewildered Indonesians after the bitter fighting for Tarakan in Borneo, 1 May 1945. Below, Australia's 9th Division lands on the Huon Peninsula, 22 September 1943 when they cleared the way for an American landing in New Britain.

Allied landings on two small islands to the east and southeast, convinced the Japanese that the attack would come from this direction. It did not. On 1 November the Americans landed at Empress Augusta Bay on the remote west of Bougainville. The following night they defeated a desperate naval counterattack and three months later smashed a belated land assault.

Meanwhile MacArthur's forces – 11 Australian and 4 American divisions, 1,000 aircraft and the 7th Fleet – had been carrying out their own part of "Cartwheel". In a complex operation, beginning on 4 September, three Australian divisions took the Japanese bases of Lae and Salamaua on New Guinea's north coast, and a month later captured the Huon Penisula which controlled the approach to southern New Britain. The JCS had already decided that heavily defended Rabaul was also to be isolated, a task which was effected by the US marines' seizure of southern New Britain in December, followed in succeeding weeks by the capture of islands which dominated the northwestern and eastern approaches to the base. By March 1944 the 140,000 Japanese in Rabaul and on the neighboring New Ireland and Bougainville were for all intents and purposes out of the war.

With Rabaul isolated MacArthur pressed ahead with the most ambitious bypass operation yet – Operation Reckless, a leap to Hollandia 800km (500mi) west of the most forward Allied base. The landing on 22 April completely surprised the Japanese. During the next three months the Americans continued to leapfrog westward until on 30 July they occupied Sansapor on the north-

Chester W. Nimitz 1885–1966

Appointed to command the US Pacific fleet shortly after Pearl Harbor, Nimitz achieved a near incredible defensive victory at Midway in June 1942. During the following 17 months, while his surface fleets were expanded, Nimitz launched a submarine offensive which within two years had crippled Japan's merchant marine. In November 1943 Nimitz's task forces went on the offensive, leapfrogging across the Central Pacific, and defeating the Combined Fleet in the Battle of the Philippine Sea and the Battle of Leyte Gulf.

William F. Halsey 1882–1959

The epitome of the aggressive "seadog", the American press christened the Admiral "Bull" Halsey after he had commanded America's first counterstrokes against Japan. Due to ill health he missed Midway but later in 1942 he was ordered to the South Pacific in which theater he cooperated closely with MacArthur and the forces of Southwest Pacific Command. Halsey was brave but impetuous. His decisions during Leyte Gulf indicated that he was not a strategist of the first rank.

Curtis Emerson Lemay 1906–

Known by his men as "Iron Ass", Lemay first saw combat as commander of the 3rd Bombardment Division based in England in June 1943. The following summer he went to China, where, with Claire Chennault, he launched a brilliantly successful low-level B-29 raid against Japanese-occupied Hankou on 18 December 1944. A month later Lemay, now commander of the Mariana-based 21st Bomber Command, introduced new low-level methods of attack which bore fruit in spectacular incendiary raids from March onward.

western tip of New Guinea's Vogelkop Peninsula, only 1,300km (800mi) southeast of Mindanao in the Philippines.

By now Japan was also reeling from the assaults of Nimitz's forces – Admiral Spruance's 5th Fleet, consisting of 7 battleships, 11 carriers with 1,000 aircraft, and 44 cruisers and destroyers and the 5th Amphibious Force and V Amphibious Corps, which together could transport 150,000 men. In mid-November 1943 this mighty force struck into the Gilberts and on the 20th the Americans stormed ashore on Makin and Tarawa. They captured Makin quickly but took four days and 3,000 casualties to eliminate Tarawa's 4,700 strongly entrenched defenders.

The consequences of Tarawa

Like New Georgia, Tarawa caused a reevaluation of strategy. When Nimitz's fleets surged into the Marshalls in January 1944 they bypassed heavily defended atolls on the rim of the group and while fast carrier task forces wreaked havoc as far apart as Eniwetok and Truk, the Americans attacked and captured Kwajalein in the center of the group, suffering relatively light casualties. Exploiting this success, between March and May a fast carrier force rampaged through the Carolines, encouraging the Japanese to think that they were the Americans' next target. But on 15 June Spruance's fleet arrived off Saipan, the main island of the Marianas, only 2,000km (1,300mi) from Japan and within bombing range of Tokyo.

The 9 carriers, 5 battleships and 41 cruisers and destroyers of Japan's Combined Fleet had been husbanded for just such an occasion. Admiral Ozawa, the operational commander, ordered it to Saipan, hoping that its 473 carrier aircraft, combined with 100 aircraft from in the Marianas, would tip the balance in Japan's favor. Disaster struck

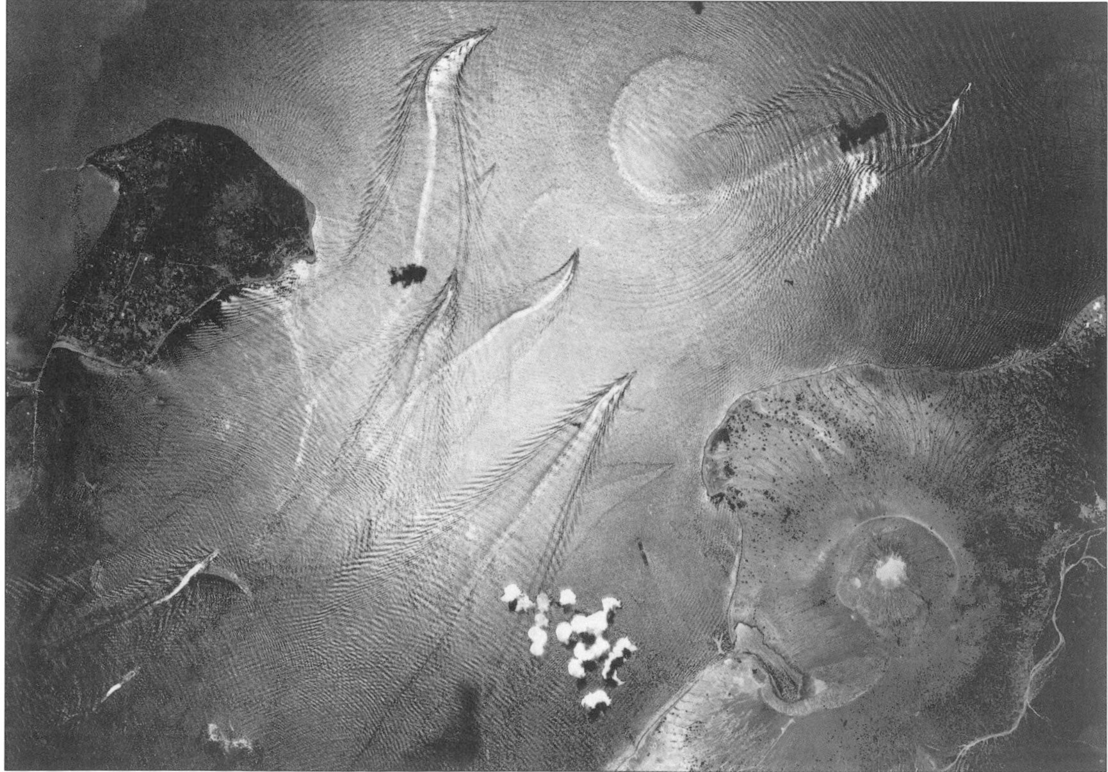

▶ The climax of Operation Cartwheel saw a series of devastating raids on Rabaul Harbor. This picture, taken by the bomb-aimer of a B-24 "Liberator" on 30 December 1943, shows bombs bursting across the anchorage, as warships desperately seek to escape the confines of the harbor.

▶▶ Beginning in June 1943, America launched twin offensives to break through Japan's defensive perimeter. In the Southwest Pacific, Operations Cartwheel and Reckless (see inset) isolated large Japanese armies in the Solomons and northern New Guinea and brought MacArthur's forces to within striking distance of the Philippines. Meanwhile Nimitz's fleets in the Central Pacific smached through the Gilberts, and leapfrogged their way through the Marshalls and the Marianas, bypassing the heavily defended Carolines to link up with MacArthur's forces for the Philippine invasion.

The Defeat of Japan 1943-45

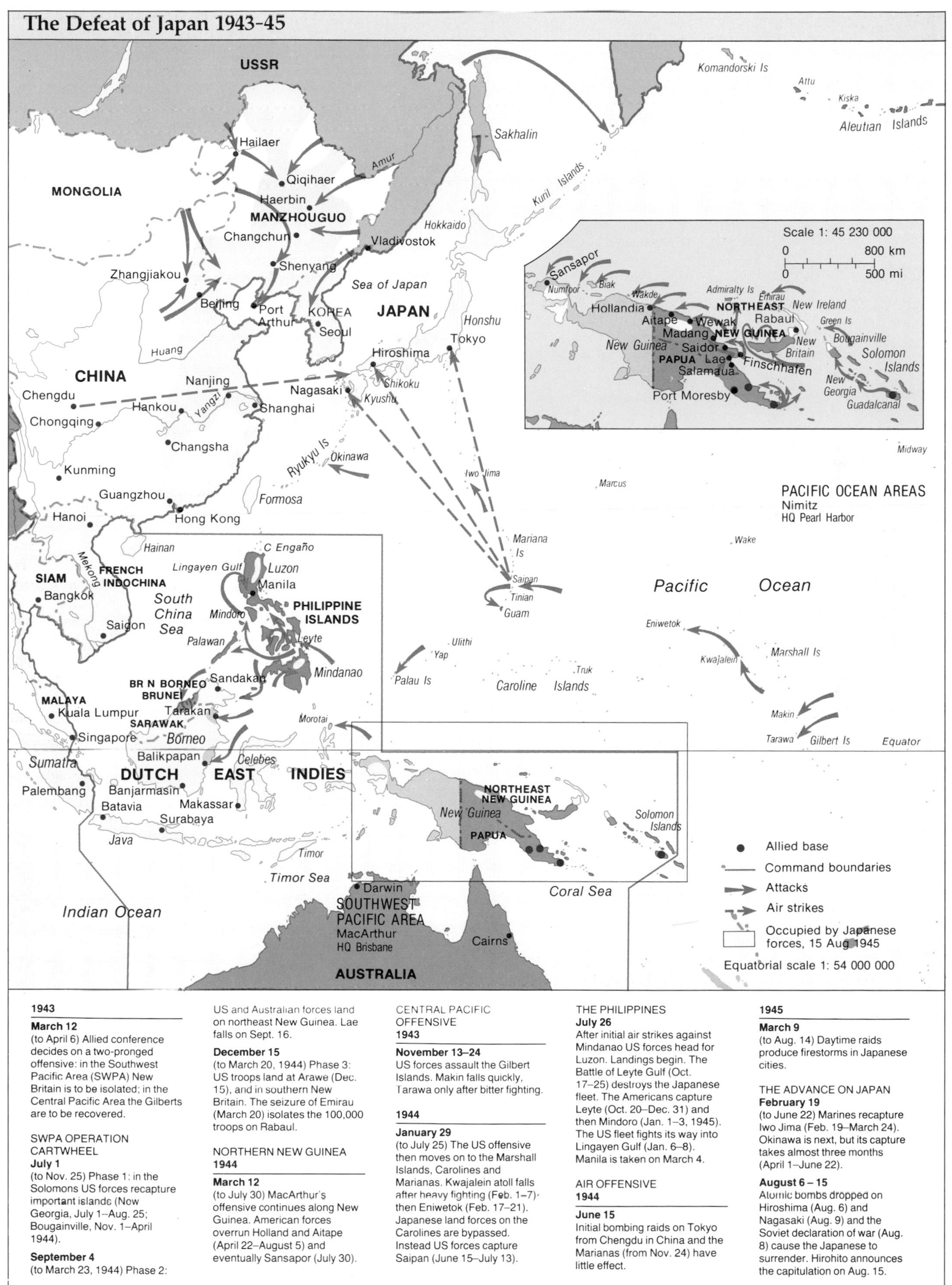

1943

March 12
(to April 6) Allied conference decides on a two-pronged offensive: in the Southwest Pacific Area (SWPA) New Britain is to be isolated; in the Central Pacific Area the Gilberts are to be recovered.

SWPA OPERATION CARTWHEEL

July 1
(to Nov. 25) Phase 1: in the Solomons US forces recapture important islands (New Georgia, July 1–Aug. 25; Bougainville, Nov. 1–April 1944).

September 4
(to March 23, 1944) Phase 2: US and Australian forces land on northeast New Guinea. Lae falls on Sept. 16.

December 15
(to March 20, 1944) Phase 3: US troops land at Arawe (Dec. 15), and in southern New Britain. The seizure of Emirau (March 20) isolates the 100,000 troops on Rabaul.

NORTHERN NEW GUINEA

1944

March 12
(to July 30) MacArthur's offensive continues along New Guinea. American forces overrun Holland and Aitape (April 22–August 5) and eventually Sansapor (July 30).

CENTRAL PACIFIC OFFENSIVE

1943

November 13–24
US forces assault the Gilbert Islands. Makin falls quickly, Tarawa only after bitter fighting.

1944

January 29
(to July 25) The US offensive then moves on to the Marshall Islands, Carolines and Marianas. Kwajalein atoll falls after heavy fighting (Feb. 1–7) then Eniwetok (Feb. 17–21). Japanese land forces on the Carolines are bypassed. Instead US forces capture Saipan (June 15–July 13).

THE PHILIPPINES

July 26
After initial air strikes against Mindanao US forces head for Luzon. Landings begin. The Battle of Leyte Gulf (Oct. 17–25) destroys the Japanese fleet. The Americans capture Leyte (Oct. 20–Dec. 31) and then Mindoro (Jan. 1–3, 1945). The US fleet fights its way into Lingayen Gulf (Jan. 6–8). Manila is taken on March 4.

AIR OFFENSIVE

1944

June 15
Initial bombing raids on Tokyo from Chengdu in China and the Marianas (from Nov. 24) have little effect.

1945

March 9
(to Aug. 14) Daytime raids produce firestorms in Japanese cities.

THE ADVANCE ON JAPAN

February 19
(to June 22) Marines recapture Iwo Jima (Feb. 19–March 24). Okinawa is next, but its capture takes almost three months (April 1–June 22).

August 6 – 15
Atomic bombs dropped on Hiroshima (Aug. 6) and Nagasaki (Aug. 9) and the Soviet declaration of war (Aug. 8) cause the Japanese to surrender. Hirohito announces the capitulation on Aug. 15.

▲ American troops use flamethrowers to incinerate Japanese in bunkers in southern Okinawa, 18 April 1945. On that day XXIV Corps Commander Major John Hodge said: "It is going to be really tough. There are sixty-five thousand fighting Japs holed up in the south end of the island and I see no way to get them out except blast them out yard by yard." He was right.

almost immediately: American submarines and aircraft sank three carriers and shot down 411 Japanese machines. It was the end of Japan's carrier aviation. By mid-July the Americans had secured Saipan and in August took Guam and Tinian, giving them firm control of the Marianas.

On 24 July 1944 MacArthur, Nimitz, and Roosevelt met in Hawaii to determine strategy. Nimitz wanted a navy-dominated thrust from Saipan to Formosa and China, bypassing the Philippines, while MacArthur urged the reconquest of Mindanao and then a drive north to Luzon. Roosevelt supported MacArthur. The liberation of the Philippines was a moral obligation; besides, it seemed unwise to leave a 350,000 strong Japanese garrison active to the American rear. He now ordered MacArthur's and Nimitz's thrusts to converge on Mindanao.

By 15 September MacArthur's forces had secured Morotai, halfway between New Guinea and Mindanao, while Nimitz's marines fought for Pelelieu, 800km (500mi) to the northwest. Feeble resistance to airstrikes against Mindanao bases now convinced the JCS that the southern Philippines could be safely bypassed and MacArthur's forces were ordered to link up with Nimitz and land at Leyte in the central Philippines.

On 17 October, as 700 Allied vessels carrying 200,000 troops, approached Leyte, Japan launched a daring counterattack. The entire Combined Fleet (64 warships) converged on Leyte from three directions. From the north came Japan's four surviving carriers, their object to lure Halsey's 3rd Fleet away from the beachhead. Meanwhile a central and a southern force, in total seven battleships, threaded their way through a maze of islands to catch the unprotected transports in Leyte Gulf. On 23 October Halsey's submarines and aircraft detected and attacked the central force, sinking a battleship and forcing it to put about. Believing he had seen off this threat and having received word of the approach of the carriers, Halsey sped northward and on 25 October intercepted and sank all four off Cape Engano, the northeastern tip of the Philippines. Meanwhile the Americans had located Japan's southern task force approaching Leyte through the Surigao Strait. All the battleships covering the landing force sped south to the entrance of the strait and on the night of 24–25 October ambushed the southern force, sinking two more battleships. During the night, however, the central force had reversed course, and at dawn debouched unopposed from the San Bernadino Strait onto the now lightly protected transports. A desperate running fight ensued. Japanese battleships sped after the Americans but at the very point of victory put about once more, in the mistaken belief that airstrikes from small escort carriers had in fact come from Halsey's 3rd Fleet. This failure of nerve turned the Battle of Leyte Gulf into an unmitigated disaster for the Japanese. They had sacrificed their carriers to no avail and had also lost 500 aircraft, three battleships, and 14 other warships. It was the end of the Imperial Navy.

Ashore resistance was fierce. The Japanese commander, General Yamashita, boosted Leyte's garrison to 70,000 but by year's end it had been overwhelmed. Yamashita determined upon a desperate expedient to prevent the capture of Luzon. During the Battle of Leyte Gulf newly formed suicide squadrons known as kamikazes (the word meant "divine wind") had crippled several American ships. Used en masse in the first week of Janauary 1945 the kamikazes badly damaged 20 ships of the invasion fleet as it approached the west coast of Luzon, but they could not prevent a landing. By 9 January 4 divisions were ashore striking south and east. In conformity with a preconceived plan the Japanese avoided pitched battles in which the Americans could bring their fire power to bear and withdrew to the Manila Bay area and the mountains of northeastern Luzon. The Americans reached the

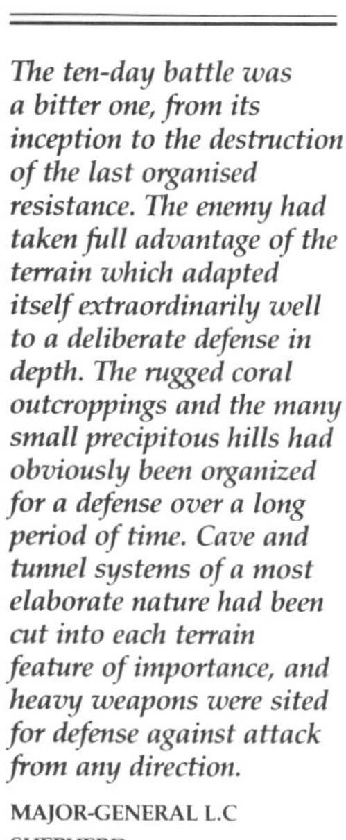

The ten-day battle was a bitter one, from its inception to the destruction of the last organised resistance. The enemy had taken full advantage of the terrain which adapted itself extraordinarily well to a deliberate defense in depth. The rugged coral outcroppings and the many small precipitous hills had obviously been organized for a defense over a long period of time. Cave and tunnel systems of a most elaborate nature had been cut into each terrain feature of importance, and heavy weapons were sited for defense against attack from any direction.

MAJOR-GENERAL L.C SHEPHERD
OKINAWA, 13 JUNE 1945

▲ The official surrender of Japan, Tokyo Bay, 2 September 1945. On the quarterdeck of USS *Missouri*, Admiral Nimitz's flagship, Japan's delegation, led by top-hatted foreign minister Manoru Shigemitsu, approaches the Allied commanders. Worried that the ceremony might be marred by an unseemly conflict between arch rivals MacArthur and Nimitz, US secretary for the navy, J.V. Forrestal, arranged a compromise. Both men would witness Japan's surrender – MacArthur as supreme commander and Nimitz as representative of the United States.

◀ 11.00 am, 9 August 1945. A photograph taken from the "Great Artiste", one of the observation B-29s. Ten kilometers (6mi) below in the inferno that was Nagasaki 35,000 are already dead and another 60,000 seriously injured.

capital on 3 February but it took them a month of bitter fighting, during which the city was destroyed and more than 100,000 Filipinos were killed, to eliminate the garrison. By the end of March Luzon had been secured except for a pocket in the northeast where Yamashita and 50,000 survivors continued to hold out until Japan's capitulation.

The approach to Japan

The battle for Manila was still raging when Nimitz's fleet sailed into the Bonins, an island group only 1,100km (700mi) southeast of Japan, to secure advanced bases for an invasion of the home islands. On 17 February two marine divisions stormed ashore on 20sq km (8sq mi) Iwo Jima, and took five weeks and 26,000 casualties to exterminate the 25,000 strong Japanese garrison. In the last week of March the largest armada ever assembled in the Pacific, 1,450 ships including the carriers and battleships of the newly formed British Pacific Fleet, sailed for Okinawa in the Ryukyu Islands only 560km (350mi) south of Kyushu. Tokyo was determined to hold it at all costs – a 130,000-man garrison and hundreds of kamikaze and conventional squadrons defended the island. During the next eight weeks Japanese airstrikes – kamikaze and conventional – sank 36 and damaged 368 Allied ships, destroyed 800 aircraft, and inflicted nearly 10,000 casualties. This, the most sustained aerial assault ever mounted against a fleet, cost the Japanese a staggering total of 8,000 aircraft. In the midst of the aerial attacks the Imperial Navy despatched the monster battleship *Yamato* on its own kamikaze mission against the Allied fleet, but swarms of American aircraft sank her well to the north of Okinawa. Ashore the fighting was the bloodiest of the Pacific War. The 300,000 troops landed in the first days of April took 11 weeks and 39,000 casualties to kill 137,000 enemy troops and 100,000 Japanese civilians – a grim foretaste of the likely cost of an invasion of the home islands.

Plans for a landing on Kyushu scheduled for 1 November (Operation Olympic) were well advanced, and while invasion forces assembled and trained pressure was kept up by an intensification of both the submarine blockade and strategic bombing raids. Since the beginning of their campaign in 1942, America's submarines had sunk more than 1,100 merchant ships amounting to 4,800,000 tons, around 56 percent of the total lost by Japan. The effect on Japanese industry had been apparent as early as 1943 and by early 1945 the import of food was also being severely disrupted, leading to widespread malnutrition and in some areas actual starvation. America's strategic bombing campaign was also becoming effective.

Japan's situation was hopeless, but on 27 July the government of Admiral Kantaro Suzuki (Tojo had resigned the previous July after the fall of Saipon) rejected the Allied demand of unconditional surrender made at Potsdam. Nothing remained but to fight it out – Japan might lose but Allied casualties would be horrific. It planned to meet the invasion fleet with 8,000 carefully husbanded aircraft (many of them kamikazes) and hundreds of explosive-packed suicide boats and "human torpedoes". Ashore two million regular soldiers and a vast "home guard" would dispose of even more invaders.

The Americans were aware of Japan's preparations. On 16 July the secret "Manhattan Project" had borne fruit with the explosion of an atomic bomb at Alamogordo in New Mexico. President Truman (Roosevelt had died on 12 April) immediately authorized its use. At 8.00 am on 6 August a single B-29 dropped an atomic bomb on Hiroshima, destroying the city and killing 80,000 instantaneously. Japan's cabinet met on 8 August but still a majority wished to fight on. On the morning of 9 August Japan reeled under two new shocks – a second atomic bomb devastated Nagasaki and Japan's ambassador in Moscow informed Tokyo of a Soviet declaration of war, which was followed within the hour by a massive Soviet invasion of Manchuria. Now at war with two superpowers, its cities being systematically destroyed by incendiary raids and a new and terrifying bomb, its industrial production at only one third of prewar levels and many of its population starving, Japan's cabinet on 10 August accepted the Potsdam Declaration, asking only that the emperor remain sovereign. It was a request to which the Americans readily acceded and on 15 August Hirohito broadcast Japan's capitulation.

PART 2

THE FIGHTING SERVICES

GLOSSARY

Air superiority
Aerial command of the battlefield, obtained by the predominance of friendly aircraft over those of the enemy.

Airborne operations
Military combined air operations in which the attacking troops are conveyed to the battlefield by air, and landed by a parachute or glider.

Aircraft carrier
Capital ship equipped with a flight deck from which aircraft operate.

Antiaircraft gun
Large ground- or ship-mounted gun for use against aircraft.

Antitank gun
Artillery piece for use against tanks and armored vehicles.

Armored forces
Mobile armored striking forces of tanks, self-propelled artillery and mechanized infantry.

Artillery
Large-caliber mounted firearms, ranging from light field guns to heavy guns mounted on railroad waggons.

Atomic bomb
Nuclear fission bomb with immense destructive force, dropped on Hiroshima and Nagasaki in Japan in August 1945.

Axis
The military alliance of Germany, Italy and Japan.

Battlegroup
Fleet of warships of various types gathered to engage an enemy force.

Battleship
Heavily armored capital ship, with main armament of heavy caliber guns.

Blitzkrieg
(Ger: "lightning war"). Offensive strategy of rapid breakthrough and exploitation by armored forces.

Bomber
Aircraft designed for delivering explosive projectiles against ground targets.

Breakthrough
The rupturing of the enemy's line of defense by an assault.

Capital ship
Large powerful ship, such as a battleship or aircraft carrier, comprising the main offensive force of the fleet.

Combined operations
Operations involving the use of a combination of arms, military, aerial and naval.

Commerce raider
Armed surface vessel for disrupting the enemy's seaborne trade by sinking merchant ships.

Convoy
A group of merchant ships sailing together under warship escort for protection against attack.

Counteroffensive/counterattack
An offensive operation to defeat an enemy attack.

Cruiser
Fast, moderately armed and armored general-purpose warship.

Deep penetration raid
Operation well behind enemy lines to disrupt lines of supply and communication.

Depthcharge
Antisubmarine weapon launched from surface vessels and designed to explode under water.

Destroyer
Small, fast, lightly armed warship designed for both escort duties and for making surface gun and torpedo attacks.

Dive-bomber
Light bomber, such as the German "Stuka", for precision air attacks of a tactical nature.

Double agent
Intelligence officer or undercover agent secretly working for the enemy.

Enigma
Complex machine employed by the Germans for enciphering military signals.

Escort duty
Employment of a warship to protect either larger warships or unarmed merchant vessels (eg in a convoy) against enemy attack.

Fighter
Machine-gun- and cannon-armed aircraft for use against other aircraft.

Firepower
The combined offensive or defensive power of weapons.

Flamethrower
Weapon which projects a jet of flaming liquid at the target, either hand-held or mounted on a tank.

Hand-held antitank weapon
Antitank rocket launcher capable of being carried and fired by a single infantryman, for example the American bazooka and the German Panzerfaust.

Incendiary attack
Attack with bombs designed to cause fires.

Infiltration
The penetration of the enemy position by stealth.

Intelligence service
Department for gathering secret political and military information from the enemy.

Kamikaze
(Jap: "divine wind") Japanese suicide aircraft employed against enemy ships.

Light tank
Fast lightly armed and armored tracked fighting vehicle for offensive or reconnaissance duties.

Maginot line
French defensive line of fortifications along the French-German frontier, built in the 1930s and designed to protect against an invasion.

Mechanized forces
Mobile forces completely equipped with motorized transport (eg motorcycles and motorized troop carriers).

Medium tank
Well-armed and armored tracked fighting vehicle with a fair turn of speed, capable of both offensive and infantry-support roles.

Morale
The mental state of troops, determining their conduct and discipline.

Offensive
Large-scale attack to break thorugh the enemy position and defeat his forces.

Outflanking maneuver
Operation to defeat the enemy by attacking his exposed flank or rear.

Pocket of resistance
Isolated area of enemy opposition bypassed by rapidly advancing attacking forces.

Preemptive strike
Attack to forestall an anticipated enemy offensive.

Preliminary bombardment
Powerful artillery bombardment before the commencement of an attack.

Radar
(Acronym: "RAdio Detection And Ranging") The use of radio waves to detect the presence of enemy aircraft or vessels.

Searchlight
Powerful ground-mounted light for spotlighting enemy planes at night.

Self-propelled artillery
Mobile artillery mounted on mechanized (and often armored) vehicles.

Siege
Extended military operations to isolate and capture a defended fortress or city.

Strategic bombing
Bombing offensive against industrial and civilian targets intended to destroy the enemy's economy and morale.

Strategy
The planning and conduct of large-scale operations of war.

Street fighting
Tactical combat in a built-up area involving infantry and supporting forces.

Submarine
Submersible vessel armed with torpedoes, designed for underwater operations against warships and merchant vessels.

Support troops
Noncombatant services for the logistical support of fighting units.

Tactical air support
The use of aircraft in close cooperation with military operations on the ground.

Tactics
The planning and conduct of small-scale military operations in the field.

Tank
Mechanized, heavily armed and armored tracked fighting vehicle.

Task force
Ad-hoc grouping of ships for a specific operation.

U-boat
German submarine, usually employed for commerce raiding.

"Ultra"
British code name for the dechipering of German Enigma codes.

Undercover agent
Clandestine spy sent into enemy territory to gather information.

War of attrition
Slow static warfare involving the wearing down of enemy forces and powers of resistance.

War of movement
Combat between mobile forces involving rapid movement over great distances.

Wolfpack
Group of U-boats collected for a coordinated attack on an Allied convoy.

Peak strength of armed forces

Australia	680,000
Belgium	650,000
Bulgaria	450,000
Canada	780,000
China	5,000,000
Denmark	25,000
Finland	250,000
France	5,000,000
Germany	10,200,000
Greece	414,000
Hungary	350,000
India	2,150,000
Italy	3,750,000
Japan	6,095,000
Netherlands	410,000
New Zealand	157,000
Norway	45,000
Poland	1,000,000
Romania	600,000
South Africa	140,000
UK	5,120,000
USA	12,300,000
USSR	12,500,000
Yugoslavia	500,000

Datafile

The land war was a massive struggle of attrition, its outcome determined by the ability to field, arm and supply huge armies, and to replace the losses they suffered in the slow process of wearing down the enemy. On the battlefield, firepower and transportation capability were the essential ingredients necessary to transform manpower into military advantage. Away from the frontline, industrial output and an intensive supply system were the driving forces which kept the armies fighting. Only the most highly mobilized industrial state could hope to stand the strain of a prolonged military campaign.

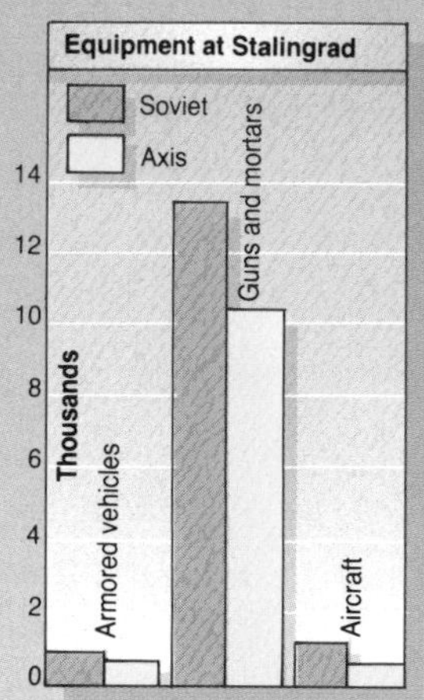

▲ The Soviet forces at Stalingrad (1941–42) enjoyed a substantial quantitative advantage in machinery, which ultimately proved too great for Germany's diminishing qualitative superiority to overcome. Strategic blunders aside, Germany was defeated at Stalingrad because it was unable to match the USSR's growing military strength or to bring new tanks, guns and transport equipment into the frontline at the same rate. German forces were outnumbered, outgunned and outsupplied.

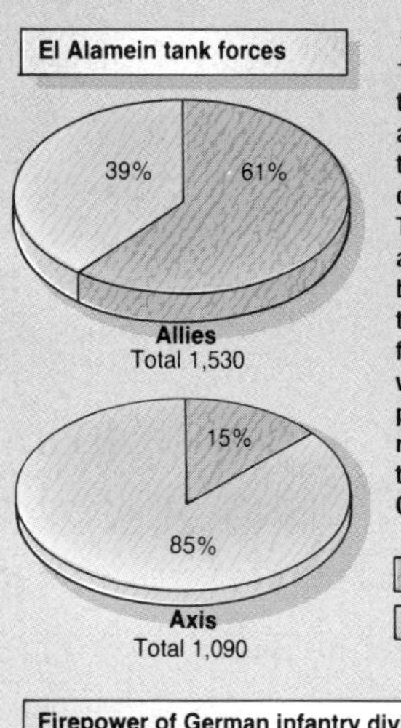

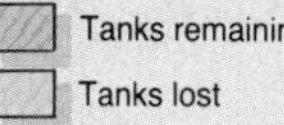

◀ At El Alamein (October 1942), Rommel's depleted tank force was destroyed by a larger force, equipped for the first time with vehicles of similar fighting power. The hard fighting in a mine- and artillery-infested battleground took a heavy toll of both sides' armored forces, but when the battle was over the Allies still possessed substantial reserves and the capability to make good losses; the Germans had neither.

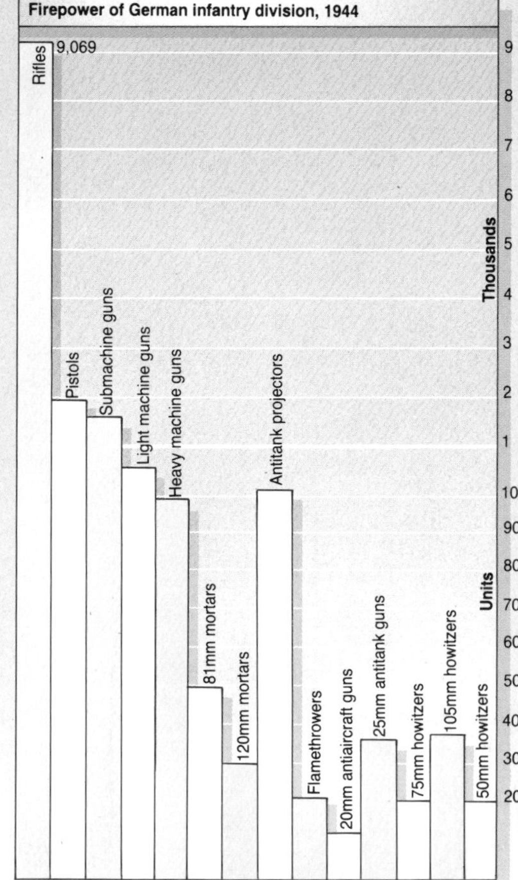

▼ Germany did not win the Battle of France (May-June 1940) as a result of numerical superiority. Even the figures for motorized and armored formations are misleading because the Allies had large numbers of vehicles in nonspecialist units. German victory was the product of a very quick and well-organized strike.

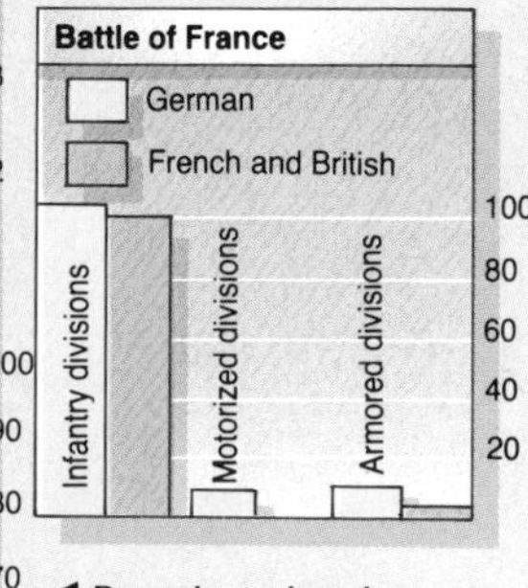

◀ Properly equipped infantry formations could deploy an impressive weight of firepower, sufficient to give them immense killing power. The large casualties of modern military engagements are ultimately the product of the greatly increased density of fire which soldiers are able to unleash on each other.

The armies of World War II retained the same basic organizational structure as that employed in World War I, itself the product of steady evolution in the 19th century. The structure of the British army can be used as an example: similar organization was found in most armies.

The smallest component was the section, composed of eight men; four sections were combined to form a platoon under the command of a subaltern. The next organizational component was the company of 120 men succeeded by the battalion of 700–800. The battalion was a permanent formation which retained the traditional regional or ethnic identity of the parent regiment, be it Highland Scottish, West Country English, Sikh or Bengali. Battalions were combined into brigades of 3,000–4,000 men, and brigades into divisions of 12,000–14,000 men. Divisions were given generic identifications, most commonly armored or infantry, although this only identified the predominant arm and all in fact included subsidiary formations of all arms. Finally divisions were combined to form corps of 30,000–60,000 men, and corps to form an army, containing 60,000–100,000 men. In most theaters, British military might was incorporated in a single army command, such as General Bernard Montgomery's 8th Army in the Western Desert in 1942, but in larger theaters, such as the Eastern Front, several armies would fight under higher organization. The German army in Russia operated as three army groups: North, Center and South.

Asian armies followed European organization closely. Indeed, most had been reorganized in the late 19th or early 20th century on European models and with the help of European advisors.

While World War II military forces were often broken up into a series of separate commands to fight in different theaters, they were at heart huge collections of men and equipment. On the eve of its defence of the Kursk salient in mid-1943, the Soviet Red Army mustered 6,442,000 officers and men, 103,085 guns and mortars, 9,918 tanks and self-propelled guns, and 8,357 aircraft. Forces of such size required a large number of high echelon formations (divisions and above) to control them. The German order of battle in May 1940 comprised some 135 divisions, including 10 armored and 9 mechanized/motorized formations. On 1 July 1944, with German ground forces stretched to the limit on the Western, Eastern and Italian Fronts, the order of battle included 139 infantry divisions, 11 air-force field divisions, 29 static infantry divisions, 11 Jäger divisions, 7 mountain divisions, 10 Panzer grenadier divisions and 24 Panzer divisions, all supplemented by 16 Waffen SS divisions.

Army formations required a massive amount of equipment, some idea of which can be gained

ARMIES AND LAND WARFARE

from a partial list of German materiel captured by the Red Army on the Don Front during the successful Stalingrad campaign: 5,762 guns, 1,312 mortars, 156,987 rifles, 10,722 automatic weapons, 10,679 motorcycles, 240 tractors, 3,569 bicycles, 933 telephone sets and 397km (248mi) of signals cable. Modern warfare made particularly heavy demands on vehicles; in 1942 alone, Germany built 4,278 tanks, the UK 8,611, the USSR 24,668 and the USA 24,997. Motorized transport was always in short supply, simply because the supply and support of fighting formations required the presence of tens of thousands of vehicles which never actually saw combat. So severe was the demand that the German army, the army most clearly associated with the development of mechanized warfare, remained heavily dependent on the horse until the very end of hostilities.

Communications and morale

Advances in communications made the World War II commander's job slightly easier than that of his World War I counterpart. Thousands of radio and telephone sets linked all components of a field force, from forward outposts all the way to general headquarters; but the sheer size and complexity of the modern battlefield, and the volume of communications engendered by the confusion of battle always made some loss of control inevitable. Under adverse conditions, brigades, divisions and even entire army corps could still collapse in complete chaos and resist all the efforts of command and technology to restore order. On the Eastern Front in particular whole formations could disappear in days, if not hours, and everywhere the confusion of defeat could produce tens of thousands of prisoners and huge hauls of perfectly functional equipment.

At the individual level, land warfare remained a matter of morale and small group cohesion. The most important unit to the common soldier was not his battalion or division, but his section or other group of close comrades. These five or ten men were the comrades with whom he shared danger, the messmates with whom he sought food and shelter, and the friends to whom he

▼ US soldiers at rest. Despite mechanization, most World War II soldiers moved and fought on foot, and, like all foot soldiers in history, were exposed to constant, wearying physical effort.

gave support and from whom he received the same in return. While World War II as a whole was probably distinguished by a higher level of identification with national cause than many wars before it, it nonetheless remained the case that small-group loyalties kept most men in the firing line. With the exception of nations, like Italy, whose general commitment to the war was never high, instances of mass surrender or abandonment of duty occurred only in militarily hopeless situations or after prolonged severe fighting. One of the most noteworthy features of the land war of 1939–45 was the ferocity and effectiveness with which the armies of the two major Axis powers (Germany and Japan) continued to fight after any rational hope of victory had long since disappeared. This durability, the product as much of sheer professionalism as of national culture or political indoctrination, ensured that there were few easy victories, even in 1944–45, and that casualties among all participants remained high to the very end.

Characteristics of land warfare

World War II on land is usually presented as an entirely different kind of conflict from World War I. It is seen as a war of movement rather than a war of static positions; as a war of tanks, aircraft and mechanized columns rather than a war of trenches, barbed wire and machine guns. These images are not entirely misleading. Technology did transform the face of war in dramatic fashion between 1939 and 1945, most obviously in the capability it gave to the general to move men and guns quickly across large distances. World War II saw the course of whole campaigns change in a matter of days as a result of rapid outflanking moves or armored breakthroughs. This notwithstanding, there was much that did not change. The vast majority of soldiers were infantrymen, who covered most of the ground they crossed on foot. Most German soldiers marched into Russia in 1941, just as Napoleon's *Grande armée* had done in 1812, and those who survived marched back out again. In most theaters rapid movement was the exception rather than the rule and much time was taken up with static or near-static warfare, the two sides holding well dug-in positions against each other.

In terms of strategy, the combination of the availability of large mechanized forces and a widespread desire to avoid the bloody stalemate of World War I (in which most senior officers would have served their military apprenticeships) placed a lot of emphasis on variations of the indirect approach – outflanking maneuvers designed to avoid head-on confrontations with strong enemy positions – and on the quick breakthrough by armored forces followed by rapid exploitation of the ensuing chaos (the idea at the heart of the German concept of Blitzkrieg or "lightning war"). Against the unprepared French and British armies of 1939–40, or in the vast interior of Russia and open sands of the Western Desert in North Africa, such approaches were sometimes viable. When operating along a coastline, the greatly improved Allied amphibious capacity of the second half of the war also allowed outflanking maneuvers to be undertaken by sea (as was tried at Anzio in Italy in 1944). In most campaigns though, the nature of the terrain or the strength of the enemies' defenses allowed little scope for such freewheeling operations. Ground had to be taken kilometer by kilometer in slow, close and often confused combined infantry, armor, and artillery operations. When a breakthrough did come, it was often only after the enemy had been ground down by weeks if not months of hard fighting and at a heavy cost in casualties. Despite air power and the tank, the German defenders of Monte Cassino were operating with many of the same advantages as the German defenders along the Somme in 1916 and, despite the technological revolution in warfare, they proved just as difficult to dislodge. Overall the machine gun, artillery piece and well-constructed entrenchment were as important as the tank and motorized troop carrier.

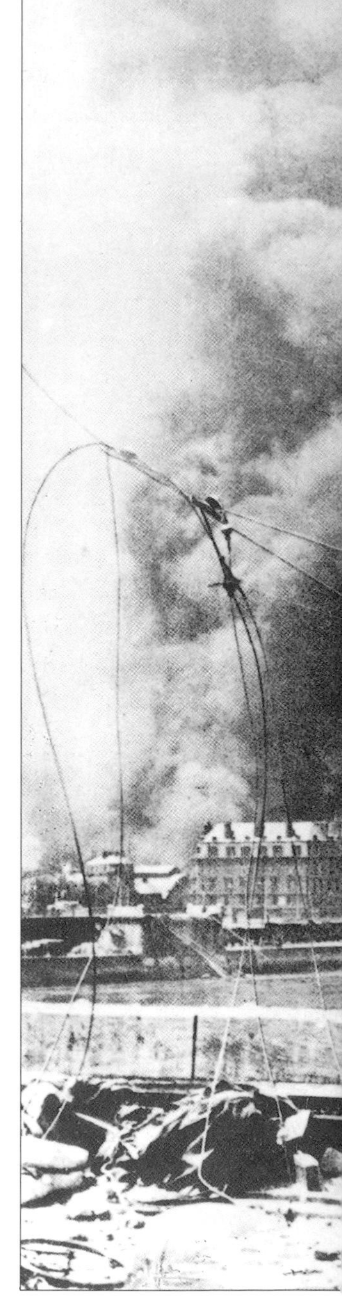

▲ Rouen in flames, May 1940. As German armies moved swiftly through northeast France, they seized one center of communications and supply after another. Air attack would disrupt efforts to restore broken lines of command and regroup retreating ground forces. The motorized leading elements of the Wehrmacht would follow swiftly to complete the process of disintegration.

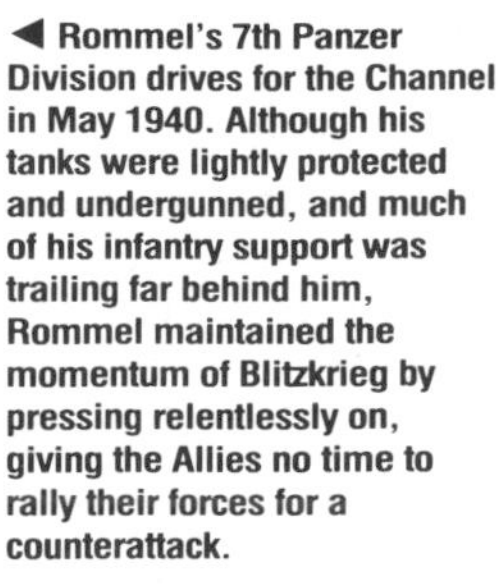

◄ Rommel's 7th Panzer Division drives for the Channel in May 1940. Although his tanks were lightly protected and undergunned, and much of his infantry support was trailing far behind him, Rommel maintained the momentum of Blitzkrieg by pressing relentlessly on, giving the Allies no time to rally their forces for a counterattack.

◀ The winged weapon of terror and disruption. The German air force or Luftwaffe played a vital role in Blitzkrieg. Close air support helped the armored spearhead to break through defended positions, while constant air attacks on rear areas disrupted efforts to reorganize resistance, destroyed communications and produced panic among the flood of refugees seeking to escape the advancing military forces. The JU87 Stuka, specially equipped with sirens to make its dive-bombing attacks even more terrifying, became the symbol of all that was irresistible in the German attack.

▲ German success depended to a considerable extent on the engineering and logistical skills of the army's support formations. The ability to construct pontoon bridges, such as the one pictured here, allowed the advancing motorized forces to press forward despite the destruction of bridges and the presence of a multitude of rivers which might otherwise have given the defenders a vital breathing space in which to regroup and reorganize.

Land warfare in Europe: Blitzkrieg

Blitzkrieg, or "lightning war", was a strategy for using new mechanized military technology. It had been developed in Germany between the wars and dominated the early years of World War II in Europe. The German victories of 1939–41 were notable for their speed and totality. They were based neither on overwhelming numbers nor on a heavy qualitative superiority in equipment. In the first instance German attacking spearheads were often dangerously overextended and vulnerable to counterattack. In the second instance German Panzer divisions were largely equipped with light tanks which were no match in armor or armament for their French or British opponents, while later experience was to prove that the much-feared JU87 Stuka dive-bomber was a mediocre machine which could not operate in the face of determined fighter opposition. The key components of the German victories, and of the doctrine of Blitzkrieg that lay behind them, were actually speed, surprise, sheer unconventionality, and the comprehensive manner in which the various parts of the military machine were integrated and employed. Blitzkrieg was less a matter of men and materiel than of how they were used.

In the Norwegian campaign of April 1940 the Germans gained the advantage by taking the risk of sailing their invasion fleets straight into Norwegian harbors, seizing as many key points as they could regardless of shortages of forces, and by deploying supporting aircraft to forward bases as soon as these became available. In France and the Low Countries a month later, the Germans attacked unexpectedly out of the supposedly impenetrable Ardennes Forest, outflanking the well-prepared French defenses of the Maginot Line and isolating French and British forces further to the north in Flanders and Belgium. After making their initial breakthrough the Germans did not stop to consolidate their position, but rather continued to push their mechanized forces forward toward the coast. While their flanks were unprotected, their supplies running low, their equipment wearing out, and their supporting

Supply on the Eastern Front

Supply has always been crucial to military success, but in World War II it assumed greater significance than ever. On the Eastern Front, the logistical problem of keeping armies moving and fighting was further complicated by the huge distances between factories and the front lines. By the beginning of 1943, the Red Army, then over 6 million strong, was using about one-fifth of its field manpower for supply and support duties. The Main Military Roads Administration, for example, employed 125,000 men to service 86,000km (54,000mi) of military highways and overhaul 185,000 lorries. The materiel demands of the war were huge as well: by the middle of 1943 the USA had shipped in 900,000 tonnes of steel and 1.5 million tonnes of food to help sustain the Russian war effort.

The Stalingrad campaign brought the problems of supply into stark contrast. When the German 6th Army was ordered to hold out after being surrounded in the city, it radioed that it would need 750 tonnes of supplies a day just to meet the minimum needs of standing on the defensive (380 tonnes of food, 250 tonnes of ammunition and 120 tonnes of fuel). The German air force could have provided only half the necessary capacity under ideal conditions. Faced with bad weather, disorganized airfield facilities and strong Russian air resistance, it never came close to meeting even that target. By mid-January the beleaguered 6th Army was living off the huge herd of horses which provided it with the largest part of its transport and had already eaten 39,000 of the unfortunate beasts. When resistance was finally crushed at the end of the month, the remnants of the 6th Army were almost completely without food, ammunition or fuel.

troops were often far behind them, the Panzer divisions so disrupted enemy forces that no effective response could be planned or executed. Having knocked the enemy off balance the Germans simply kept pressing until he collapsed in complete confusion.

The vital tactical device behind these lightning assaults was close cooperation, not just between the assorted units of the German army, but also between the army itself and the air force. The Panzer divisions did not consist entirely of tanks: rather they combined tanks with motorized infantry, artillery, and other supporting services to form semiautonomous strike forces capable of dealing with almost any situation that might confront them. Thus, for example, they were able to blunt an attempted British armored attack on their flank, not just with tanks, but with artillery, antitank guns, and even antiaircraft guns (the 88mm AA gun provided a particularly effective tank destroyer). As the Panzer divisions incorporated their own infantry formations, they would not have to launch unsupported tank attacks against defended positions. They were not, however, expected to deal with pockets of resistance and isolated enemy detachments left behind by their advance – these were the responsibility of the infantry divisions which followed along more slowly, in the wake of the armored spearhead. Finally, Blitzkrieg depended not just on aerial superiority but also on direct air intervention in the land engagement. The German air force generally succeeded in gaining the advantage in the sky by launching preemptive strikes to destroy enemy air power on the ground. It then carried out a comprehensive program of tactical air support – a role for which it had been specifically designed and trained. Enemy communications and command structures were disabled by attacks on bridges, railways, roads and population centers well ahead of the advancing German armor while that armor itself could call on close air support to help it push through any obstacle that blocked its path. Overall, Blitzkrieg was as destructive to enemy morale as it was to the physical fabric of his fighting forces, and, as was the case with France in 1940, it was the loss of the will to fight rather than the absolute destruction of the means that produced surrender.

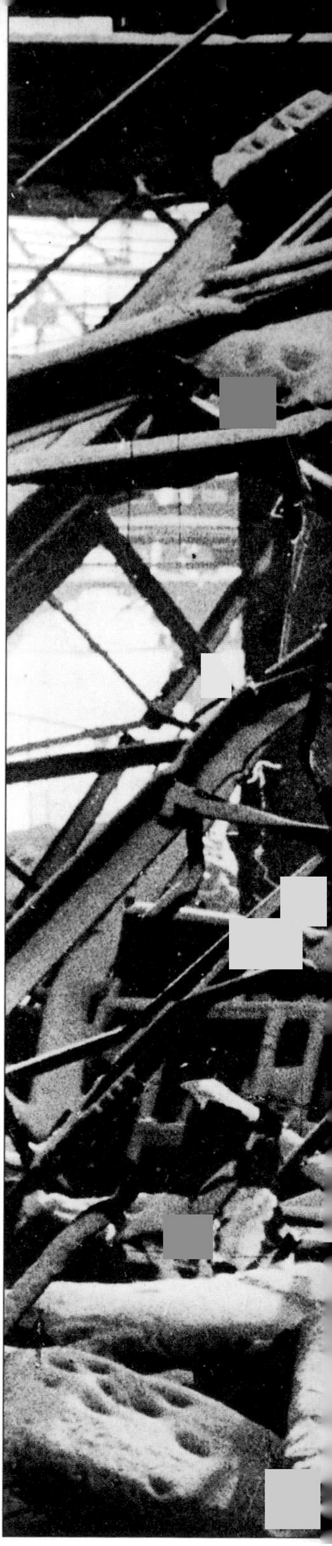

War of attrition: the example of Stalingrad

The Stalingrad campaign was almost certainly the key turning point in the war on the Eastern Front. It began with the Germans driving deep into southern Russia, reaching as far east as the Volga and as far south as the foothills of the Caucasus. It ended with the Germans reeling back beyond their start line leaving an entire army cut off and doomed to surrender far to the east in the ruins of Stalingrad. The campaign itself can be broken into three distinct phases: the initial German offensive, the battle for the city of Stalingrad itself, and the Russian counteroffensive with its attendant siege of the trapped German 6th Army.

The Russians enjoyed a decided advantage in manpower throughout the campaign, but the German army still enjoyed qualitative superiority

◀ **A German soldier dug in at Stalingrad. "General Winter" was the great ally of the Soviet resistance. Weather played a crucial role in the German defeat: snow and freezing temperatures played havoc with the inadequately equipped German forces. German morale and fighting effectiveness declined steeply as men froze and weapons failed to function.**

▲ **Fighting in the Red October Tractor Factory, Stalingrad. The German army in Stalingrad was swallowed up in confused street fighting among the ruins of a devastated industrial city. Unit cohesion disappeared in a maze of broken buildings and rubble-filled streets. Von Paulus's forces were gradually worn down in a series of confused and vicious local actions which destroyed central command, made resupply impossible and broke elite fighting units into isolated groups of exhausted and desperate street fighters.**

at the start of the Stalingrad drive. This advantage was lost through the overextension of German forces in pursuit of distant geographical objectives and a growing strategic obsession with the capture of Stalingrad itself. Eventually, an entire army of 22 divisions was sucked into the ruined city, leaving its long and vulnerable flanks protected mainly by second-rate Hungarian, Romanian and Italian forces. As the German advance ground slowly forward, the Russians had time to build up massive forces to the north and south of the city. On 19 November 1942 the Russians attacked, cutting off Stalingrad completely in two days. The Germans failed either to break out from the trap or to push a relieving force through. After Luftwaffe attempts to supply from the air had failed, the 6th Army was overrun and forced to surrender between 30 January and 2 February 1943. Of a force originally numbering more than a quarter of a million, only 80,000 men were left alive to march into captivity.

The fighting around Stalingrad followed a pattern common to much of the Russian war. Each side counterattacked in turn as the other overextended itself; the Russians used mass infantry assaults to overwhelm enemy positions by sheer weight of numbers; and armored columns repeatedly broke through, causing massive confusion and precipitate retreat.

The fighting inside the city itself was entirely different in nature. Stalingrad was a modern, heavily industrialized center which stretched for

▶ **Free French armored cars in the Western Desert. Desert warfare was a mixture of set-piece engagements in which armor, infantry and artillery were used to attack heavily defended positions, and mobile pursuit or outflanking operations in which light armored forces such as the column pictured here played a vital role. The requirement for fast-moving light forces made armored cars and small reconnaissance tanks a far more important component of the desert armies than of military formations in other theaters of war.**

18km (11mi) along the high west bank of the River Volga. As two German armies – 6th and 4th Panzer – fought their way in from the west, and as the Russians reinforced the defenders from the east to a strength of five armies, the whole battle degenerated into confused street fighting. Units lost all cohesion as they were fed into the ruined buildings and rubble-choked streets of the inner city. Combat revolved around small group struggles for individual houses and factory buildings, and the morale of both sides was stretched to breaking point by the daily hazards of snipers, boobytraps and ambushes.

As the weather deteriorated, snow, wind and cold were added to the soldiers' problems. After they were cut off and thrown on the defensive, the Germans gradually ran out of food, ammunition and fuel. Russian reinforcements were ordered across the Volga and instructed not to come back, while the Germans missed whatever chance they might have had of an early breakthrough by waiting for definite orders from Hitler. The beleaguered 6th Army was slowly pushed back within an ever-shrinking perimeter. By late January 1943 the enclave was only 16km (10mi) across and the last airfield had been captured. Shortly afterward the Germans were cut into two pockets and overrun.

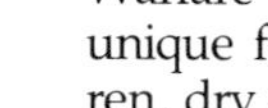

Desert warfare

Warfare in North Africa's Western Desert was unique for several reasons. The terrain was barren, dry and hostile, and provided little by way of either supplies or natural obstacles to movement. Supply was the critical factor and the strength of the rival armies was generally determined by their distance from their main bases (Tripoli for the Axis and Alexandria for the Allies). Rapid advances tended to outstrip supply capability, leaving the attacker vulnerable to a counterthrust from an enemy who had retired closer to his own supplies. As a result the campaign seesawed back and forth several times across the entire length of the North African desert.

The terrain exercised a decisive impact on the fighting. In the absence of major natural obstacles or easily defensible positions, both sides relied heavily on barbed wire and mines to provide defensive boxes in which they could group their infantry and artillery. Around these the armored forces of both sides enjoyed great freedom of movement, and there was always scope for outflanking moves or surprise attacks on lightly defended areas of the front. For most of the campaign the Germans got more out of their tanks than the Allies did from theirs. Until mid-1942, when American-built Grant and Sherman tanks began to arrive in numbers, British armor was generally slow and undergunned. In Rommel they were confronted by an enemy commander with an instinctive flair for tank warfare. Finally, the Germans possessed the most effective anti-tank gun in their dual-purpose 88mm weapon. They had no hesitation in deploying these weapons at the very front of positions where they were often able to destroy British tanks before the latter could get close enough to do any damage.

▼ **Italian infantry move forward in a sandstorm. The Western Desert presented unique problems to the men expected to fight across its arid terrain. Blowing sand reduced visibility, clogged equipment and produced severe personal discomfort. Temperatures varied widely, ranging from baking heat by day to bitter cold at night, and supply was a constant problem in a land which offered little by way of natural resources and rendered transportation extremely difficult. As was the case in so many other theaters, the infantryman's lot was grindingly hard and unpleasant.**

Overall the campaign has been justly compared to sea warfare with both sides operating from fixed bases in an almost featureless and generally hostile environment. The campaign also involved polyglot forces on both sides. While the German Afrika Korps attracted most of the attention on the Axis side, it was in fact supported by a large Italian contingent. This force was inferior in equipment and had suffered severely in the opening stages of the campaign, but under Rommel's direction it could not be completely written off by its opponents. On the Allied side almost all of the British Commonwealth and Empire was represented, the 8th Army comprising Australian, New Zealand, South African and Indian formations as well as native British ones. It also included a Free French detachment and by 1942 was looking to the USA for some of its equipment, most importantly Grant and Sherman medium tanks to allow it to take on Rommel's Panzer MkIIIs and MkIVs on equal terms. In the final stages of the campaign, after Rommel's defeat at El Alamein and the Allied landings in French North Africa, American troops also became involved, picking up hard-earned combat skills at the hands of their outnumbered but much more experienced opponents.

All armies coped well with the peculiar demands of the desert climate, and despite the profound swings in temperature and the eye- and engine-clogging clouds of sand, the standard of military performance was generally very high.

Rommel has been criticized for attempting too ambitious a campaign with the forces at his disposal, but, whatever the strategic ramifications, there can be no doubting the fact that his units achieved as high and as consistent a level of battlefield performance as any German military formation during the entire war. His victories were triumphs of improvisation based on the excellent training and spirit of his men. In defeat, the Afrika Korps maintained its morale and fighting capability to the very end, delaying the final Allied triumph for several valuable months. For British and Imperial forces, the campaign was significant because it allowed them the opportunity to recover their morale and develop their fighting skills in preparation for the reconquest of mainland Europe. After the disasters of 1940–41, North Africa was the only theater in which the land forces of the western Allies were in regular combat with the German army. While they were frequently outmaneuvered and outfought by their highly professional opponents, they emerged from the campaign with a far better knowledge of the requirements of modern mechanized warfare than they had been able to draw upon in Norway, France or Greece. Furthermore, the heroic defense of Tobruk, the stout performance of men and machines in the seesaw campaigns of 1941–42, and the final triumph of El Alamein gave the units of the 8th Army a faith in their ability to defeat the enemy which had been sadly lacking by the end of the first campaigns in northern Europe. Overall, there can hardly have been another theater of military operations which saw the opposing sides perform so well and so consistently over such a prolonged period of time.

▼ The critical engagements of the desert war were set-piece battles fought around entrenched defensive positions. Mines (below), dug-in artillery and infantry emplacements equipped with periscopes (bottom), barbed wire, etc were employed to stop armored assaults. Only when such barriers had been destroyed or outflanked did the war assume its more famous mobile character.

◀ A raiding party of the British Long Range Desert Group. The open flanks of the desert armies and the long, vulnerable lines of communication behind them allowed great scope for light mobile operations. Fighting in the tradition of Lawrence of Arabia, small formations like the LRDG caused disruption well out of proportion to their size, but the issue was always ultimately settled in a set-piece battle between the main armies.

▲ The RAF drops supplies to Slim's 14th Army on the River Irrawaddy in Burma. The fighting forces' need for constant supply was all the more critical in the jungle where the terrain made traditional modes of transport impossibly difficult, and the appalling climate rendered mass disease the inevitable consequence of lack of good food and medical supplies. British success in the Burma campaign of 1944–45 was firmly based on air superiority and the massive air supply effort which command of the skies and possession of a large transport fleet allowed. While the British were able in this way to feed the fit and evacuate the sick, their Japanese opponents were ravaged by sickness and malnutrition.

▶ The sheer volatility of the jungle climate produced problems of its own. In the monsoon season, whole areas were inundated and normal patrolling became a matter of amphibious operation, with all manner of craft being employed to move men to areas of suspected enemy activity. In many cases, the volume of water pouring down from the skies was such as to make military movements on any but the smallest scale a complete impossibility.

Jungle warfare: the case of Burma

In their Burma campaign of 1944–45, Allied forces succeeded by utilizing advantages and tactics similar to those employed by the Japanese in the preceding two years. They were also aided by a preemptive Japanese attack (March 1944), the initial success of which overextended General Mutaguchi's 15th Army and exposed it to a devastating counterthrust. Initially Allied attacks met stiff resistance, but when the defensive Japanese barrier finally cracked, the advance gathered momentum. The Japanese lost first Mandalay (20 March 1945) and then Rangoon (3 May 1945), falling back toward the Thai border in complete disarray, and leaving the Burmese army to defect to the Allies.

Japan's successful conquest of Burma in 1942 (see p. 46) had been based on superior air power and better jungle-fighting techniques. British positions were consistently infiltrated and outflanked, and position after position had to be abandoned to prevent encirclement. By 1944 air superiority had passed to the Allies. This enabled them to support deep penetration raids which were aimed at disrupting enemy supplies and communications.

On the British sector of the front, along the Indian frontier, these raids were mounted by General Orde Wingate's Chindits. On the Chinese-American sector further inland, the raiders were a US formation known as Merrill's Marauders. The efforts of these semiguerrilla forces were supported by conventional offensives. General Joseph Stilwell's Chinese-American forces in the north eventually managed to reopen the land link to China along the Burma road, while Slim's British 14th Army advanced on Akyab through the coastal mountains and on the Chindwin river in central Burma.

The second of these offensives was disrupted by a massive Japanese counterattack which broke through the mountains into India. The Japanese invaded the Manipur plain, encircling the strategic towns of Imphal and Kohima. In the case of both the coastal and Imphal campaigns, Slim did not allow his troops to retreat when outflanked and threatened with encirclement. Rather he ordered his garrisons to hold firm and depended on his superior air resources to supply and support them. He then ordered his own relieving forces into the jungle to outflank and encircle the Japanese besiegers. The critical battle took place at Imphal where the siege lasted 88 days (29 March–22 June 1944). When it was finally broken, the Japanese 15th Army had been all but destroyed, severely reducing the defender's ability to resist a renewed offensive in 1945.

For both sides Burma was a remote theater of war, girdled by mountains and cluttered with difficult jungle terrain. The key to success for both sides was their ability to supply and support frontline forces fighting kilometers away from prewar lines of communication. The Japanese managed to complete the Burma railway, running from Moulmein on the coast of Burma to their supply bases in Thailand – its construction the Japanese sacrificed more than 50,000 lives, the bulk of them Asian laborers and Allied prisoners of war – and to conquer terrain. The Allies built a whole series of airfields and a new mountain road from northeast India across the Naga Hills into Burma. Altogether, it required several hundred thousand laborers and support troops to keep the Allied fighting effort in Burma going. In terms of terrain difficulties and the quality of enemy opposition, it was one of the most exacting campaigns of the war. At the tactical level, Slim's 14th Army might have won the war in Burma because it successfully adopted the flexible techniques of jungle fighting, but at the strategic level the outcome was undoubtedly decided by the organization of a massive supply and support service making very heavy use of transport aircraft. Slim's generalship of the 14th Army was particularly distinguished, and his conduct of the Imphal-Kohima battle alone would be sufficient to place him in the front rank of World War II commanders.

Jungle Warfare

War in the jungle was dominated by terrain and climate to a greater extent than war in any other landscape. Much of the jungle terrain fought over by the Japanese and Allied forces between 1942 and 1945 was a cluttered nightmare of dense undergrowth, stagnant water and overgrown tracks. The climate, alternately arid and extremely wet, often almost unbearably hot, made things much worse, not only through the sheer physical discomfort it brought, but also because it bred disease at an army-destroying rate. Success in jungle warfare was thus largely a matter of limiting the impact of natural hindrances on operations and the maintenance of effective fighting capacity.

In the Burma battles both the Japanese in the early years of the campaign and the British and Allied forces in the battles of 1944–45 succeeded by ignoring the conventions of normal strategy and tactics. Armies which stuck to the established road network or attempted to hold fixed positions in open terrain usually found themselves outflanked through the surrounding jungle. Operating in the jungle itself placed a heavy emphasis on independent action by small units, and as a result only troops with high morale and good training could be expected to survive. It was essentially an infantryman's war; tanks operated in small numbers tightly constricted by terrain, while only light artillery could be moved around with sufficient ease to be of any use.

Jungle casualty lists resembled those of the wars of the preindustrial age with far heavier losses through sickness and disease than through enemy action. This was partly a result of climate and partly a result of difficulties in maintaining adequate supplies of food and medicine. The army that could fight best in the jungle was the one which could keep its sick list down by maintaining the flow of food and medical supplies.

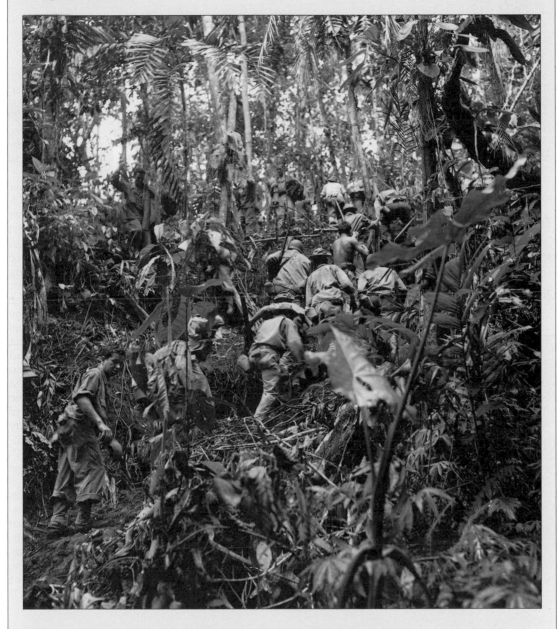

▲ Near-impenetrable terrain on Guadalcanal, 1943.

Then one day we had our first shower. This was the end. The rains. The steep sides of the ridge became a bog and a treacherous foothood. The water seeped into our foxholes and ran in great gushes down the slope, carrying the mud with it. The track from the airstrip was an adventure, but for those who carried the stretchers down to the strip it was a nightmare. The strip itself was soggy. The shower soon passed off, but it left a feeling that we could not get rid of. Rains, misery and sickness, vistas of mud stretching into the endless future, tearing at our boots, covering our bodies, filling our bivouacs and folding over us as we slept.

RICHARD RHODES JAMES
BURMA, 1944

▲ Well-trained and tightly disciplined, the Japanese soldier was all the more fearsome for his willingness to surrender his own life to destroy the enemy. This soldier was crouching in a foxhole in Burma with an explosive device between his legs which he hoped to detonate when an enemy tank passed over his position. Such suicidal bravery ensured that Allied victory could only be completed by the systematic destruction of the entire defending force – a slow, dangerous and bloody business, however inevitable the final result.

THE WEAPONS OF LAND WARFARE

World War II was notable for the dramatic increases it brought to military firepower and mobility. The most famous contributor to this process was the tank, but artillery of all sorts, infantry weapons from the submachine gun to the bazooka, and a multiplicity of motorized transport and light vehicles also played a part.

The tank, a tracked armored vehicle with a turret-mounted main armament, had made a hesitant combat debut in World War I, but in World War II it came of age as the queen of the battlefield, a valuable tactical weapon in traditional infantry engagements and a potential campaign winner when deployed en masse to break through or outflank the enemy's lines. The German Panzer divisions, which first brought the tank to prominence in 1939–41, were largely equipped with small, lightly armed vehicles, but as the war progressed, armored divisions were progressively strengthened by the arrival of larger, more heavily armed and armored tanks like the American Sherman, the Soviet T-34 and the German Panther.

However, important the tank, it never achieved total dominance of the battlefield. In broken terrain, or against well dug-in organized resistance, armor could suffer heavy losses without achieving any significant breakthrough.

The basic infantry weapon of the war was the same bolt-action magazine rifle carried in 1914–18, of which the British Lee-Enfield of 1903, capable of placing five shots in a 10cm (4in) circle at 180m (220yd), is perhaps the most famous example. Beyond the rifle, however, World War II infantry formations could call on a far more formidable arsenal than their predecessors. Large numbers of submachine guns added immense close-range killing power, while two-man weapons from the well-established Vickers machine gun to the new German MG42, a weapon capable of the terrifying rate of fire of 1,200 rpm, provided the capability to saturate the enemy with bullets at longer distances.

The infantry could also deploy mortars and flamethrowers but throughout the conflict they operated under the shadow of the artillery barrage. Field guns and howitzers formed the backbone of every army's firepower. They were backed up by larger weapons, usually of 150mm (6in) caliber, self-propelled guns mounted on tracked, armored chassis, and even rockets like the Soviet Katyusha. The weight of firepower which could be deployed was as terrifying as any deployed in the far more famous artillery barrages of World War I.

◄ German armor on the move in difficult terrain. One advantage of tracked fighting vehicles was their ability to operate off roads or other hardened surfaces. Cross-country performance was often critical to the success of mobile operations.

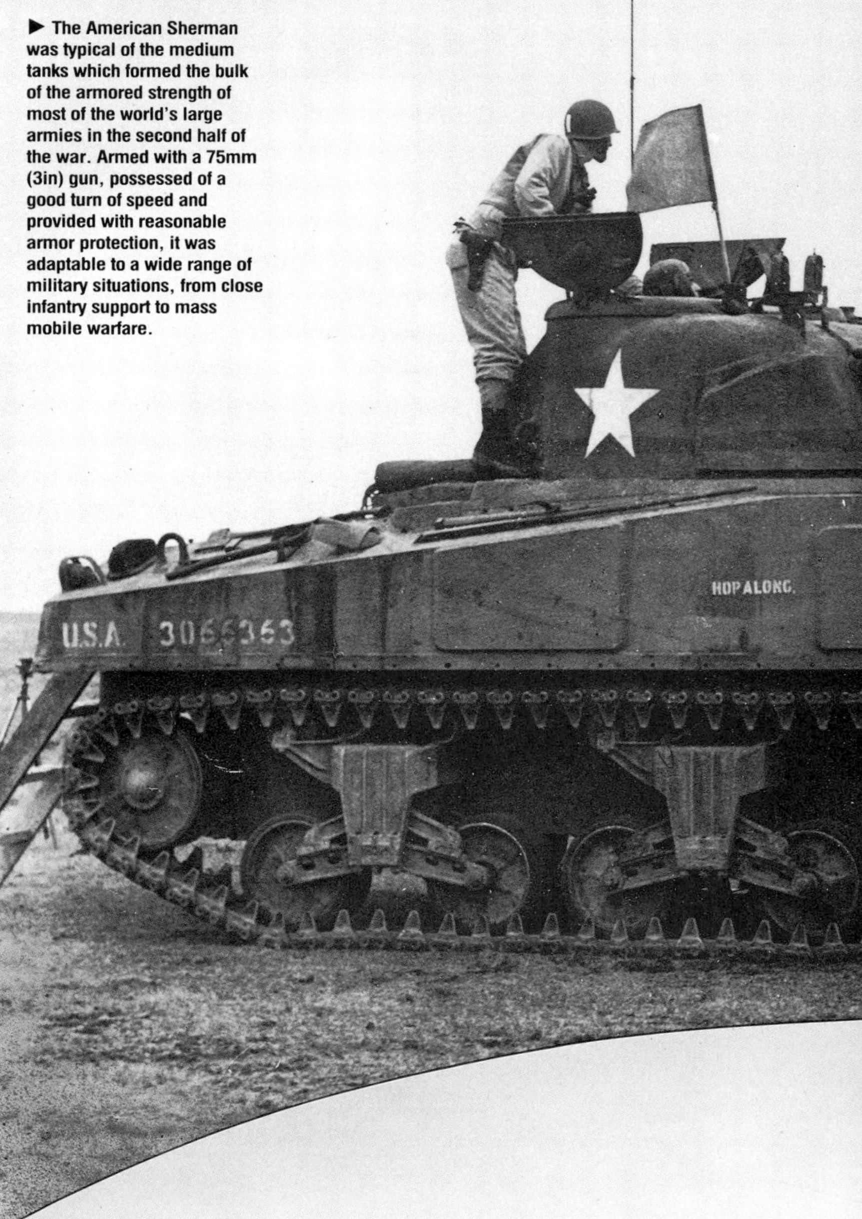

► The American Sherman was typical of the medium tanks which formed the bulk of the armored strength of most of the world's large armies in the second half of the war. Armed with a 75mm (3in) gun, possessed of a good turn of speed and provided with reasonable armor protection, it was adaptable to a wide range of military situations, from close infantry support to mass mobile warfare.

Rocket launcher

► The only rocket launcher of World War II was the Soviet Katyusha. It fired 16 132mm missiles with a range of 8,460m. Katyusha batteries were massed for salvo firing. They were manned by special troops, designated "Guards Mortar Units" to conceal their true function.

Howitzer

► Howitzers provided armies with the bulk of their artillery support, firing medium- and heavy-caliber shells in high trajectories against distant targets. Antitank guns were specialized weapons for use against armor. They fired special armor-piercing ammunition at high velocities.

◀ A British soldier with a submachine gun. This weapon was neither accurate nor useful at long ranges, but in close-quarters fighting its high rate of fire was very useful. All of the armies introduced great numbers into their ranks.

▶ By mid-1943 the Soviet army had four divisions of Katyusha rocket-launchers. Each division could fire 3,840 projectiles, to provide saturation bombardment before an attack.

▼ An American 8in (200mm) howitzer in action against Japanese positions. Most artillery strength consisted of lighter pieces; such heavy guns were deployed against well dug-in positions.

◀ Mortars were usually relatively small and portable weapons used to provide infantry with close artillery support. The mortar was particularly useful against an entrenched enemy. The bazooka was widely used against bunkers and pill-boxes as well as tanks.

Datafile

The sea war involved tens of thousands of fighting and merchant vessels – more than any other campaign before or since. The introduction of the aircraft carrier, and the impact of airpower generally, added a new dimension to seaborne conflict. Once again, industrial might played a crucial role. Nations incapable of replacing losses to torpedoes, bombs, mines and shells were swept completely from the seas.

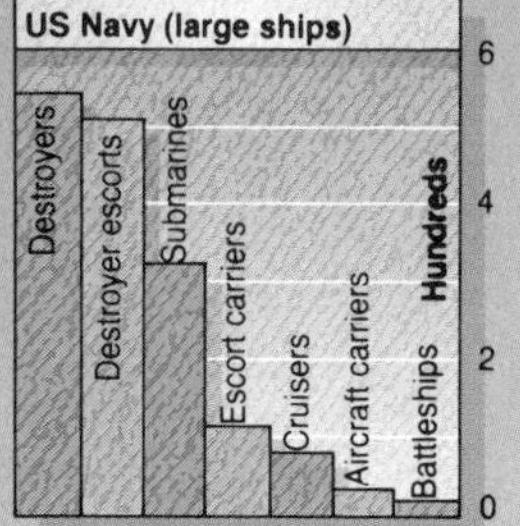

U-boats

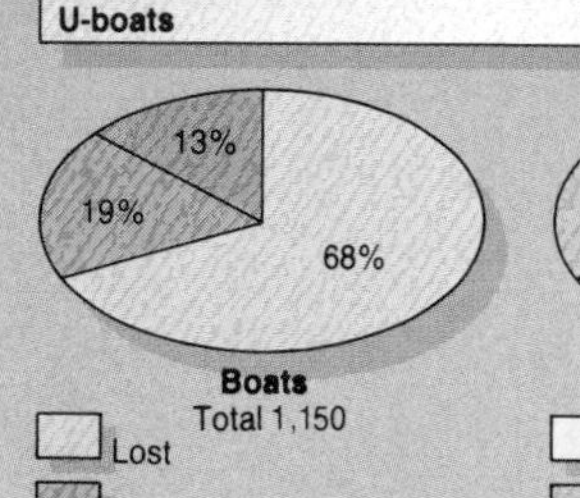

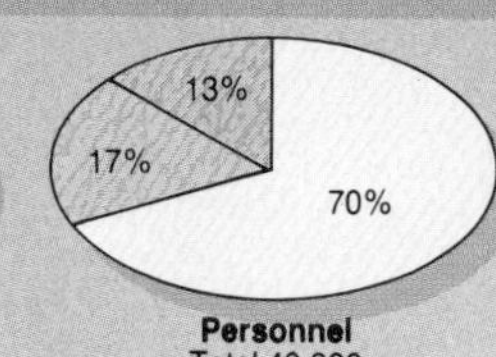

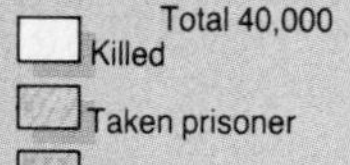

▲ The German U-boat service paid a heavy price for the damage inflicted on Allied shipping. In the second half of the war, improved air and surface antisubmarine techniques took such a heavy toll of the German submarine force that by the last year of the war it was powerless to interfere with the enemy's supply effort.

▼ In the first half of the war, Allied merchant shipping losses were horrendous, but from 1943 new building programs, particularly that launched in the USA, were more than compensating for losses, and the faltering German U-boat campaign could not prevent a progressive growth in Allied seaborne carrying capacity.

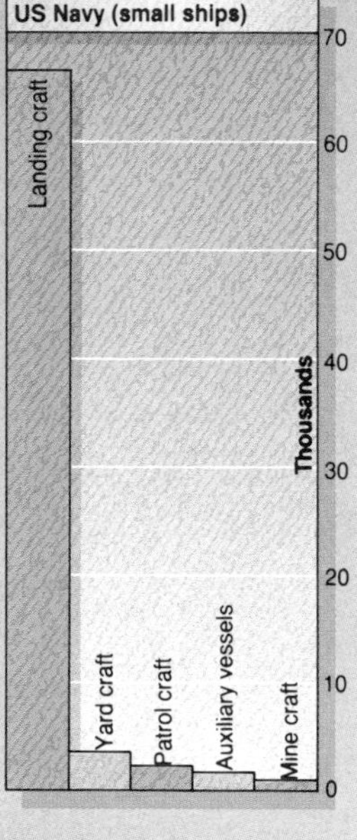

▼ The Western Allies enjoyed a substantial numerical superiority over Axis naval forces at the beginning of the war, but the early removal of the French navy from the scene, the diversion of British forces to meet threats away from home waters, and the growth of the German U-boat force dramatically changed the situation. Between 1940 and 1942, Allied naval forces in European waters were stretched very thin, and even isolated Axis surface units could pose a dangerous threat.

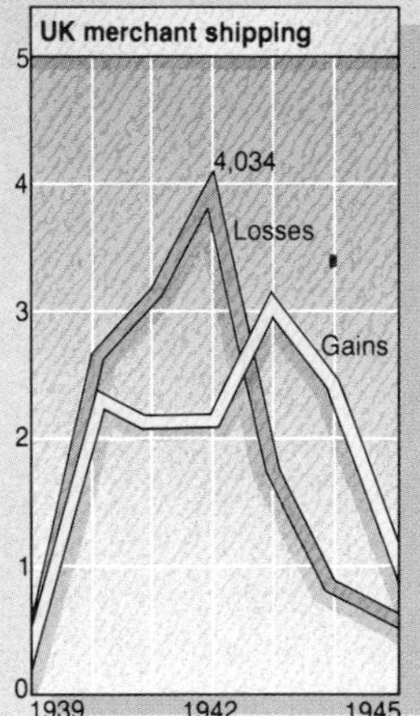

▲ The US navy emerged from the war as the greatest fighting armada ever seen. A huge shipbuilding program, based on the country's dynamic heavy industrial sector, produced not only an impressive fleet of modern carriers, battleships, cruisers, destroyers and submarines, but also tens of thousands of small craft to carry the amphibious landing forces which reconquered the islands of the Pacific and opened the second front in Europe leading to the defeat of Germany.

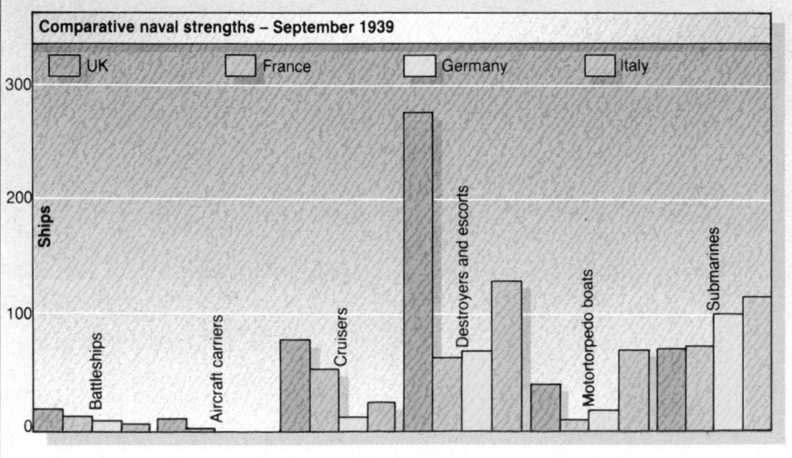

The navies of World War II were fighting services caught in a period of transition. The navies of the late 19th and early 20th centuries had been built up around the battleship, as a result of the introduction of steampower, armor-plating and the long-range rifled gun firing high explosive ordnance. Naval thinking revolved around the employment of battleships in fleet-to-fleet gun duels, with the result that the entire navy would be built around the battleline.

By World War II the introduction of submarines and aircraft had been changing this pattern for several decades, but in none of the combatant forces was the transition to a more balanced fighting fleet complete. Submarine forces were still expanding and in many cases had yet to evolve a coherent strategic doctrine for use in war. Antisubmarine technology was still in its infancy and very few purpose-designed antisubmarine vessels were actually in service. Battleships were still being designed and built, and indeed the last such example of the type was not to enter service until World War II was over. Finally, carrier aviation was in an advanced state of development only in Japan and the USA, and even there naval high commands had not completely rid themselves of a belief in the battleship as the ultimate arbiter of seaborne warfare.

The organization of naval forces

Naval organization was far more flexible than that of land and aerial forces. Vessels were grouped in permanent squadrons and flotillas, but this was mainly for administrative convenience rather than operations. Fighting fleets were usually put together as situations required and the availability of units allowed. Thus a modern light cruiser might find itself operating alone searching for a commerce raider, or helping to screen a battle-group of capital ships, or leading a mixed task force of cruisers and destroyers on a raid against enemy coastal forces, or on escort duty with a convoy fighting its way through the aircraft-dominated Mediterranean. The multiplicity of roles and the frequent unavailability of purpose-built types often exposed warships to challenges they had not been designed to meet.

The war itself saw a tremendous change in the nature of the naval balance. In 1937 the Imperial Japanese Navy (IJN) was the world's third largest force, operating the entire range of aircraft carriers, battleships, cruisers, destroyers and submarines. By 1945 the IJN had been all but annihilated and its place as number three had been taken by the Royal Canadian Navy. This force had entered the war with nothing but a handful of destroyers, but had expanded as one of the major participants in the Battle of the Atlantic. Designed almost entirely as an antisubmarine

NAVIES AND SEA WARFARE

Changing features of naval warfare at the time of World War II

The organization of naval forces

The varieties of naval warfare

The Battle of the Atlantic

Convoy organization

Carrier warfare in the Pacific

force, it controlled hundreds of frigates and corvettes but nothing resembling a modern capital ship. The navy that experienced the most dramatic expansion, however, was that of the USA. With the full resources of a massive shipbuilding industry behind it, it left the war with between two and three times the number of modern capital ships and cruisers that it had begun with, to say nothing of an armada of amphibious warfare vessels capable of launching major operations simultaneously in the South Pacific and Northern Europe.

The roles of technology and morale

Technology played a crucial role in naval warfare, and the outcome of battle was often determined by the quality of radar or communications equipment, the intensity and accuracy of antiaircraft fire, and the ability to use established systems to limit the impact of fire, explosion and other battle damage. Training and coordination were crucial if the complicated machinery of the modern warship was to be used to full effect in combat.

While morale is always important in war, it did not exercise a decisive influence over the sea campaigns of World War II – simply because most navies almost always fought to the best of their technical abilities. Both the German U-boat force and the Imperial Japanese Navy suffered appalling losses in the second half of the war, due to industrial and technological advances on the part of their enemies, but both forces continued to fight hard to the bitter end. Neither British forces dive-bombed by the Axis out of the waters of Crete and the Aegean in mid-1941, nor US forces, swept from the island waters of the South Pacific six months later, suffered any decline in fighting spirit. Only the Italian navy demonstrated a poor appetite for fighting the enemy. A partial explanation for this may be its low level of politicization: it was disinclined to fight because it saw the war more as a fascist adventure than a matter of national survival. The French navy was denied the chance to distinguish itself. Equipped with some of the most modern warships in the world, it was first attacked in harbor by its ex-ally, Britain, and then scuttled in harbor to avoid capture by its rather dubious friend, Germany.

Trade, supply and the sea war

Beyond the drama of fleet engagements and amphibious operations, naval warfare in the modern age has always centered around the control of sea lanes for the purposes of supply and trade. The blockade of enemy coasts, the marshalling of huge escorted convoys of merchantmen and the employment of the free-ranging surface commerce raider had all been essential features of naval campaigns for centuries before the aircraft and the submarine appeared on the scene. World War I had introduced the submarine as the most potent threat to a warring nation's seaborne trade. World War II saw the underwater threat intensify, with German submarines in the Atlantic and US submarines in the Pacific sending millions of tonnes of enemy merchant shipping to the bottom. The coming of age of the aircraft added a whole new dimension to the situation. Merchantmen had to be protected against a threat from above as well as one from below, and the achievement of air supremacy over the sea could allow the destruction of trade and supply on a scale of which not even the submarine was capable. In World War II the navies of the world refought the age-old battle for control of the sea lanes with new weapons as well as old on a scale unimaginable in any earlier conflict.

The UK and Japan were the major combatants

▼ Funeral at sea: American casualties of the naval war in the Pacific are prepared for a watery grave.

most dependent on seaborne supply for survival and suffered most at the hands of these new weapons. The victory of the first and the defeat of the second were inextricably linked to their performance in the war between merchant shipping, submarines and aircraft. Beyond the convoy battles on the major sea lanes, however, many other naval engagements were also fought over trade and supply. Perhaps the best example of this phenomenon was the German surface commerce raiding campaign, which involved operations in every major ocean.

Beyond its participation in the invasion of Scandinavia, the only major role of Germany's small surface fleet was to pose a threat to British commerce. Either by operating against independently routed vessels in distant waters (a role left largely to converted merchantmen with concealed armaments) or by descending suddenly in force on a single north Atlantic convoy, German surface raiders could cause heavy damage and even more serious disruption. Perhaps most importantly, they could tie down large numbers of British warships needed in other theaters.

▲ A U-boat commander uses his periscope to survey the surface. The secret of the submarine's effectiveness was its ability to approach its target undetected. But the necessity of constructing a vessel small enough to avoid easy detection yet capable of carrying a mass of special technology, a full complement of torpedoes, and two different engines produced a cramped and uncomfortable environment for the crew. Submariners spent long hours trapped in a metal tube which a single depth-charge would transform into their coffin.

Convoy Organization

Asdic range
Destroyer
Destroyer
Destroyer
Corvette
Corvette
Merchant Ships
Corvette
Corvette
Corvette
Corvette
Corvette
4.6km

▲ Typical deployment of antisubmarine escort forces.

The basic principle of the convoy is to concentrate merchant ships in large groups so that they can be protected by warships. A typical convoy of 45 ships in 9 five-ship columns would cover an area of 13sq km (5 sq mi), and with the underwater detection device Asdic normally only capable of picking up submerged submarines at distances shorter than a kilometer and able to sweep in an arc of 80 degrees ahead of the vessel, only a strong escort could expect to prevent a determined mass U-boat attack from getting through to the merchant vessels.

Until the mid-way point of the war, overworked escorts could rarely take enough time away from screening duties, shepherding straying vessels back into formation and rescuing survivors to depth-charge a submarine to destruction. Only after sufficient vessels became available to provide large screens and independent groups to cooperate with aircraft in a hunter-killer role, did U-boat losses begin to climb and merchant ship sinkings decline. Before that the convoy only served to keep losses within manageable proportions. When U-boats concentrated in numbers a massacre could ensue.

German surface raiders caused intermittent trade crises in areas as far from their homeland as the South Atlantic, the Indian Ocean and the coastal waters of Australia until the middle of 1942. Germany's fleet in Norwegian waters posed a serious threat to the Arctic Convoys from the USA and UK to the USSR between 1942 and 1944, playing a central if indirect role in the destruction of the ill-fated convoy PQ17 and tying down the best of Britain's handful of modern capital ships until the two main German vessels, the battleship *Tirpitz* and the battlecruiser *Scharnhorst* were destroyed. The most famous action of the campaign, however, was fought in the North Atlantic in 1941 when Germany's first modern battleship, the *Bismarck,* was hunted down and destroyed. This campaign revealed the strengths and weaknesses of the German position. At a time of the war when it was severely stretched in other theaters, the Royal Navy was forced to deploy a huge force of capital ships, cruisers and destroyers to prevent a single enemy capital ship (with only one heavy cruiser as a consort) destroying entire convoys. On the other hand, by committing this formidable vessel on what was always likely to become a one-way mission, the Germans were throwing away a valuable asset, which could have been better deployed.

Attrition at sea: the Battle of the Atlantic

The Battle of the Atlantic was essentially a struggle of attrition, in which the German U-boat force attempted to sink Allied merchant vessels at a faster rate than they could be built, while not themselves accepting losses too high to allow the growth of submarine strength. It was also a struggle of technology, in which complicated machinery was continually under development either to conceal or detect the movements of the submarine; in which new weapons were added to both surface and underwater arsenals; and in which the rival intelligence services struggled to break each other's codes while keeping their own

impenetrable. Ultimately the U-boat force was defeated because it failed in both struggles.

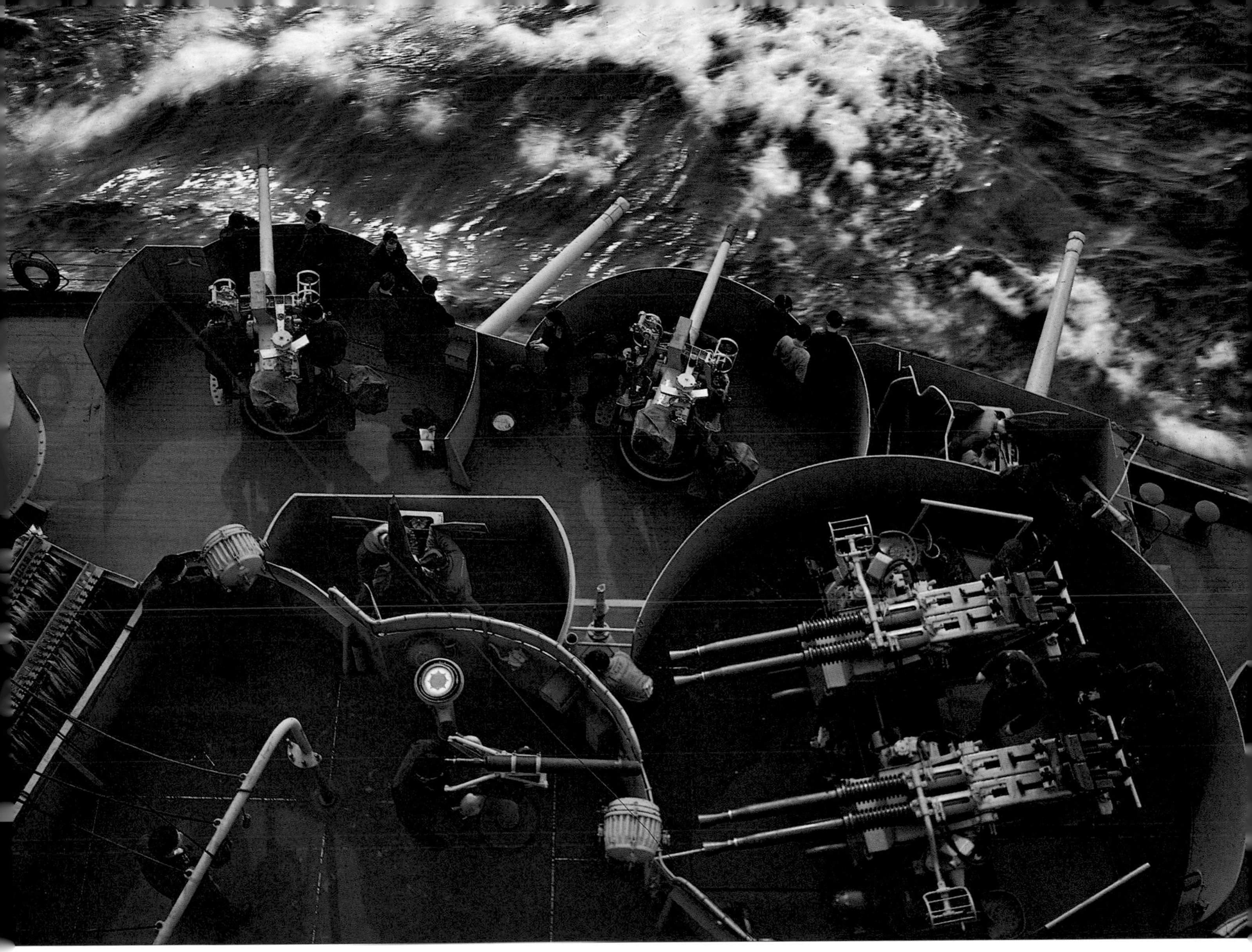

While U-boats in other theaters had the luxury of operating against single unescorted merchant vessels, their counterparts in the North Atlantic had to contend with escorted convoys from the beginning of the war. Because of this fact, U-boats were concentrated in ad-hoc wolf packs under the general coordination of their headquarters in France whether a convoy was detected or not. While the U-boats themselves could not cooperate closely without risking detection of their radio signals, the general intention was to overwhelm the escort by force of numbers and sink as many merchant vessels as possible while they remained a concentrated target.

Both the geographical focus and the technological balance of the U-boat war changed several times during the war. In the early months German submarine activity was concentrated in the western approaches to the United Kingdom, but increasing patrol activity gradually drove them out into the Atlantic, where, in late 1940 and early 1941, they enjoyed a period of great success. Improved antisubmarine techniques brought this period to an end in spring 1941, but when the USA entered the war at the end of the year, the U-boats were able to find easy pickings along the

▲ Overhead view of part of the secondary armament of the US battleship *Arkansas*, taken while the vessel was on convoy duty in the Atlantic. Such vessels had a limited role in protecting convoys against German surface raiders, but it was the far smaller destroyers and corvettes, usually operating in far worse weather conditions than those portrayed here, which represented the last line of defence against the far more dangerous German U-boat force.

◀ An Atlantic convoy at sea. The escort in the foreground is a British destroyer of World War I vintage modified for antisubmarine duties. Convoys of 40–60 merchant ships occupied sea space of several square kilometers, presenting U-boats with a huge target and escorts with a dangerously overextended perimeter to defend.

▲ USS *Bunker Hill* on fire after being hit by two kamikazes off Okinawa in April 1945. Kamikaze attacks often did spectacular damage to the plane-crowded, wooden flight decks of US carriers, but the effect was rarely decisive. The huge Japanese losses incurred in such suicidal assaults failed to destroy the American capability to dominate the Pacific battlefield with naval airpower.

American coast for several months, until routine wartime measures like blackouts and convoys were introduced.

The convoy battles reached their peak in the winter of 1942–43 far out in the mid-Atlantic away from the ever-increasing patrol areas of land-based aircraft. After a final burst of success in the early spring, U-boat losses rose sharply, and merchant ship sinkings fell by a corresponding margin. Under almost constant threat on the surface from long-range aircraft, not only in the Atlantic but also on their outward and inward passages through the Bay of Biscay, and hunted down by free-roving groups of anti-submarine vessels (often accompanied by a small escort aircraft carrier), the U-boats had difficulty even reaching attack positions near convoys. New devices like the schnorkel and new submarines of advanced design came too late to change this situation, and although U-boats remained at sea until the end of the war, their low success rate made them a nuisance rather than a menace.

The Battle of the Atlantic was fought at great cost on both sides. Many torpedoed merchantmen sank so quickly as to allow no chance of crew survival, and many of those who survived the actual sinking were never rescued, either because they were not found or because escort vessels were too busy trying to deal with enemy submarines. The U-boat force itself suffered the unprecedented casualty rate of 90 percent, and in most cases the entire crew would perish with the vessel. The outcome of the battle, however, was determined by ship-building capacity and technological development. After the introduction of extensive Anglo-American merchant-ship building programs, the U-boats could no longer sink vessels at a rate sufficient to reduce the pool of available tonnage. With aircraft able to patrol almost the entire ocean, with newly built escorts

▶ Royal Navy deckhands push a strike aircraft into position on the flight deck of HMS *Indomitable* in the Pacific in 1945. By this time a British carrier force had joined the already huge American fleet. The aircraft carrier was a floating airfield as well as a warship, and the need to maintain fully equipped hangars with all the necessary repair, maintenance and refueling machinery produced crews of over a thousand per vessel – a floating community of specialists in the complicated art of naval air warfare.

appearing in sufficient numbers not only to protect convoys but to form extra support groups as well, and with the technology of detection and destruction developing heavily in favor of the anti-submarine forces,the U-boats of the last two years of the war were more often the hunted than the hunters.

The impact of carriers

The naval war in the Pacific was dominated by the use of aircraft operating from carriers. Fought around the North Pacific island of Midway on 4–5 June 1942, the Battle of Midway marked the turning point in the naval war between the USA and Japan. In it, Japan lost two-thirds of its main carrier strike force, and a similar proportion of highly-trained flight crew. Neither the ships nor the men could ever be replaced, and, stripped at one blow of its most valuable asset, the IJN was doomed to eventual defeat by superior American air power.

The Japanese defeat was due largely to poor planning, confusion of objectives, and over confidence, although sheer bad luck also played a role. Admiral Yamamoto dispersed his forces too widely, sending a small but valuable detachment (including two light carriers) on a futile diversionary attack on the Aleutian Islands, and splitting his main body into separate invasion, battle, and carrier groups, thus denying the carriers themselves adequate escort. While the successful occupation of Midway Island would have been a valuable gain, this mission could only be successfully completed if the opposing US carrier force was first destroyed. In fact, the latter objective was of far greater strategic importance. Despite this, Japanese attention remained focused on the island until it was too late, with the result that the Americans were able to get in the first and, as it turned out, decisive blow.

The battle began with Japanese air strikes on Midway itself. Aircraft were being readied for a second strike, when aerial reconnaissance reported the presence of the American carriers to the northeast. By this time an American strike force was already on its way and, when this arrived overhead on the morning of 4 June, Japanese

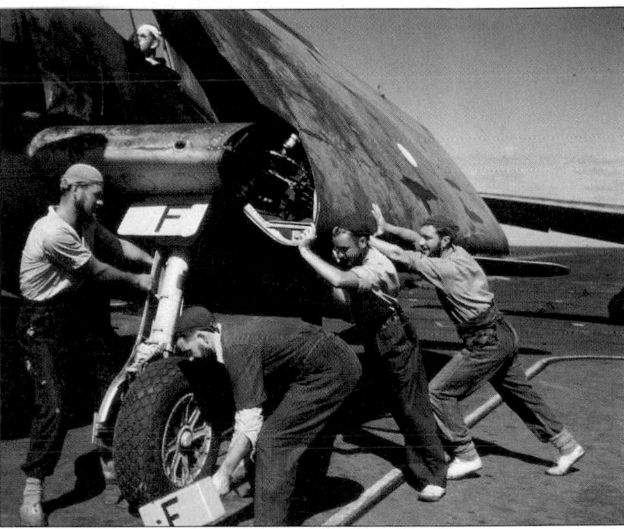

deck crews were still frantically at work rearming aircraft for antiship operations. By sheer coincidence low-flying American torpedo bombers found the enemy first, and, although these were all but wiped out by Japanese fighters, they did pull the latter down to sea level, leaving the way entirely clear for the higher-flying dive-bombers which followed. At about 10.30 am these began their attacks, and within five minutes the outcome of the battle had been decided. The carriers *Akagi*, *Soryu*, and *Kaga* received two, three, and four bomb hits respectively. These, by themselves, might not have been fatal but, with flight decks and hangars crowded with fuel lines, unstowed ordnance, and flight-ready aircraft, a series of seconary explosions and uncontrollable fires soon reduced all three vessels to wrecks.

The one undamaged carrier, *Hiryu*, managed to launch two airstrikes against the enemy, but, again by coincidence, these both found the same target, the carrier *Yorktown*, leaving the other two US carriers, *Hornet* and *Enterprise*, (which were sailing in a separate detachment) free to find and disable the *Hiryu* at 5.00 pm. All four Japanese carriers proved unsalvageable, and those which remained afloat were eventually scuttled to prevent capture. The sole US carrier casualty, the *Yorktown*, was badly damaged, but remained afloat until 6 June, when it was sunk by a Japanese submarine in an attack which also accounted for one of its escorting destroyers. By this time Japanese surface forces, bereft of air cover, had begun to withdraw. In the confusion of the retreat two Japanese cruisers, *Mogami* and *Mikuma*, collided, and were then subjected to heavy air attack. While the former eventually limped home, the latter was sunk by a hail of bombs, drawing the IJN's disastrous battle to a close.

Although the IJN was not so superior in carrier airpower as the overall size of its fleet at Midway would indicate, the Japanese should have been able to win the battle, or at least to have avoided complete disaster, had the operation been more clearly conceived and better planned. Their failure to do so left them with insuffiient naval airpower to fight the remainder of the war with any chance of success.

▼ Mealtime at sea on a US aircraft carrier. A large warship was an entirely self-contained fighting unit. An important requirement for efficient operation was the ability to feed the crew from on-board resources – a major exercise in supply, preparation and organization in a vessel which might be expected to operate for weeks without resupply.

WARSHIPS OF WORLD WAR II

World War II saw the metal-hulled, steam-powered fighting vessel reach its highest and most varied level of development. The largest of the world's naval services deployed the full range of modern fighting vessels, from the smallest coastal craft, through the submarine, destroyer and cruiser, to the last of the great battleships and the first of the purpose-built fast fleet carriers. The naval war was fought not only on the surface of the world's oceans, but in the skies above and in the silent depths below, and warships evolved in a bewildering variety of forms to combat the threats they had to face.

The British Royal Navy, for so long the premier seaborne fighting force in the world, was the most heavily and widely engaged service of the war. By the end of the conflict, it had been far outstripped in size by its US counterpart, but from the opening skirmishes in northern European waters through to the last assault on Japan, its vessels compiled a fighting record second to none, engaging in every possible maritime role, deploying every type of warship,and fighting in every theater save those centered on the Soviet Union's inland seas – all at a cost in vessels sunk and crew lost which none of the Royal Navy's earlier fleets had ever been forced to face.

In common with all of its major contemporaries, the Royal Navy entered the conflict with a force centered on the heavily armed and armored battleship. As the war developed, the aircraft carrier received increased attention, playing a vital role in the destruction of the German battleship *Bismarck*, crippling the Italian fleet in a daring surprise attack on its base at Taranto, helping to fight through the Malta convoys, and finally participating in the kamikaze-threatened naval air campaign against mainland Japan. It was the RN's smaller vessels, however, which bore the heaviest brunt of the sea war. The cruiser and destroyer force fought a long hard campaign in the coastal waters of north Europe and the Mediterranean, suffering heavy losses in the early years from German U-boats and Germany's highly dangerous air force. Along the vital convoy routes of the North Atlantic, these forces were joined by hundreds of far smaller vessels, frigates, corvettes and other anti-submarine escort vessels, all of which were forced to keep the seas for long periods in the worst of weather conditions to protect the UK's supply lines from the potentially fatal threat of the largest underwater campaign against trade the world has ever seen.

► As the war developed, aircraft carriers (above) came to be seen as the most important element of the battlefleet, their strike aircraft capable of destroying the enemy well out of gun range and their fighters of defending friendly vessels from a similar fate. Here aircraft fly off the British carrier HMS *Indomitable* to help defend a convoy to Malta.

► The battleship reached its final stage of development during the war. Modern vessels such as HMS *King George V* were fast, heavily armored and possessed of massive firepower. In a traditional gun engagement they could be very effective, but they could be sunk by aircraft long before they could bring their main batteries into action.

◀ Destroyers, with their high speed, good maneuverability, and mixed gun, torpedo and anti-submarine armament, were the most versatile and most heavily used components of all the fighting fleets. These British craft were photographed escorting an Arctic convoy, providing close protection against aircraft, submarines and surface raiders alike, but they could as easily have been employed on screening duties for a battlefleet or in independent strike forces.

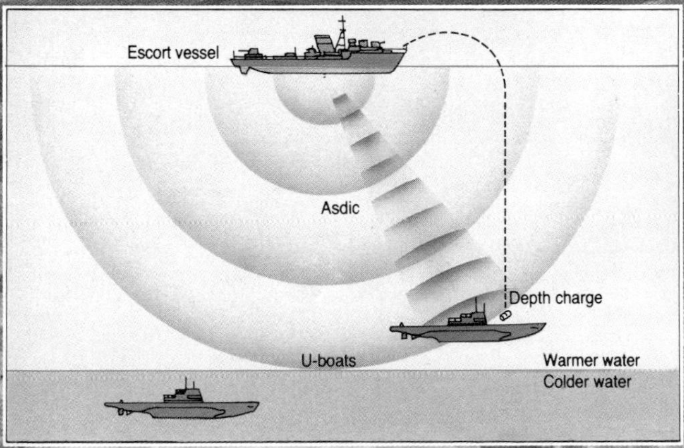

Escort vessel
Asdic
Depth charge
U-boats
Warmer water
Colder water

◀▲ The underwater war of the submarine required a unique collection of equipment and techniques. As this photograph of British vessels tied up alongside their depot ship makes clear (above), the submarine itself was a tiny craft, but its torpedoes could sink even the largest surface vessel. All navies devoted great efforts to improving methods for detecting and destroying submarines underwater. The most common combination was the use of Asdic to produce a sound echo revealing the enemy's postion and the dropping of depth charges, large explosive devices set to explode at the submarine's estimated depth, to crush his pressure hull (left). A wide variety of other equipment was deployed by both sides. Sophisticated listening equipment could pick up engine noises at great distances.

Datafile

The air war swallowed up aircraft and aircrew by the thousands. The greatly increased firepower of the aircraft themselves, the massive growth of ground- and sea-based antiaircraft weaponry and on occasion the use of aircraft as suicide missiles all contributed to increase the high wastage rates normally associated with aerial operations. All of the combatants had to launch massive industrial efforts to keep abreast of the constant need for expansion and replacement. They were also faced with the problem of training enough personnel. Toward the end, the Axis powers could no longer sustain their replacement programs and were all but driven from the skies.

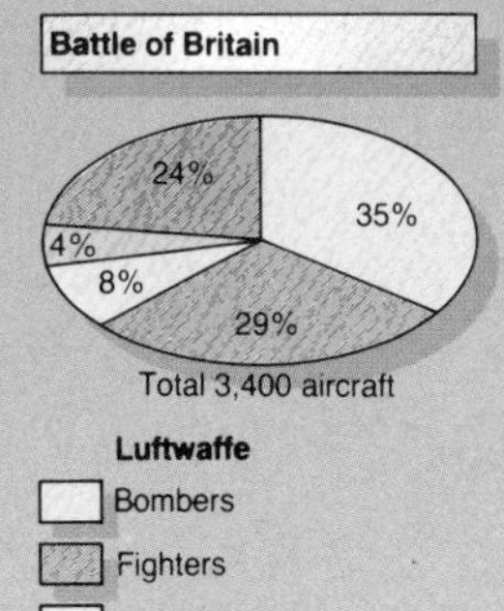

▲ The RAF entered the Battle of Britain heavily outnumbered, but a large part of the attacking German force was made up of lightly armed bombing aircraft ill-suited for their task. The opposing fighter forces were closely matched in equipment, training and numbers, and the British aviation industry was actually able to outbuild its opposite number. With the advantage of fighting over its own territory, the RAF was not in quite as severe a position as has often been thought.

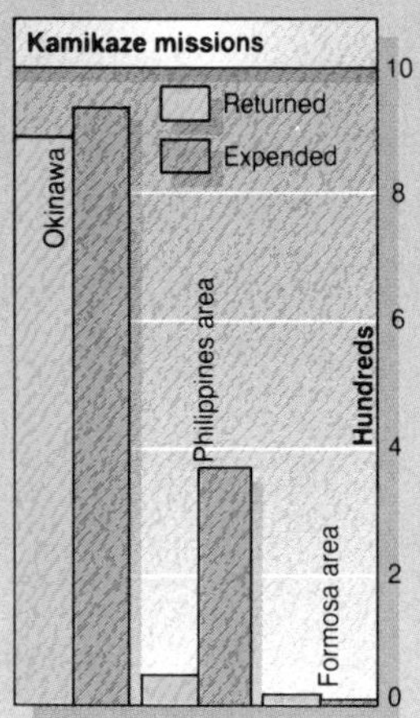

◀ In the last year of the war, the Japanese expended hundreds of planes and pilots in desperate one-way missions aimed at reversing the course of the war by using aircrafts as flying bombs. Most kamikazes did not survive the flak and defending fighters sent up to meet them. Those which did reach their targets could not do enough damage even to delay the Allied advance, and the cost in irreplaceable aircraft and aircrew was such as to destroy Japanese airpower.

Kamikaze damage

11%

89%

Total 222 ships

Warships damaged

Warships sunk

▲ While the kamikazes did not inflict damage on a large number of vessels, much of the effort was expended on destroyers and other small vessels on screening duties. No major Allied capital ships were actually lost to kamikaze attacks, and those which did suffer damage were quickly replaced or repaired.

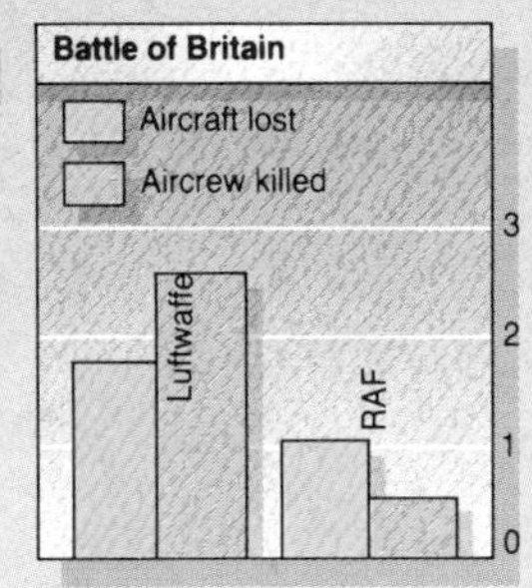

◀ Enjoying all the benefits of fighting over its own territory, RAF Fighter Command suffered appreciably lighter losses than the Luftwaffe. While the balance of losses was not as favorable as portrayed by wartime propagandists, it was sufficient to force the cancellation of the German offensive.

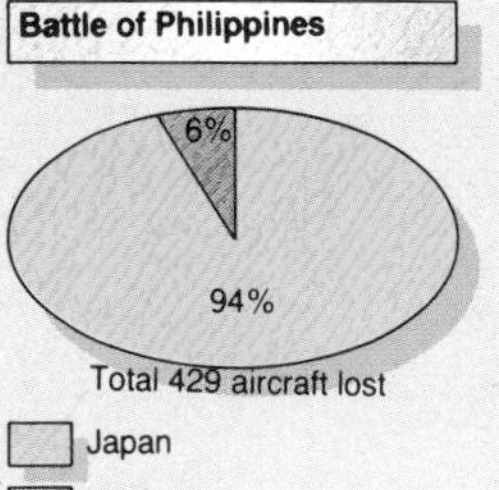

▶ The air battle of the Philippines Sea was almost completely one-sided. Ill-trained Japanese pilots in obsolescent aircraft proved no match for US carrier fighters and Japan's naval air capacity was destroyed for good in a few hours of hectic action without a single US naval vessel receiving either bomb or torpedo damage.

Aircraft exercised a profound influence on the course of hostilities, but at the start of the conflict Europe's air forces probably mustered fewer frontline machines than their predecessors had at the end of World War I in 1918. The French air force, for example, had over 2,000 aircraft in November 1918, but in early 1940 could place only 1,147 machines in the front line, and of these only 678 could be described as modern designs. Only the German air force (Luftwaffe), could really be described as a fully effective striking force. It took 3,530 operational aircraft into the Battle of France in May 1940.

Aircraft construction programs, however, had already become items of urgent priority all over the world before the first shots were fired. As the war developed, the major combatant air forces swelled dramatically in size. By the end of hostilities, the largest of them rivaled armies in the numbers of men employed. The British Royal Air Force (RAF), which had a personnel strength of 175,692 on 3 September 1939, reached a peak of 1,185,833 men on 1 July 1944. The striking might of the aircraft themselves reached levels scarcely dreamed of six years before. In September 1939 RAF Bomber Command could muster 352 aircraft (none of them heavy four-engined types); in the first four months of 1945 it had an average daily strength of 1,420 aircraft (1,305 of them four-engined) and during this period flew 67,483 sorties and dropped over 180,000 tonnes of bombs.

Organization and efficiency

Most air forces employed similar organizational structures: aircraft were grouped into squadrons, wings, groups and finally semiautonomous commands like the US 8th Air Force, responsible for the strategic bombing offensive launched against Germany from bases in the UK. The German air force, which employed a tightly structured command system right through the war, was fairly typical in the way it organized its frontline formations. The smallest tactical unit was the *Staffel* of 12 aircraft (for combat purposes alone the *Staffel* was broken down into three *Schwarmen*, each composed of a pair of two-plane *Rotten*). Three *Staffeln* were combined to form a *Gruppe*, and three or four *Gruppen* in turn comprised a *Geschwader*. Organization up to this level was fixed, but *Geschwadern*, and on occasions detached *Gruppen*, were assigned to *Luftflotten* (air fleets – effectively area air commands) to meet the needs of the operational situation.

The performance of the aircraft was naturally one of the key determinants of combat effectiveness. Pilots unfortunate enough to be sent aloft in inferior equipment, such as those of the Russian air force in 1941 or the Japanese aviators of the last two years of the war in the Pacific, were

AIR FORCES AND AIR WARFARE

bound to suffer crippling losses. Beyond equipment, however, it was the training, experience and élan of a unit which really shaped its fighting efficiency. Flying high-performance aircraft under combat conditions required a degree of individual skill far above that expected of most fighting men. Improved aircraft were of little use without a long and intensive training program and the operational experience necessary to hone individual and group flying skills.

Tactics and resistance

The air war in World War II varied tremendously in nature and intensity. Aircraft were not completely absent from any campaign, and they played an important if not decisive role in most, ranging from Japanese tactical bombing in 1937 to the final bombing attacks on Japan in 1945. Much of the air war was directly related to the land and sea wars below it. The German Blitzkrieg of 1939–41 depended heavily on support from the air and provided the model for later tactical air operations by all the combatants. At sea, aircraft proved a greater danger to warships than other warships, and, even in campaigns as different as those fought in the Mediterranean between 1940 and 1943 and in the Pacific between 1941 and 1945, it was land- or carrier-based air power which exercised the dominant influence.

This was largely the case because the aircraft had developed greatly in range, speed, reliability, and destructive capability since the previous war. The introduction of the metal-framed monoplane with enclosed cockpit and engine/engines of vastly improved power allowed the aircraft far greater scope for interference in land or sea campaigns, enabling it to deliver a large bomb or torpedo payload at great range with considerable accuracy. It also allowed the aircraft to operate as a strategic weapon in its own right, squadrons of bombers being deployed against industrial and population targets hundreds of kilometers from

▼ RAF ground crew gathered in front of a Lancaster bomber in 1944. Keeping a large bombing force operational required a massive force of fitters, riggers, electricians and armorers.

the Front in an effort to cut off the enemy war-effort at its roots.

The tactics of the air war varied with its nature. In the classic fighter-against-bomber confrontations of the Battle of Britain and the Allied strategic bombing campaign against Germany the introduction of radar, radio-control, and a new generation of fast monoplane fighters such as the Spitfire and the Messerschmitt Bf109 disproved the prewar adage that the bomber would always get through. Even large, concentrated formations of bombers with strong defensive armaments would suffer unacceptably high losses until provided with fighter escorts of their own. In operations of a tactical nature, usually undertaken at lower altitudes, the roles of the fighter and the bomber were often combined in one aircraft, and heavy casualties had to be accepted due to antiaircraft fire from the ground.

Ground-based antiaircraft (AA) weaponry grew in strength as the war progressed. At high altitudes aircraft could be subjected to constant danger from time- or proximity-fused shells fired from large guns like the German 88mm. Lower down they could encounter a wall of automatic and semiautomatic gunfire from heavy machine guns, 20mm Oerlikons, 40mm Bofors, and the like. Major cities and industrial targets were protected by tightly integrated rings of searchlights, guns and control stations, all guided by radar, while both warships at sea and military formations on land carried large complements of AA guns with them wherever they went.

Overall, the air war increased in intensity as the conflict progressed, with Allied forces recovering from early losses to gain gradual ascendancy in a long war of attrition. When the breaking point came, the decline of Axis airpower was swift, and the closing months of the war in both the east and the west were characterized by almost unchallenged Allied air superiority over Axis soil. Axis targets came under attack not only from long-range multiengined bombers, but also from smaller aircraft flying from advanced landing

▲ RAF pilots rush toward their Hurricane fighters in the summer of 1940. Advances in radar and communications technology allowed fighter forces to be deployed rapidly from their bases to make last-minute interceptions of incoming raiders. As a result, it was often possible to break up bomber formations before they reached their targets.

► A German bomb-aimer mans the machine gun in the nose of his machine while other aircraft fly in formation beneath him. In 1940 Germany's bomber force was composed entirely of lightly armed two-engined aircraft. Even in tight formation its squadrons could not generate enough defensive firepower to protect them from heavy losses at the hands of attacking RAF fighters.

◀ The heart of the air–sea war: British naval and air force officers control operations in the eastern Atlantic. The combination of advanced detection and communications equipment allowed senior officers to monitor and control complicated multiservice operations in far-flung regions from control complexes a long way from the dangers of bombs and gunfire. Ultimate success was the product as much of radio/radar control as of frontline bravery and firepower.

strips just beyond the front or from carrier flight decks just off the coast.

Debate still rages about airpower's claims to be a war-winner in its own right, but there can be no doubt that it made a major contribution. It employed tens of thousands of aircraft to kill hundreds of thousands of people and inflict millions of dollars worth of damage. The dropping of two atomic bombs on Japan in August 1945 gave a sinister warning that in any future war the destruction would be even worse.

Attrition in the air: the Battle of Britain

The Battle of Britain, fought in the skies over southern England in the late summer and autumn of 1940, began as an attempt by the German air force to eliminate Britain's Royal Air Force as a prelude to the planned invasion of the island. As this objective proved elusive, and the invasion itself became a more distant prospect, the German campaign degenerated into an assault on the morale of the British people through the bombing of major population centers. At each stage of the campaign the Germans failed, not so much because of poor planning or bad intelligence (although these were both factors of some importance), but because the forces at their disposal were not designed for the tasks in hand. They were also severely hampered by recent advances in aviation technology, which had greatly increased the capabilities of the British defenders.

The German air force had been designed first and foremost as a tactical air-support force for military operations on the Continent. It possessed no heavy bombers, and its medium bomber force of JU88s, Heinkel 111s, and Dornier 17s carried small bomb loads and light defensive armaments: they were in no way suited for long-range strategic operations in the face of heavy fighter opposition. The much vaunted JU87 Stuka dive-bomber proved almost useless over Britain and had to be withdrawn after heavy losses in early raids. Finally, the German fighter force could not give its bombers adequate support. The two-engined Messerschmitt Bf110 proved no match for its single-engined opponents, and while the excellent Messerschmitt Bf109 was in every way a match for the British Spitfires and Hurricanes, it lacked the fuel capacity to linger long in the combat zone.

On the other hand the near-simultaneous introduction of radar, integrated radio control of airborne units from the ground, and the fast, low-wing monoplane fighter allowed the British to mount a far more effective defense than would have been possible even a few years before. German air attacks could be detected as their aircraft took up formation over France, and fighter resources could be allocated and moved rapidly to contend with each threat as it developed. As a result, German raids could often be disrupted, if not completely broken up, before they reached their target. The British also enjoyed the added advantage of fighting over their own territory. This allowed pilots, who managed to escape unharmed from disabled aircraft by taking to their parachutes, to return to operations rather than languish in prisoner-of-war camps. It also exposed German aircraft to loss through antiaircraft fire.

Kamikaze Pilots

◄▲ The contest between man and mass machinery: a Japanese pilot (above) ties his nation's flag around his head before departure on a one-way journey to attack the enemy, and the US light carrier *Belleau Wood* (left) burns after being hit by a kamikaze off the Philippines in October 1944. Mass attacks by suicidally brave Japanese pilots exposed US carrier forces to constant danger and heavy damage, but even the mass sacrifice of men and aircraft could not offset the USA's huge material superiority. While damaged carriers were quickly repaired and replaced, the Japanese lost pilots and machines at a rate which their war-effort could not sustain.

German air commanders have been criticized for failing to pursue a coherent targeting policy, but in truth, few of the targets at their disposal were easily neutralized. Radar stations were small and notoriously difficult to put out of commission. Air fields near the coast could be made untenable, but defending fighters could be moved to bases further inland with only a marginal loss of effectiveness. Also, the British aircraft industry was well dispersed, with many of its factories located far to the north in the industrial Midlands. German intelligence persistently underestimated the combat strength and reserves of the RAF, but the more important factor here was simply that the British aircraft industry, more completely mobilized for war, was producing machines faster than its German equivalent.

German losses made their air force the real loser in the war of attrition. It takes nothing away from the RAF's reputation for gallant defense in the Battle of Britain to indicate that the odds against it were not as heavy as has usually been thought.

Air power in the ascendant: the Pacific, 1944–45

In 1944, after a year of steady but unspectacular progress in 1943, the American war-effort against Japan's Pacific empire accelerated. This was most clearly the case in the air war. With the arrival of a new generation of aircraft, a massive expansion of the carrier force, and the continuing growth and success of pilot-training programs, American aviation went from strength to strength. The Japanese case was the reverse. Air services remained heavily dependent on prewar designs of light construction that could not stand up to enemy firepower. Some new aircraft did appear, but were not manufactured in sufficient numbers to make any impact. The carrier force losses of 1942 were never made up, and 1944 saw the extinction of the remainder of the fleet. Worst of all, Japanese pilot-training programs, originally geared to produce small numbers of highly skilled aviators, never adjusted to the demands of the war. The flying skills of aircrews went into such steep decline that the quality of aircraft being flown was scarcely material.

American superiority was graphically demonstrated during the Battle of the Philippines Sea in June 1944. While American carrier strike planes were mauling enemy surface units, their fighter aircraft all but annihilated the Japanese attempt to counterattack, in an air battle so one-sided as to earn it the nickname "the Marianas Turkey Shoot". This massacre destroyed Japanese carrier aviation for good. When the navy launched its massive effort to regain the initiative at Leyte Gulf in October 1944, it was so short of planes and pilots for its remaining carriers that they were used merely as a sacrificial decoy force to draw US carrier forces away from the battle area. This battle resulted in further heavy aircraft losses for the Japanese; it also witnessed the first systematic use of aircraft in the kamikaze role – a role in which the aircraft itself became a projectile to be crashed deliberately on its warship target.

In the Iwo Jima and Okinawa campaigns of 1945 the Japanese turned wholeheartedly to the kamikazes in a last-ditch effort to save the final barrier between the home islands and the enemy. Significant damage was inflicted on American naval forces, but it was never sufficient to stem the tide, and Japanese aircraft losses were so high as to leave few resources for more conventional missions, such as the provision of air cover for the giant battleship *Yamato* on its own one-way mission against enemy forces off Okinawa. In the last weeks of the war US and British carrier forces cruised with near-impunity just off the Japanese coast, striking at shore and naval targets in the face of only light and ineffective resistance.

While the naval air war was making its way inexorably toward Japan's shores, Japanese cities were coming under heavy attack from another direction. Using air bases in China and on captured Pacific islands, the Americans launched a strategic bombing offensive against enemy morale and industrial capacity. Strengthened by the arrival of the new B-29 Superfortress, this air assault inflicted massive damage. Under incendiary attack, Japan's lightly built cities proved veritable tinder-boxes. Japanese industry and transport disintegrated far more rapidly than had those of its ally, Germany, under similar attack. Unlike the Germans, the Japanese failed to develop an adequate home air defense system, partially through the sheer unavailability of aircraft and pilots. By the last months of the war aerial resistance to these raids had virtually ceased and the centers of many of Japan's cities had been razed to the ground. The final blow from the air came in early August 1945 when first Hiroshima and then Nagasaki were devastated by single atomic bombs dropped by B-29s. Convinced that further resistance was pointless, the Japanese surrendered, giving American air power the somewhat tainted distinction of being the first to force a modern industrial power's surrender without the necessity of an invasion. The bomber had played a more decisive role in the defeat of Japan than it had in Europe. The crucial difference (notwithstanding the atomic bomb) was that Japan, unlike Germany, did not possess the resilient infrastructure of the modern industrial state, and could not survive a prolonged aerial assault.

Named "divine wind" after the typhoon that had destroyed a Mongol invasion fleet on its way to Japan in the Middle Ages, Japan's kamikaze pilots were the agents of a last-ditch effort to turn the tide of the Pacific War and save the homeland, yet again, from invasion. The kamikaze, drawing on self-sacrificial ideals, set out from base with the explicit intention of using his aircraft as a guided missile and immolating himself, to cause maximum destruction to the enemy.

The kamikaze campaign began on a small scale on the initiative of a local commander in late October 1944 during the closing stages of the Battle of Leyte Gulf. On their first mission the kamikazes successfully destroyed an American escort carrier, and a number of further successes (and exaggerated reports of others) seemed to provide a desperate Japanese high command with a solution to the US naval threat. In 1945, particularly during the long struggle for Okinawa – the last island barrier between the enemy and the homeland – the kamikaze campaign became the dominant strain in Japanese air policy, with pilots and planes being sacrificed in their hundreds.

The kamikazes inflicted heavy damage and severely shook US morale. They failed, however, to delay or deflect the Americans. Some carriers were temporarily put out of action, but the kamikazes failed to sink any major warships.

There were severe constraints on their effectiveness. The pilots were usually inexperienced volunteers. Their aircraft were not as efficient at destroying ships as the more conventional bomb or torpedo. Much of the damage done affected the superstructure or deck of the target vessel, not its hull or vital machinery. Finally, the inexperienced kamikaze pilots tended to attack the first enemy vessel they saw regardless of size or importance. In many cases this would be a destroyer or similar small craft on picket duty in the outward screen of the invasion force.

This pilot knew his job thoroughly and all those who watched him make his approach felt their mouths go dry. In less than a minute he would have attained his goal; there could be little doubt that he was to crash his machine on the deck of the Enterprise. *All the batteries were firing... but it came on, clearly visible, hardly moving, the line of its wings as straight as a sword.*

FROM USS ENTERPRISE
13 MAY 1945

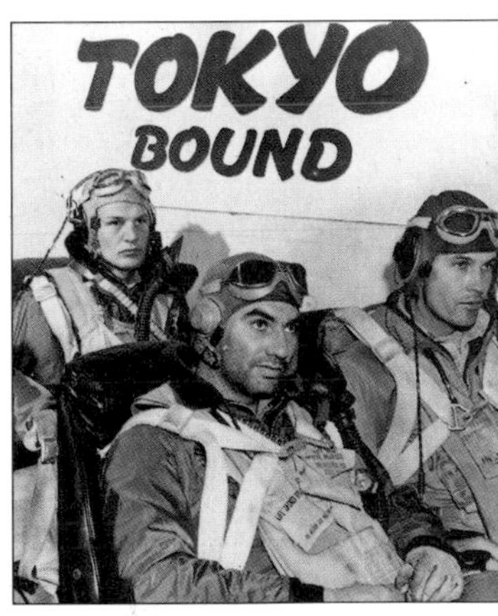

▲ US naval airmen are briefed for a carrier air attack on Japan in early 1945. Although Japanese air units could offer little effective resistance, the strain of even victorious air operations is clearly etched on the faces of these highly trained pilots.

◀ US B-29 bombers on a forward air base in the Mariana Islands. American successes in the naval air war in the Pacific brought the massive might of land-based airpower closer to Japan's shores. Operating from specially built bases, constructed and supplied on captured island groups through a massive logistical exercise, B-29s carried huge loads of high explosive and incendiary bombs directly to Japan's industrial heartland. As one city after another was razed by fire, Japanese air resistance weakened to the point that the bombers' defensive armament was reduced to allow the carriage of even heavier bomb loads.

AIRCRAFT OF WORLD WAR II

The technological development of the aircraft has been characterized by alternating periods of rapid progress and relative quiescence. In the 1920s and early 1930s, design advanced only slowly from the short-range lightly constructed biplane configuration which had emerged from World War I, but from the mid-30s on, the introduction of the all-metal, low-wing monoplane, powered by engines of increased power, transformed the aircraft into a potent fighting machine. By the middle years of the war, tactical aircraft were capable of dominating any land or sea battle area while long-range, multiengined bombers could bring mass destruction on strategic targets hundreds of kilometers behind the front.

Fighter aircraft such as the British Spitfire, the American Mustang and the Japanese Zero were capable of far greater speeds and carried aloft far more potent armaments than their biplane predecessors. They transformed the nature of aerial warfare because their performance advantage over bombing aircraft was such that the latter could only operate with fighter escort of their own if they were to avoid crippling losses. The major problem with early war fighters was that they lacked the range to operate for long periods over enemy territory. Mid-war improvements solved this problem, allowing Allied fighter forces to take the war to their beleaguered Axis opponents, driving them from the air and leaving Germany and Japan exposed to crippling attack by long-range bombers.

At the other end of the design specturm, the war saw the emergence of long-range, multiengined bombers capable of carrying bombloads measured in tons. While such aircraft as the British Lancaster and Halifax, and the American B-17, B-24 and B-29 carried heavy defensive armaments, they continued to suffer heavy losses from defending fighters until their own fighter forces developed the range to suppress such opposition. With or without escort, however, the huge formations the Allies were able to throw at the enemy heartlands in 1943–45 wreaked tremendous havoc, gutting entire cities and obliterating factories and transportation networks. Debate still rages on the exact contribution of strategic bombing to the outcome of the war, but there can be no doubt that the failure of the Axis air forces to develop long-range bombing forces of their own placed their countries at a severe disadvantage in the struggle.

Between the fighter and the heavy bomber, was a wide-range of tactical aircraft whose basic purpose was to intervene directly in the ground and sea wars. Germany made the most important early contribution in this area, with its JU87 divebombers and HR111, JU88 and DO17 medium bombers contributing significantly to the success of the Blitzkrieg and all but driving naval vessels from European waters. By the middle of the war, however, Axis armies and navies were being savaged by a new generation of Allied tactical aircraft, including the powerful Soviet Shturmovik ground-attack machine.

◄▼ Allied bombers used in 1942–45, such as the B-17 Flying Fortress (bottom left), were heavily armed. British aircraft generally operated at night in long, loose streams (left). US aircraft flew by day in tight formations. They waged a bloody battle with enemy fighters, but by the time the larger B-29 (center left) appeared in 1944–45, enemy fighters had been driven from the sky.

▼ Successful aerial operations ultimately depended on the ability of fighter forces to maintain air superiority. The war produced a new breed of adaptable high-performance fighters, of which the British Spitfire was perhaps the most famous. Entering service in 1939, it was modernized and remained capable of vanquishing any foe until the end of hostilities.

◄▼ A German HE111 medium bomber being "bombed-up" for a raid. The bombs used in World War II were capable of causing great destruction. The normal technique was to mix high explosive and incendiary bombs – the former would cause widespread structural damage and disruption, while the latter would turn the resulting shambles into a caldron of fire.

▼▼ The offensive and defensive capabilities of aircraft in aerial combat were considerably enhanced by the high-powered metal monoplane's ability to mount heavy machine guns and light automatic cannon. The latter were particularly effective. One or two hits could usually cause fatal damage to any airframe, but concentrated machine-gun fire could have a similar effect.

RADAR: THE SCIENTISTS' CONTRIBUTION

Among the many wartime technical advances, that of radar was crucial to the Allied victory. Its development greatly increased the effectiveness of night fighters, it enabled long-distance bombers to navigate and locate their targets with accuracy, and was equally crucial in the war at sea in the detection and tracking of submarines.

Britain, Germany and the United States had all conducted research prior to the war into the system of locating distant objects by using pulses of radiowaves. A transmitter generates a short burst of radio-frequency energy, which is then radiated from an aerial. This pulse is reflected off solid objects, and the "echos" received at the source. From the time elapsed between transmission and reception, the distance of the object can be calculated. If a directional aerial is used, rotation of the aerial enables the direction of the object to be determined.

By 1939 the British and German systems were the most advanced. The German Freya was a ground-based system for detecting aircraft; in 1940 the Germans introduced Würzburg, another ground-based system, for detecting fast-flying aircraft at a range of 32km (20mi).

British radar had first been demonstrated in 1935 by Robert Watson-Watt, and in 1938 a chain of ground-based radar stations was set up to protect the British coastline, known as CH stations (Chain Home). The same year an airborne system was demonstrated to detect surface vessels on the sea; in 1939 AI (Aircraft Interception) was demonstrated.

The CH stations and AI were crucial in the RAF's defeat of the Luftwaffe in the Battle of Britain. Their effectiveness was heightened by the introduction of IFF equipment (Identification Friend or Foe), a system whereby every friendly aircraft or ship could be identified on the operator's radar screen.

Equally crucial to the survival of Britain was the role played by radar in the Battle of the Atlantic (1939–45). Radar gave the Allied aircraft their first reliable means of detecting a U-boat. Its effectiveness was hugely improved by the introduction of a new kind of transmitting valve known as the cavity magnetron. This gave greatly improved definition and accuracy to the radar and enabled the detection by ships of even periscopes of partially submerged U-boats. In 1943, the year of its introduction, U-boat sinkings rose from 84 to 238.

American radar, which in 1939 had lagged behind British and German, was improved by close technical collaboration with British scientists from September 1940. Japan, on the other hand, had greatly inferior radar, mostly deriving from German designs.

◀ The threat posed by German U-boats to convoys was met by the ship-borne radar system known as Type 271. This could detect a submerged submarine's periscope. Lightweight Air to Surface Vessel radar (ASV), carried on the aircraft supporting the convoy, was almost as sensitive.

▶ The RAF's defeat of the Luftwaffe in 1940 owed much to Ground Control Interception. Radar on the ground tracked all the aircraft in the vicinity; their positions, minute by minute, were logged at the control room. The location of the enemy was relayed to the RAF fighter pilots by radio telephone, who were then guided onto the tail of the German aircraft within range of their own AI systems.

◀ The cavity magnetron, which enabled the radar wavelength to be reduced from 100cm to 10cm, brought a dramatic increase in definition. It was described by Hitler in January 1944 as the single invention that had done the most to hold up the German war-effort – in the U-boat war against Allied convoys. It was also the key to postwar development.

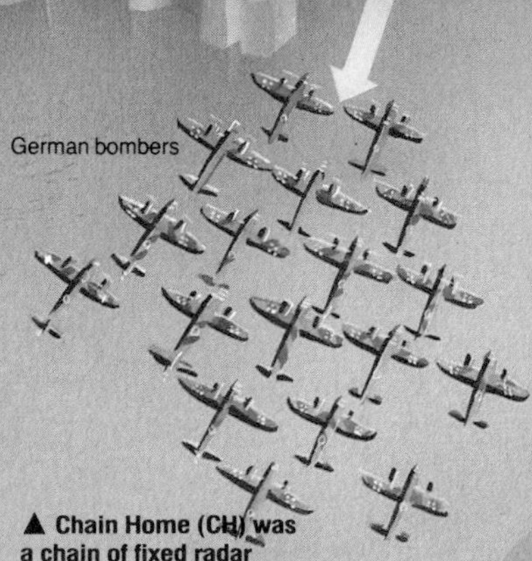

▶▶ Radar assisted the war effort of all major participants in World War II, but especially the British. Its uses were defensive, in the air war and at sea, and offensive, helping fighter aircraft and bombers to navigate and locate their targets.

▶ Window, known as "chaff" in the United States, consisted of small strips of aluminum foil dropped by aircraft. This foil was detected by enemy radar, and meant that the position of the aircraft itself could not be detected with certainty.

▲ Chain Home (CH) was a chain of fixed radar installations around the coast of Britain, giving long-range warning of the approach of enemy aircraft in daytime. CHL stations performed a similar function but were rotating.

▼▼ Intersecting radar beams were used to guide bombers to their target. The bomber flew along an arc at a fixed distance from one station (the "cat station"); the interception of a beam from the "mouse" station gave the signal to release its bombs.

▼ H2X radar, and its later refinement, was an airborne system that swept the area beneath a bomber to build up a picture of the terrain over which it was passing. In this photograph, ships and the coastline can be clearly identified.

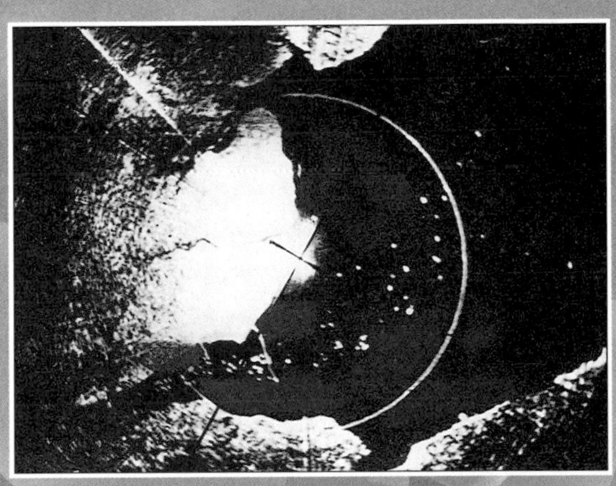

◀ The movements of formations of fighters over occupied territory were monitored at home by the Fighter Direction radar; navigational information could then be relayed to the aircraft by means of a radio telephone link.

◀ Homecoming bombers navigated by GEE, whereby a signal was beamed out by three widely dispersed ground-based beacons. From the compass bearing and distance of these stations, the position and course of the aircraft could be calculated.

▲ Airborne radar, as on the Messerschmitt shown here, used aerials on each side of the aircraft's nose. Each one sent out a lobe-shaped beam and picked up its own signals. The signals from the two aerials were displayed side by side on a screen.

Datafile

Several millions of the men who donned the uniforms of their countries during World War II ended their active service in enemy prisoner of war camps, many as the result of mass surrenders such as Singapore or Stalingrad, others in the day-to-day struggle of battlefield or bombing campaign. For some, imprisonment did at least mean an end to the danger of death or dismemberment – in the west, in particular, only a handful of POWs suffered anything worse than boredom and prolonged separation from friends and families. For others, the act of surrender could lead to maltreatment, exploitation and almost certain death. German attitudes to slavs and Japanese contempt for all prisoners meant that on the Eastern Front and in the Japanese camps in Asia, respecively, thousands of prisoners were deliberately murdered while even more were allowed to die as a result of malnutrition or overworking under appaling conditions.

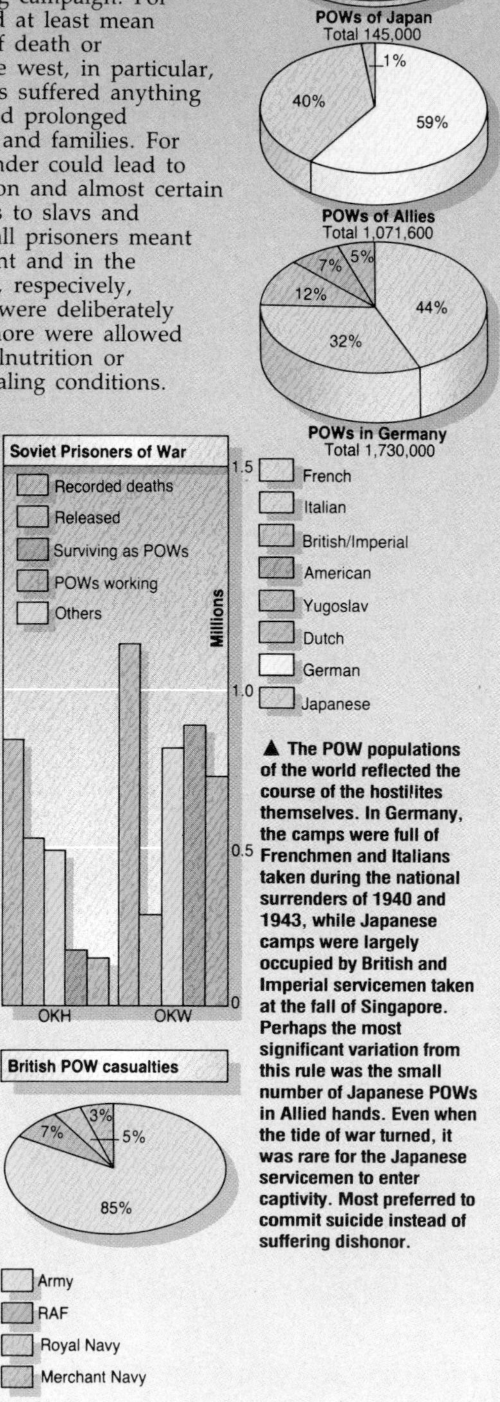

▶ By May 1944 German forces operating in the USSR had captured just over 5 million Soviet soldiers. Of these, substantially less than half were still alive – and even of these the majority were either serving under arms or working as slave labor for their captors. Almost 2 million had died of starvation or disease in prison camps, and a further million had been killed out of hand, many of them at the Front soon after they had surrendered. This sort of treatment, and that meted out to German POWs in retaliation, reduced the war on the Eastern Front to barbaric savagery, a state of war far removed from the normal conventions of modern warfare.

▶ Soldiers normally make up the bulk of POW populations. As a result of its long bombing campaign over Germany, the UK probably lost more airmen to the enemy's prison camps than any other country while its sailors were similarly exposed to capture on a more prolonged basis. Nevertheless, the majority of the British population in POW camps were from the army.

▲ The POW populations of the world reflected the course of the hostilites themselves. In Germany, the camps were full of Frenchmen and Italians taken during the national surrenders of 1940 and 1943, while Japanese camps were largely occupied by British and Imperial servicemen taken at the fall of Singapore. Perhaps the most significant variation from this rule was the small number of Japanese POWs in Allied hands. Even when the tide of war turned, it was rare for the Japanese servicemen to enter captivity. Most preferred to commit suicide instead of suffering dishonor.

Warfare on a large scale always produces large numbers of casualties and, if the relatively humane conventions of the modern age are being observed, of prisoners. Precise figures for prisoners of war (POWs) in World War II are difficult to obtain, but rough figures for servicemen recorded as missing in action (a very large percentage of whom would have been captured) were as follows: USA 139,000; UK 214,000; Italy 350,000; Germany 3,400,000; and USSR 5,750,000.

POWs fall into two categories: those taken in small numbers over long periods of time, such as aircrews of the RAF and USAAF who parachuted

Colditz

Colditz Castle was originally built as a hunting lodge for the kings of Saxony. Sited on a steep hill overlooking the Mulde river, the building was ideal as a place of confinement. The German prison administration decided to use Colditz for particularly important prisoners and for prisoners who habitually attempted to escape from regular camps. Major redesign work was undertaken to make it escape-proof.

It was designed to hold a maximum of 200 men, but as the war went on the number increased well above this limit. The multinational population included Poles, Dutch, Belgians, French, and Britons.

From 12 April 1941, when the first prisoner (a Frenchman) went missing, the prisoners of Colditz continually pitted their courage and ingenuity against the castle and its guards in an attempt to find freedom. Attempts to escape ranged from simple bluff or disguise to the most ingenious utilization of technical expertise. Foremost among the latter was a glider, built in sections in the upper loft above the prison chapel. Never used, it was found only when Colditz was liberated. All told, 130 prisoners actually managed to escape from Colditz, but of these only 32 evaded death or recapture to cross the frontiers of Hitler's Reich to a safe neutral haven.

By the last months of the war conditions inside Colditz deteriorated due to the collapse of the whole German war economy. The inmates of Colditz were freed by American troops in mid-April 1945. While they constituted only a tiny fraction of the POW population, the Colditz prisoners have rightly attracted a huge share of historical attention. Many servicemen staged daring escapes from other camps, but only the men in Colditz had to contend with the high security of a prison which had been constructed to make escape impossible.

▶ The place of confinement: Colditz courtyard.

▶ One way out (inset): a hidden escape tunnel.

PRISONERS OF WAR

Quantities of prisoners in World War II

Prewar conventions for the treatment of POWs

Prison camps in western Europe and North America

The special case of Colditz

Barbarous treatment of prisoners in Eastern Europe and in Asia

The notorious Bataan death march

from crippled aircraft during the strategic bombing campaign against Germany; and those taken together in one large group as a result of a mass surrender, such as the 80,000 German survivors of the Battle of Stalingrad who surrendered on 2 February 1943. Most of the combatant nations had adequate facilities to deal with the first type of POW, but all had severe problems when confronted with the second.

Conventions for the humane treatment of POWs had been established by the Hague Conventions of 1899 and 1907. These attempted to ensure that the prisoner would be adequately fed, clothed and sheltered, that he would be treated in accordance with his rank, and that he would not be overexploited as a laborer, forced to contribute to the enemy war-effort or brutalized in any way. At the start of the war in western Europe in 1939, all combatants made reasonable efforts to conform with international standards. As the war developed, however, the mushrooming of the POW population, shortages of food, manpower and accommodation, and the barbarization of warfare itself in certain theaters produced a wide variety of standards, ranging from the completely correct to the overtly murderous.

POWs in western Europe and America

Generally speaking, the western nations treated their prisoners of war as correctly and humanely as resources would allow. There were isolated cases of groups of prisoners being shot in cold blood by both Allies and Axis, but the vast majority were safely transported out of the combat zones and confined under tolerable conditions. The USA, with vastly superior resources, compiled the best record and the 425,000 POWs (mostly German) transferred to camps in the USA fared as well as any, save that geography made escape and return to Germany a virtual impossibility. The UK generally maintained similar standards, although lack of space and general shortages did produce problems, and many prisoners had to be shipped to Canada, running the risk of having their transport torpedoed by one of their own submarines on the way. Finally, the Germans were as a rule correct in their treatment of western European and American prisoners, but severe problems occurred in the last year of the war, and SS and Gestapo units did perpetrate isolated atrocities.

It is difficult to generalize about POW camps themselves. They varied in construction from converted castles and country houses to specially built compounds incorporating a large number of buildings, and ranged in capacity from less than 100 inmates to many thousand. Officers and men were generally incarcerated in separate institutions, and it was normal, but by no means universal, practice to segregate prisoners of different nationalities. Security was provided by barbed-wire fences, searchlight towers and armed guards (the latter normally drawn from the ranks of those classed unfit for combat), but efforts were usually made to employ nature itself to make escape difficult: isolated locations distant from any frontier were particularly common.

While most of the literature on POWs concentrates on daring escapes, most prisoners made no attempt to break free. They lived a regimented life, generally under the discipline of both captors and superiors in the prison population, and while some were employed on a variety of labor tasks

▼ Mute testament to tragedy, the last message of an imprisoned British-Chinese member of the former Hong Kong garrison, executed by his Japanese captors. Japanese disregard for POW life resulted in a steady string of executions, often without any real pretext, throughout the war.

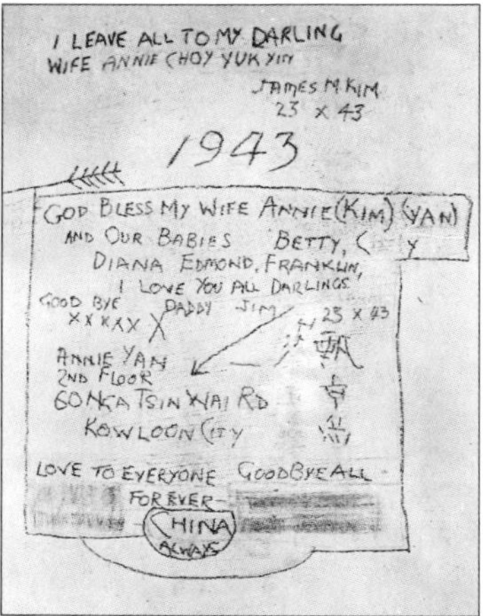

I LEAVE ALL TO MY DARLING
WIFE ANNIE CHOY YUK YIN
JAMES M KIM
23 X 43
1943
GOD BLESS MY WIFE ANNIE (KIM) (YAN)
AND OUR BABIES BETTY, C[illegible]Y
DIANA EDMOND, FRANKLIN,
I LOVE YOU ALL DARLINGS
GOOD BYE DADDY JIM
XXXXX X
23 X 43
ANNIE YAN
2ND FLOOR
60 NGA TSIN WAI RD
KOWLOON CITY
LOVE TO EVERYONE GOODBYE ALL
FOREVER
CHINA
ALWAYS

the normal day of the typical prisoner was characterized by long periods of boredom and inactivity. Food was adequate if neither plentiful nor varied, and living conditions were spartan but rarely harsh or dangerous. The biggest problem facing most of the hundreds of thousands of servicemen who spent part of the war in Europe in a POW camp was enduring separation from home, friends and family while waiting for freedom.

Two kinds of barbarity

The situation on the Eastern Front was entirely different. The German conduct of a race war against the USSR transformed the war on the Russian Front into something almost primeval in its savagery. Of nearly 6 million Russians captured by German forces, only about 1 million were still alive at the war's end. The rest were either killed at or near the Front, or allowed to die of exposure, starvation or exhaustion in overcrowded compounds or on slave gangs in factories and mines. The Russians, who had set an early precedent by murdering 4,443 Polish officer prisoners in the Katyn Forest in 1940, responded in kind to German barbarities, treating POWs with similar cruelty and disregard for life.

The Japanese case was entirely different, and possibly even worse. There was no place in the Japanese military code of behavior for the prisoner of war. A fighting man was expected to seek death rather than accept the dishonor of surrender. Most Japanese servicemen deliberately chose death. Western soldiers who gave up their arms voluntarily were treated simply as if they had sacrificed any right to life. A minority were deliberately killed and most were worked as slave laborers, denied adequate food, shelter or medical supplies and allowed to die in their thousands. For many POWs this meant slow starvation in a neglected prison camp.

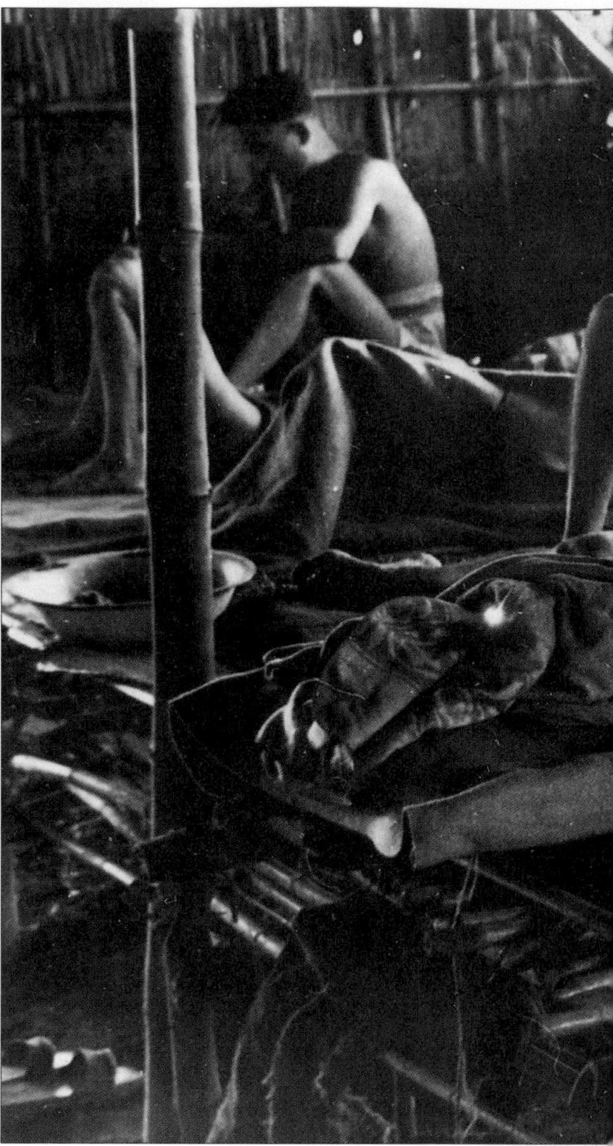

▼ The boredom of captivity: British POWs in Stalag Luft III, Sagan, Germany. Western Allied servicemen in German hands rarely suffered severely at their captors' hands, but the forced inactivity of life in the camps and the long separation from friends and family imposed a burden of their own.

The Bataan death march

Another significant minority, however, did not survive long enough to reach camp or slave gang. They died from brutalization and maltreatment on their way into captivity. One of the most infamous incidents of this type was the Bataan death march of April 1942. In the first week of that month, after several months of resistance against ever-increasing Japanese forces, American and Filipino troops were forced down to the tip of the Bataan Peninsula. Weakened by disease and lack of supplies, the 35,000 survivors surrendered unconditionally to avoid annihilation. On 15 April, six days after the capitulation, the prisoners were sent off on a forced march from the south of Bataan to the railroad sidings at San Fernando 100km (60mi) to the north. Already in poor physical condition, they were driven along in columns of four through fierce heat and choking dust, without anything approaching an adequate supply of food and water. Anyone who dropped out was murdered by clean-up squads moving behind the column; anyone found with Japanese currency or any article presumed to have been looted from Japanese dead was beaten to death; and all the marchers were systematically robbed

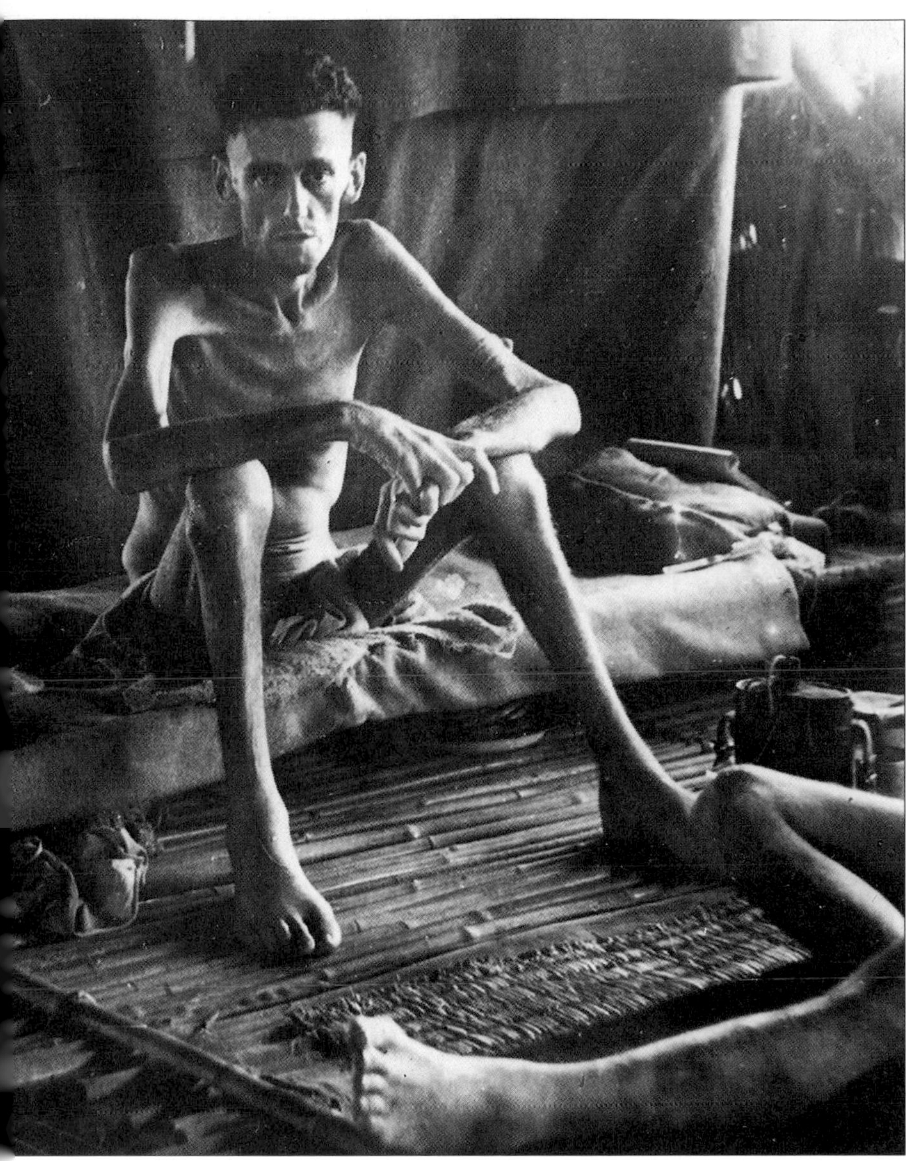

of their possessions. This ordeal lasted six days, but when the survivors did reach San Fernando, they were first packed into poorly ventilated boxcars for a three-hour journey and then forced to march another 11km (7mi) to Camp O'Donnell in the jungle province of Arlac.

More than 10,000 prisoners (approximately one-third of those who had surrendered two weeks before) did not survive the Bataan death march. Sickness, starvation, brutalization and summary execution accounted for almost all the deaths, and only a handful of men managed to escape into the jungle. Although it is usually portrayed as a crime against American prisoners, more than three-quarters of the dead were actually Filipino soldiers (US casualties on the march totaled 2,300). It was therefore fitting that Lt-Gen. Masaharu Homma, who had commanded the Japanese forces in the Philippines and was thus ultimately responsible for the outrage, was brought to justice in the islands. Arrested at the end of the war, he was tried in Manila and executed by firing squad in April 1946.

Whatever Homma's personal culpability, it is only fair to point out that he was something of a scapegoat. To Western eyes the Bataan death march was an atrocity, a sadistic and completely unnecessary exercise in brutality and mass murder. Homma and his men, however, were functioning within a code of behavior which did not so much degrade the prisoner of war as deny the possibility of his existence. For much of the war Japanese fighting men behaved in a fashion quite consistent with this code, often going to extraordinary lengths to find death rather than accept imprisonment. It is a moot point whether someone behaving within the terms of one code should be found guilty within the terms of another, but whatever the verdict on Homma himself, it remains impossible to see the Bataan death march as anything less than an unnecessary act of savagery.

▲ The ravages of neglect, a liberated British POW in a Japanese hospital camp in Thailand. Underfed, denied proper medical attention and often worked until they were too sick to continue, servicemen captured by the Japanese suffered an appalling ordeal which killed tens of thousands and reduced survivors to mere skeletons.

► Chinese POWs march toward an uncertain fate. The Japanese campaign in China was waged with particular savagery, and it was common for prisoners to be killed out of hand. Those who did succeed in surrendering could look forward to nothing but brutalization and the constant possibility of execution.

Datafile

Intelligence plays a vital role in any war. It works on various levels and includes the use of spies and agents as well as technically gathered intelligence, especially from radio signals. Counterintelligence units and domestic security agencies are also essential to protect one's own secrets. Every major combatant scored intelligence successes, but in the end the intelligence war was conclusively won by the Allies.

Principal intelligence organizations	
Japan	
	army general staff 2nd (intelligence) division
	navy general staff 3rd (intelligence) division
Owada Tsushin-tai	Owada signals intelligence unit (army)
Kempei	military security police
Germany	
Abwehr	military intelligence
RSHA	Reichssicherheitshauptamt (state security apparatus), includes SD (Sicherheitsdienst – security police) and Gestapo
B-Dienst	navy sigint
Italy	
SIM	Servizio Informazione Militare – army intelligence
SIS	Servizio Informazione Segreto – navy intelligence
Pubblica Sicurezza	domestic security agency
UK	
MI6/SIS	secret intelligence service – foreign intelligence
MI5	security service – counterintelligence
GCHQ	Government Communications Headquarters – sigint (Government Code and Cipher School to 1942)
SOE	Special Operations Executive – covert action
USA	
OSS	Office of Strategic Services – intelligence and covert operations
SSA	Signal Security Agency – army sigint
FBI	Federal Bureau of Investigation
USSR	
NKVD	Narodnyi Kommissariat Vnutrennik Del – domestic/foreign intelligence, covert operations
GRU	Glavnoe Razvedyavatelnoe Upravlenie – military intelligence

▼ **The British became increasingly efficient at cracking German codes. Among their major breakthroughs were the U-boat keys "Heimisch/Hydra" (read from August 1941) and "Triton" (December 1942). Some codes, however, remained unbroken.**

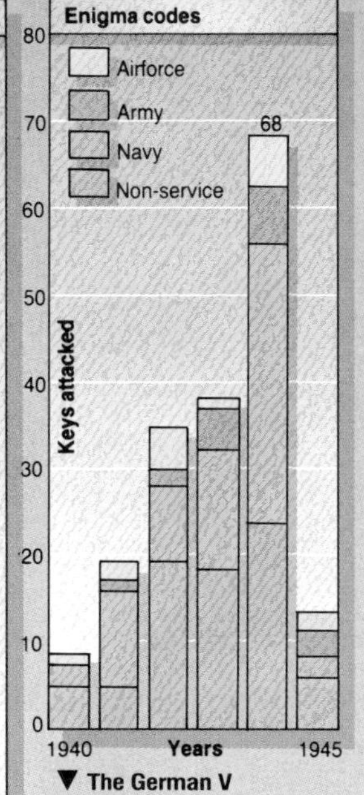

▼ **The German V (Vergeltung, or retribution) weapons well illustrate the importance of good intelligence. The 5,800km/hr (3,600mi/hr) V2 rocket was so technologically advanced that many on the Allied side at first doubted its existence and were only convinced by a combination of intelligence information.**

Chronology: British intelligence and the V2 rocket

1939

November
British intelligence receives the "Oslo Report", which mentions a secret research establishment at Peenemünde on the German Baltic coast due north of Berlin.

1942

October
First successful launch of a V2.

December
Intelligence receives report of the testing of a "large rocket".

1943

March
Captured German airmen inadvertently confirm the existence of long-range rockets.

April
Aerial photographs of Peenemünde indicate that site might be used for firing rockets.

June
Photographs and Enigma intercepts, convince intelligence that rockets are being developed at Peenemünde. The British cabinet orders the site to be bombed.

August
Peenemünde is bombed heavily and progress on the rocket set back several months. Existence of rockets at Peenemünde is confirmed by a detailed intelligence report, which also reveals launch sites near Le Havre and Cherbourg.

October
Intelligence receives report of "secret weapon sites" near Abbeville in northeast France.

1944

April
Intelligence receives a report of the testing of "aerial torpedoes" at Blizna in Poland.

May
Intelligence receives report of an unexploded V2 in the hands of Polish partisans.

June
First V1 lands in London on June 13; on the same day, a German V2 rocket crashes in Sweden and is examined by British intelligence officers.

July
Blizna abandoned by Germans as Soviets advance; Polish V2 remains flown to Britain.

August
No V2 launching-sites can be located with certainty, but suspected areas are bombed.

September
First V2 lands in London; the launch site remains unknown.

December
V2 sites located at The Hague in the occupied Netherlands; heavy bombing begins.

During World War II the Allied powers enjoyed a marked advantage in intelligence over their Axis opponents, which contributed substantially to the Allied victory in 1945. The sorts of intelligence which were most useful to the Allies, moreover, reflected the increasing importance of technology in warfare. Human intelligence, known as "humint" – the classic style of espionage, involving undercover agents, "moles" and traitors – was significant, but its value was largely superseded by technically gathered information, especially signals intelligence, known as "sigint". Indeed, some historians believe that the Allied command of sigint shortened the war by two or three years.

Problems of intelligence information

At the start of the war in Europe the advantage mostly lay with the Axis powers. British, French, Soviet and American intelligence failed to give much warning of enemy attacks. Although the British and French, for example, were well aware in the late 1930s that German policy aimed at dominating Europe, and that the Germans were quickly building up the resources to sustain a large-scale war, they were less sure about when and where precisely they might strike.

Even when good information was available, intelligence chiefs still faced the problem of convincing their military and political masters of its quality and reliability. French intelligence reports of growing German strength were largely ignored by the politicians, in part because they did not want to believe them. Dealing with as unpredictable and opportunistic a leader as Hitler, moreover, made intelligence assessment even more difficult. Sometimes even *German* generals did not know what Hitler was going to do. For the most part the Axis side had better intelligence about Allied intentions in the period just before war broke out in Europe. Both the Italians and the Germans had broken British diplomatic ciphers, which helped them to assess how far the British could be pushed before declaring war.

Political and diplomatic intelligence is an area where human sources can be useful. Lax security at the British embassy in Rome enabled a local employee, Secondo Constantini, to pass valuable secret documents to Italian intelligence. Despite a security investigation in 1937, Constantini was not exposed as a spy until after the Italian surrender in 1944. In 1943–44 the British suffered similarly from the activities of "Cicero", Elyesa Bazna, in Ankara. Baszna was valet to the British ambassador and photographed top secret papers in his safe, which he passed on to the Sicherheitsdienst or SD (the German security service). But the material which Cicero provided was so good that, paradoxically, it became suspect in Berlin.

The two main types of intelligence: "humint" and "sigint"

The varied experience of wartime secret agents

Codebreaking in Europe and Asia

The Allies' greatest wartime triumph: the breaking of Enigma codes

Intelligence and technological advances

◀ Good intelligence – vital for victory – means keeping your own secrets as well as discovering the enemy's. The British ran a campaign which depended heavily on humorous posters with catchy slogans. It was amazingly successful. Despite the huge number of people who worked in sigint, the secret of British wartime codebreaking was kept not only during the war but for nearly 30 years after.

Human intelligence in wartime

In terms of humint the work of the British secret intelligence agency, MI6, got off to a poor start. In November 1939 the head of the MI6 organization in Holland and another officer were lured to a meeting with alleged anti-Nazi German dissidents at a place called Venlo on the Dutch–German frontier. But the meeting had been set up by the SD; the two British officers were captured by an SS snatch-squad and bundled over the border.

During the war MI6 had to share responsibility for clandestine operations with a new agency, the Special Operations Executive (SOE), which was set up in July 1940. Although SOE, like MI6, gathered intelligence, its main function was, in Churchill's words "to set Europe ablaze" by means of sabotage and subversion. There was considerable rivalry between the agencies, with MI6 especially resenting the diversion of men and materiel to SOE. One MI6 recruit dropped into Greece found that his predecessor had been shot by an SOE man who suspected him (rightly as it turned out) of being a double agent.

The experience of agents working in occupied Europe varied considerably from place to place. In France and the Low Countries they worked covertly in the classic espionage mold. They were either brave patriots – often civilians – determined to do what they could to resist German

A Soviet Spy in Japan

One of the most successful spies of the modern age was Richard Sorge (1895–1944), a Soviet agent who worked in Tokyo. The son of a German father and a Russian mother, raised in Berlin, he became a communist at university and was recruited in the 1920s by the GRU (Soviet military intelligence). In 1933 Moscow decided to send him to Japan.

He worked in Tokyo as a journalist and was so successful that he was named correspondent for the prestigious *Frankfurter Zeitung*. This position helped him gain the confidence of the German embassy in Tokyo. He became a trusted adviser, was actually given an office in the embassy and sometimes drafted the ambassador's reports to Berlin. He also built up a small but very impressive network of agents, of whom the most important was Ozaki Hotsumi, another journalist, who had first met Sorge in China. Ozaki had very good lines into senior government circles and for a while was even a consultant to the Japanese Cabinet. For eight years up to October 1941 Sorge's ring of agents provided Moscow with top-level intelligence about both Germany (from Sorge's embassy connections) and Japan. Most of all, the Soviets wanted to know if the Japanese were likely to invade Siberia at any stage, especially after the German-Japanese Anti-Comintern Pact of 1936. This concern reached its highest level after the German invasion of Russia in June 1941.

Drawing particularly on Ozaki's sources, Sorge reported in early October 1941 that there would be no Japanese strike against the Soviet Union, thus enabling Moscow to transfer precious troops from the Far East to fight against the Germans in Russia. This was Sorge's last and greatest intelligence coup. Shortly afterwards, let down by poor security on the fringes of his ring, he was arrested, convicted and executed (with Ozaki) in November 1944.

Sorge fitted the popular image of what a spy should be. He was glamorous, high-living, a hard-drinking womanizer, with a splendid cover, who gained the full confidence of his targets and provided his masters with very valuable high-grade political and military intelligence. He was also an agent who worked from ideological commitment, not for money, and that sort of spy is the very hardest to catch.

The Breaking of Enigma Codes

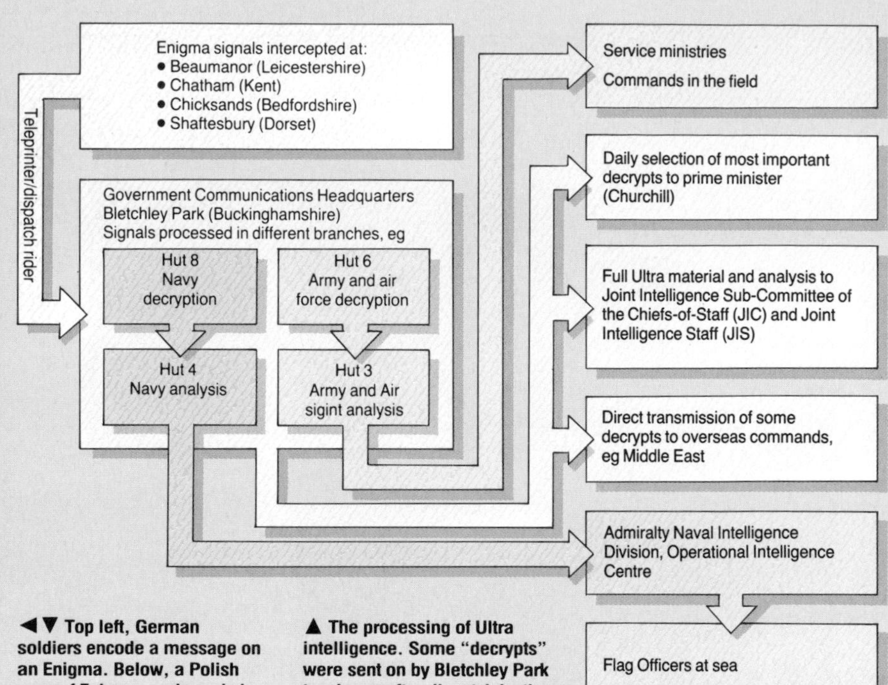

◄▼ Top left, German soldiers encode a message on an Enigma. Below, a Polish copy of Enigma made early in the war.

▲ The processing of Ultra intelligence. Some "decrypts" were sent on by Bletchley Park two hours after dispatch by the Germans.

The main machine used by the German forces in World War II for encoding messages for onward transmission by radio was the "Enigma". Through a complex arrangement of plugs and rotors the possible number of encoding positions for each letter was immensely large: 5,000 billion trillion trillion trillion trillion (5 followed by 87 zeros). Each branch of the German armed services used different keys for their machines and the settings were changed every 24 hours. Since they took good care to protect their daily settings, issuing them only for short periods at a time, the Germans believed that the Enigma machine was absolutely secure. They were wrong.

From May 1940, when the German air-force key was mastered, British cryptanalysts working at Bletchley Park in Buckinghamshire, England, were increasingly able to read enemy signals. There were three reasons for this success. In the summer of 1939 Polish intelligence officers, who had attacked Enigma with some success in the 1930s, decided to share their knowledge with the British and the French. This gave the British a useful start. Secondly, the British poured enormous resources into the signals intelligence effort, which eventually involved over 10,000 people. The head of the so-called "Government Code and Cipher School", Commander Alistair Denniston, was given virtually a free hand to recruit whomever he wanted. He combed the universities for the most intelligent people, men and women, professors and students. Classical scholars were often favored, especially if they played chess or bridge. The third reason for the British breakthrough lay in scientific and technological advances. Two brilliant mathematicians, Alan Turing and Gordon Welchman (originally recruited not because of mathematical skill but for ability at chess), led research which produced an electromagnetic calculating machine called a bombe. It enabled possible Enigma settings to be tested at high speed. In order to cope with improved German cipher methods later in the war, Turing and his colleagues also developed the "Colossus", effectively the world's first computer.

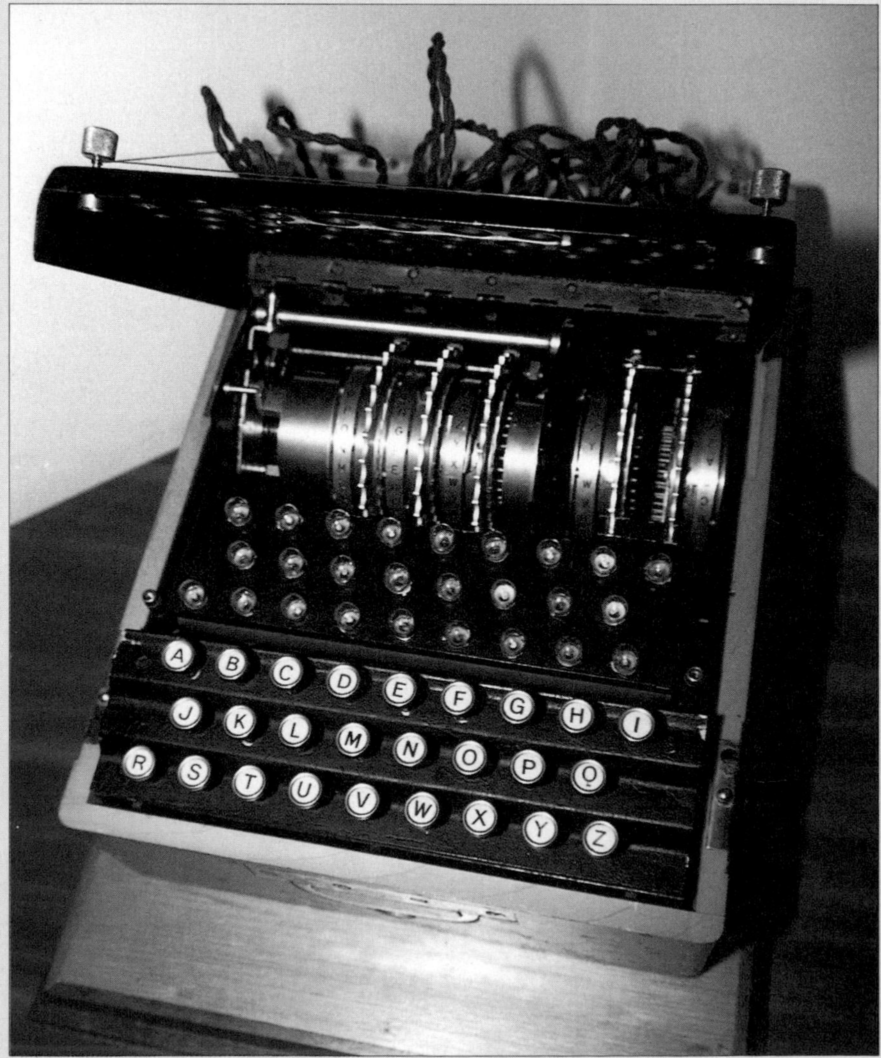

domination, or Allied officers dropped in by parachute or quietly landed by boat. In the Balkans, by contrast, intelligence officers who were attached to local resistance groups frequently wore uniform and worked quite openly with the partisans.

German humint also had a mixed war. Attempts to penetrate the UK and the USA were notably unsuccessful. Espionage in North America, based mostly within the German-American community, was effectively neutralized within a year of Germany declaring war on the United States in December 1941. In Britain the domestic security service, MI5, managed to "turn" most of the German agents sent during the war and used them to feed back misleading information. An important early success was with Wulf Schmidt, a 26-year-old Dane, who was dropped into East Anglia in September 1940. He was quickly arrested by the Home Guard, broke down under interrogation and worked as a double agent for the rest of the war under the codename "Tate". Since his German masters in the Abwehr (military intelligence) continued to trust him, Tate was able to make arrangements for subsequent agents, who were picked up on arrival in the UK and usually also persuaded to work for the British.

Another key double agent was Dusko Popov ("Tricycle"), a Yugoslav actually working for the Abwehr who offered himself to the MI6 station in Belgrade. At the end of 1940 he traveled on Abwehr instructions to London where he became the center of a significant network of double agents. In the summer of 1941 the Germans ordered Tricycle to travel to the United States and set up an espionage network there. They also gave him a detailed questionnaire, about a third of which dealt with Hawaii and the American naval base at Pearl Harbor. The American authorities were kept informed about Tricycle's mission, but the head of the Federal Bureau of Investigation, J. Edgar Hoover, refused to cooperate with any arrangement run by a self-confessed German spy, and Tricycle returned to Britain without having recruited any more double agents. The significance of his instructions to investigate Pearl Harbor, moreover, was entirely missed by the Americans.

The contribution of codebreaking

The importance of signals intelligence during the war can scarcely be overestimated. It played a crucial role in the North Atlantic. Until 1943 the German Navy's interception and codebreaking service, the *B-Dienst*, was able to break enough British codes to provide vital information about convoy routes. In consequence U-boats sank large amounts of British shipping. For a time it looked as though they might bring the UK to its knees. But the British were working hard on German naval codes and began to read their signals from May 1941 until February 1942 when the Germans adopted a more secure U-boat cipher. This defeated the Allied codebreakers until mid-December 1942, during which time shipping losses mounted catastrophically. But the decisive advantage now lay with the Allies. A combination of the decoded U-boat signals, radio direction-finding, and improved weaponry, such as very-long-range escort planes, won the battle of the Atlantic in the spring of 1943.

The importance of "Ultra" (the British codename for these decrypted intercepts) extended to every theater of the war. The Axis supply line across the Mediterranean to North Africa was effectively shut off with the aid of intercepts. Ultra, too, helped to make the double-cross system of bogus enemy agents more credible, since the intercepts revealed precisely which bait the Germans were taking. The deception operations relating to the Allied invasions of Sicily in July 1943 or Normandy in June 1944 could not have worked as well as they did without the intelligence provided by Ultra. The Germans also profited from sigint. Between late 1941 and early 1944 a team from the German Post Office Research Bureau intermittently intercepted and unscrambled one of the radiotelephone links between the UK and the USA. On two occasions they actually listened in to conversations between Churchill and Roosevelt. At a tactical level the intercept service of the Afrika Korps provided Rommel with very valuable information until the summer of 1942 when the unit was eliminated. On the Eastern Front communications intelligence – direction-finding and code-breaking – provided the Germans with a great deal of information about Soviet forces.

The Soviets have revealed very little information about their intelligence during the war. Claims that spy networks within Germany and occupied Europe – the "Lucy" ring and the "Red Orchestra" – provided high-grade material seem greatly exaggerated, although the spy Richard Sorge in Tokyo did provide vital information about Japan. Like the other combatants, the Soviets' main source of good intelligence was probably sigint. They had expertise in the area and the evidence suggests that they were able to break at least some of the German air force and army Enigma keys by the spring of 1943.

▲ Bicycles dropped by SOE (Special Operations Executive) into Denmark as part of supplies for resistance groups. During the war SOE sent in some 50 agents who were able to stiffen local resistance and provide London with intelligence about the enemy. This pattern was repeated in other places, especially France, the Low Countries, Yugoslavia and Greece. The organization was set up in 1940 to run sabotage and subversion teams throughout occupied Europe. Churchill recognized that "covert action" was in 1940 one of the few ways that the British could hit back at the Axis. By 1945, with an estimated strength of 10,000 men and 3,000 women, SOE had extended operations to Burma and Malaya.

▼ Both Allied and Axis forces depended on enemy civilians for information. Here an alleged fifth columnist is held by French marines at Dunkirk in June 1940.

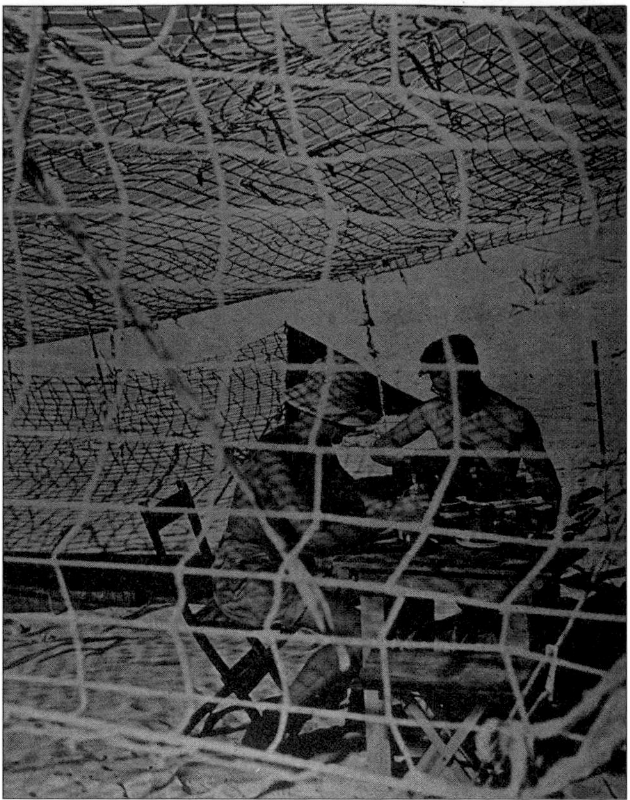

◀ A camouflaged military post in North Africa. Netting such as this could be used either to disguise installations or specifically to deceive enemy air reconnaissance. Before the Allied invasion of Sicily in July 1943 bogus army units were established in Egypt to try to convince the Germans that the main assault would be against Greece.

The Americans had considerable success in breaking Japanese codes. In summer 1940 US army and navy cryptanalysts began to read a diplomatic cipher which they called "Purple" and which enabled them to penetrate the communications of Japanese diplomats in Berlin, who reported on all aspects of the German war. Late in 1943, for example, the Japanese ambassador and the military attaché toured the German fortifications of the Atlantic Wall in France. The information they sent back to Tokyo was intercepted and provided valuable intelligence for the invasion of Northwest Europe in 1944. In the Pacific, US navy codebreakers broke the main Japanese fleet cipher – code-named "JN25b" – by early 1942 and were able to follow the Japanese Admiral Yamamoto's plan for a decisive battle to capture the key island of Midway in June 1942. The subsequent US victory was the turning-point of the Pacific war, and in the US Admiral Nimitz's own words "essentially a victory of intelligence."

The American signals intelligence capability, however, had not been able to prevent the surprise Japanese attack on Pearl Harbor in December 1941. Decrypts from the "Purple" cipher did give Washington a few hours' warning that something was planned which would cause a breach between Japan and the USA. At the time, too, other pieces of intelligence were available which could have pointed toward the possibility of an attack on Pearl Harbor, but the main problem in this case was appreciating that the Japanese would even contemplate such an strategically illogical operation. Few people believed that the Japanese would be so foolish, and the best intelligence in the world might not have been enough to persuade them otherwise.

Japanese intelligence work

Like the other major combatants, the Japanese also expanded their sigint effort during the war. They seem to have had the most success with codebreaking in the field. British army and USAAF signals were decrypted in Burma and China, and claims have been made that a very high proportion of all Chinese codes were broken. Luck played a part in this. In early 1944, due to a navigational error, a Chinese plane landed at a Japanese airfield in southwest China. On it was an officer carrying details of the complete code for a Chinese army preparing a major offensive on the Yunnan Front. Thus, the Japanese 56th Division had ample advance warning of Chinese plans. Indeed, the divisional intelligence chief reported that some 80 percent of his information came from wireless intercepts.

Another stroke of luck for the Japanese occurred when the German raider *Atlantis* captured a British liner, the *Automedon*, off Singapore in November 1940. Among the top-secret mail for the British commander in chief, Far East, being carried by the liner was a British chiefs of staff paper on the situation in the Far East if Japan were to attack British imperial territories. The memorandum revealed not only how stretched British resources were generally but also that the British would not necessarily go to war if the Japanese invaded French Indochina. It was passed on to Tokyo, certainly influencing the decision to occupy southern Indochina in July 1941 and making it much more likely than hitherto that the Japanese would eventually attack British possessions in Asia.

Japanese intelligence also used human sources. One notable Japanese spy was a navy sublieutenant, Takeo Yoshikawa, who in March 1941 was posted under the name "Tadasi Morimura" to be a secretary in the Japanese consulate in Honolulu. Yoshikawa was able himself to gather much of the intelligence used in the attack on

▼ Transmitting to the enemy? Much Chinese battlefield radio communication was compromised. This operator in northern Burma may unintentionally have been providing intelligence for the Japanese.

◄▼ **Interpreting "photint" (photographic intelligence) was not easy, as the photograph (left) shows. It was taken in June 1944 of the V2 experimental area at Peenemünde on the Baltic coast. A was interpreted as light flak positions; B as transport cradles; C as two rockets. Such information was supplemented by information from Polish agents on the ground (one of whom even got out to London with fragments from a rocket test-firing, as below) and sigint (codenamed "Corncrake") from the Enigma traffic between Peenemünde and Blizna (the test site in Poland).**

Pearl Harbor. This was partly due to Yoshikawa's ability to gather intelligence from open and freely available sources. In the cockpit of each of the attacking Japanese planes on 7 December 1941, for example, was a panoramic view of Pearl Harbor provided by Yoshikawa. It was a postcard view of the Harbor which he had bought in a Honolulu souvenir shop.

Intelligence and technical advances

An important side of intelligence in World War II involved keeping tabs on the enemy's technical advances. Here sigint was not always the most important source. The first warning that Germany was developing the V-weapons (the V1 flying bomb and the V2 rocket) came from a Danish engineer who told MI6 in Stockholm. Subsequent aerial reconnaissance provided more information. The German scientific research station at Peenemünde in the Baltic was repeatedly photographed and rocket launch sites were identified. When photographic intelligence revealed similar installations in France and the Low Countries, it was clear that V-weapons were becoming operational. Within occupied Europe local volunteers provided additional intelligence. In northern France, agent "Amniarix" – Jeanne Rousseau – used her position as an interpreter liaising between the Germans and French industrialists to gather information about the construction of V1 launch sites.

Another concern was the possibility of the Axis developing an atomic bomb. In 1941 British intelligence learned that the Germans had stepped up production of heavy water – a clear indication that they were interested in atomic research. Athough the Germans' heavy-water plant was destroyed by an SOE team in February 1943, the knowledge that the Germans were interested in the subject was a factor in the Allied decision to work on the bomb. In January 1945 an intercepted Japanese cable from Hanoi to Tokyo revealed the discovery of rich uranium ore deposits in Indochina. There was also evidence in the spring of 1945 that atomic-related materials were being delivered to Japan by U-boat. Although the Japanese were nowhere near developing atomic power, this intelligence helped justify the American decision to continue development of the bomb, and, in the end, to use it.

Some of us were told to collect pedestrian facts; others were given specific instructions; others still were asked to keep ears and eyes open for the unusual, the improbable; all were working in small compartmented fields in almost complete isolation – the price to pay for lessening danger – and yet saw friend after friend, comrade after comrade fall into the ever gaping trap.

JEANNE ROUSSEAU
"AMNIARIX"

PART 3

THE MOBILIZATION OF PEOPLES

Datafile

The impact of the war varied from belligerent to belligerent, but few countries, including neutrals, remained untouched. Disruption occurred at all levels. International political and economic relations were among the first nonhuman casualties of the war as diplomatic relations were broken, currencies were controlled and trading relationships were severed. Within countries, the powers of governments over the citizen were strengthened, while in some cases internal government collapsed for a time. Personal expectations and relations were also transformed, as families were broken up, populations migrated and women assumed a public role.

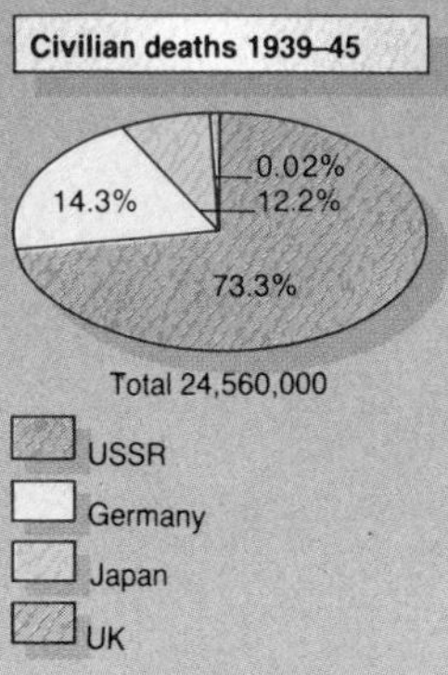

▲ In many of the belligerents, civillians were very much in the front line, with the US being a notable exception (except for Hawaii). The figure for USSR deaths is an educated guess, as the exact figure is not known.

◀ The divorce rates for the USA and UK are interesting for both their differences and their similarites. They increased more rapidly in the UK during the war, possibly because of visiting US servicemen, but both shot up in the postwar period.

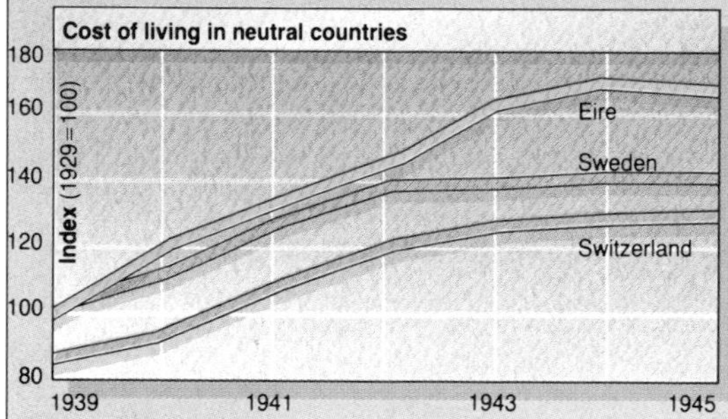

◀ Neutrals as well as belligerents suffered from sharp increases in the cost of living, as shortages pushed up prices. Wages, too, went up, as industries competed for labor to produce for the lucrative export markets. The poorest of the three neutrals in the chart, Eire, suffered the largest increase in living costs.

▼▶ Grain production shot up in the neutral countries, since there was a ready market in the belligerent countries. In 1943–44 the UK, for example, had to feed not only its own citizens, but the American troops awaiting D Day. The production of barley and oats showed very great increases in Switzerland.

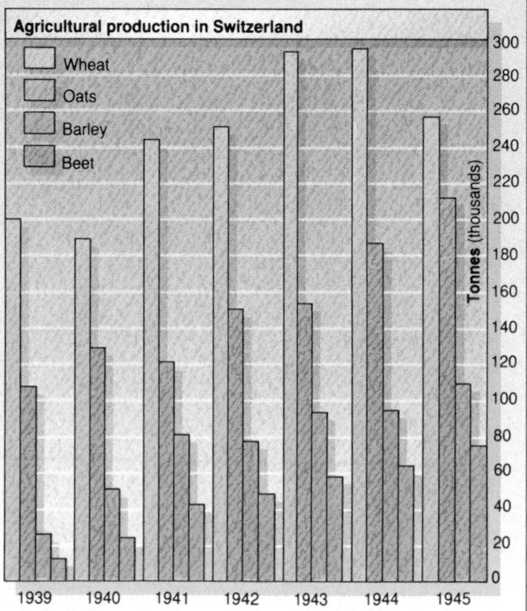

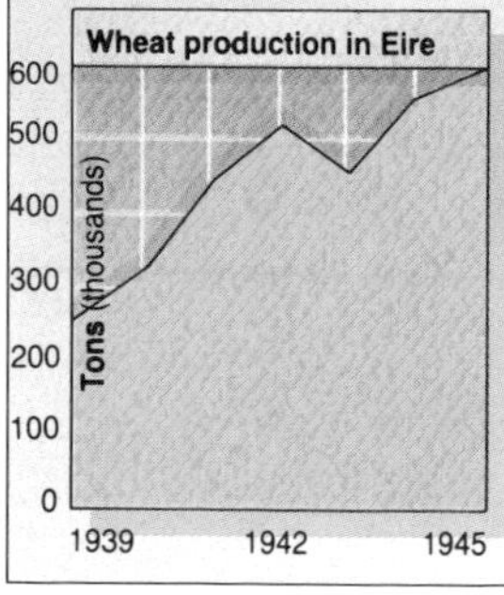

Both World Wars have popularly been termed "total wars", the implication being that not only the fighting services but the whole of society in the affected countries was directly involved. In this sense World War II was even more "total" than World War I: economic mobilization went much further in many countries, as did mobilization of the female section of the population. The number of civilians killed was also much higher, 17–18 million out of a total toll of 40 million. The strategic bombing of civilian populations was a notorious feature of the war, accounting for 500,000 Japanese, 300,000 German, and 60,500 British deaths. But far more significant was the widespread policy of intimidation and reprisal practiced upon civilians by invading and occupying armies, which resulted in, for example, the deaths of 7 million Russians and 6 million Poles. One statistic has an especial horror: of the 8.5 million Jews living in occupied Europe in 1941, only 3 million survived the German policy of racial extermination.

Beyond the numbers killed, World War II was more "total" than World War I because of its wider geographical range, both in Europe and in the world beyond. World War I had involved six main European powers (Germany, Austria-Hungary, Russia, France, the UK and Italy) plus their associated possessions, four minor European powers, and the USA. In World War II, the *only* European states not caught up in the fighting, either as invader or invaded, were Sweden, Switzerland, Spain, Portugal and Eire. There were in addition the USA and Japan, whose activities emphasized the global nature of the conflict. During 1914–18 fighting had, for the most part, been confined to Europe, the Near East and Africa; naval battles were largely a matter of submarine attacks and attempts to sink them, with the Battles of the Falklands Islands and of Jutland notable as the exceptions. During 1937–45, by contrast, fighting spread over most of the world, on sea as well as on land: only inhabitants of the Americas remained safe – as long as they stayed on dry land. The one exception here were the citizens of Brazil: although fundamentally a fascist regime itself, Brazil declared war on the Axis powers in 1942, and in 1944 sent an expeditionary force to Italy, becoming the first South American state to send fighting troops to Europe.

The end of international relations

The first implication of the involvement of most of the world in total war was the breakdown in international relationships, both political and economic. One casualty was the League of Nations, or what remained of it (Brazil, Turkey, Japan, Germany, Italy and the USSR had all withdrawn between 1926 and 1939, while the USA

THE IMPACT OF TOTAL WAR

The cost of World War II

The breakdown of the international economy

The impact of war on ordinary people

Neutrality: four case studies

Argentina, Switzerland, Eire and Sweden

had never been a member). Nevertheless, it was the world's only international political organization, and it had been instrumental in helping to solve certain problems relating to the smaller European states. However, during the months of crisis in 1938–39 the great powers had virtually ignored the League, and with the outbreak of war, its political usefulness effectively ceased. The League's permanent organization remained in being in Geneva, and it attempted to continue its nonpolitical activities as far as possible during the war until it was formally dissolved in April 1946. Its remaining responsibilities were then assumed by the new United Nations organization.

Bilateral political relationships, of course, broke down as well. Sometimes there were formal breaks of diplomatic relations, but sometimes communications simply collapsed when a country was invaded. Belligerents tried to retain links with neutral powers, in particular with the USA before it entered the war, but this frequently depended upon communications links remaining open and usable. A recurring pattern was the invasion of a country, followed by a cloud of confusion settling over that country, as links with the outside world were broken and patched together only with difficulty.

International economic relationships were equally the victims of total war. Most currencies instantly became inconvertible, or convertible only under the most stringent controls. In the UK, for example, private citizens found it very difficult indeed to convince the treasury to allow the purchase of dollars. But at least the pound sterling largely retained its value: other currencies collapsed. This fed into the wider problem of the rending of the web of international trading relationships: if the currency of country A cannot be converted into the currency of country B, then it becomes more difficult for the citizens of country A to purchase the goods produced by country B.

▶ Individual banks could perhaps protect themselves, as in Shanghai, China, in September 1937, but international economic relationships were one of the first casualties of the outbreak of war. At this point, Japanese troops had already invaded Chinese soil, and intense fighting was continuing.

▼ German snipers scatter women and children in Paris, 1944, underlining the fact that civilians were very much in the front line. Horror and fear are overwhelming.

At this point it is worth noting that for a decade before the outbreak of war, more and more countries had begun to practice autarky, or economic self-sufficiency: the prime examples were the USSR, out of necessity, and Germany, out of choice. Also an increasing number of countries (such as the USA) had retreated behind high tariff barriers, or had developed preferential trading links, such as the members of the British Empire and Commonwealth. Therefore, the outbreak of war hardly destroyed a system of international free trade. What it did do was to wipe out a large number of private trading relationships, as countries concentrated on war-related exports and imports. Even those neutrals who retained trading links with both sides found that the exigencies of war and geography tended to direct their trade predominantly into one channel or the other.

If the first implication of the involvement of most countries in total war was the breakdown of international relationships, the second was the destruction or diversion of the executive power and the relegation of the legislative power for the duration. Elections were a victim in most countries other than the USA, which was constitutionally required to hold them: the UK held no general election between November 1935 and July 1945 (after the defeat of Germany), although by-elections were held to fill vacant seats. The dictatorships were hardly likely to bother with such fripperies, but France, which had been used to them, was occupied for most of the war.

That, indeed, was the most obvious cause of breakdown in national patterns, and with the exception of the five countries already noted, and the UK, every European country was invaded and occupied. Even those which were not invaded suffered the very tight controls of centrally directed war economies, with direction of investment, production and labor. All, of course, experienced great internal shifts in population, as millions left home to join the forces or to work in war-related industries.

The impact of war on human relations

This upheaval of populations was a major factor in the third implication of total war: the breakdown in every country of family and social relationships, and of social mores. Men went off to war and yearned for the women left behind while chasing those who were immediately available. But this time, women went off to war as well. The USSR was an exception in having women who were official combatants: they included snipers, and the "Night Witches", the female pilots who were instrumental in helping to defeat the Germans around Stalingrad. Women were also prominent in the resistance forces in the various occupied countries. But in most others women in the armed forces filled administrative, training and other support roles. Germany was notable in not having women even in these noncombatant positions until the very end of the war, because Hitler believed that their place was in the home, rearing children.

Many women, however, were unable because of family commitments, or unwilling, to join the forces: they made their contribution by working in munitions factories or on the land instead. Others stepped into the breach in civilian occupations left by those who had departed. Many of these women had paid employment – or decently paid employment – for the first time, and very much enjoyed the economic advantages as well as the new experiences and new friends provided by such opportunities. Their husbands, fathers and boyfriends were frequently less sure about the advantages, but there was considerably less grumbling in the UK, which was in the front line, than in the USA, where either inclination or external pressure kept 71 percent of working-age women at home.

The bare fact that women stayed at home did not, however, render them immune to the new experiences available. The stress of war heightens the flow of adrenalin: women and men in all belligerent countries shared the dominant mood of "eat, drink and be merry, for tomorrow we die". In both the USA and the UK – countries at war but not invaded – the rates for births, both legitimate and illegitimate, and for divorces all increased, while in the USA the venereal disease rate shot up strikingly. Marriage rates also increased in both countries; many of these marriages were hastily contracted and therefore fragile. Premarital and extramarital affairs became much more common, as the threat of death loosened conventional restraints, and the presence in many cases of foreign troops provided the allure of the unfamiliar. The well-paid American troops stationed in Britain before D Day provided the prime example of this syndrome: at the end of the war, 60,000 British women followed their American husbands to the USA as war brides.

▲ Members of the Chinese youth comfort corps about to set out on a clearing-up operation in 1937, when their country was at war with Japan. In all combatant countries children were caught up in the work of the war. This might take the form of replacing female work in the home, working in the fields, or, in some cases, taking up a gun.

◀▶ These Wrens (members of the British Women's Royal Naval Service) fitting smoke floats to a trainer aircraft exemplify the importance of women in the forces in two of the major front-line countries, the UK and the USSR. In the UK, however, their roles were noncombatant, whereas in the USSR they saw front-line service and shone particularly as snipers and pilots. But the general view of women was ambivalent, as the poster (right) warning the soldier against ladies of easy virtue, demonstrates. In World War II the incidence of venereal disease rocketed.

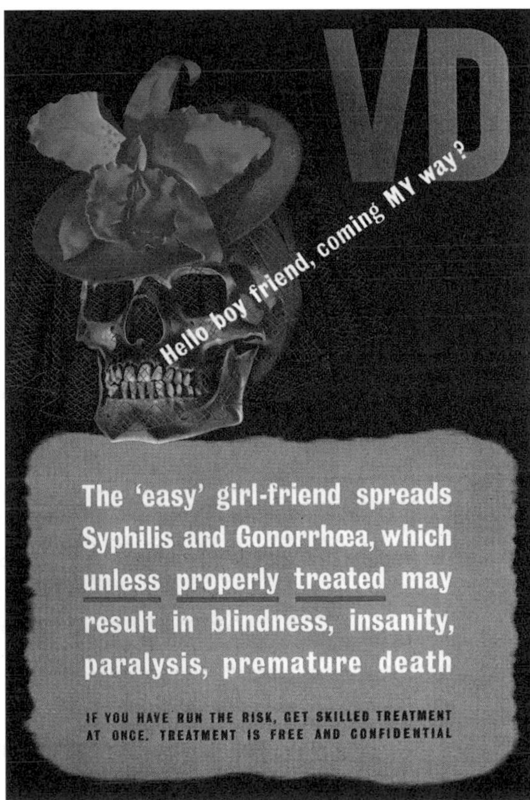

VD
Hello boy friend, coming MY way?
The 'easy' girl-friend spreads Syphilis and Gonorrhœa, which unless properly treated may result in blindness, insanity, paralysis, premature death
IF YOU HAVE RUN THE RISK, GET SKILLED TREATMENT AT ONCE. TREATMENT IS FREE AND CONFIDENTIAL

Neutrality: four case studies

The impact of total war on the belligerent nations, then, was thorough upheaval – of international, national and personal relations alike. But not all nations were at war. Many countries declared themselves neutral, but even those situated far from the fighting could not escape effects of the war. Argentina's declaration of neutrality in 1939 and its refusal to join the US-led Pan-American defense alliance meant that in 1941 the USA viewed it as virtually fascist. The USA supplied arms to other Latin American countries, especially to Brazil, and put economic pressure on Argentina to break relations with Germany. Argentina's response was to become increasingly stubborn and, by 1943, fearing an invasion by Brazil, to look to Germany for arms. This, however, cost it the protection of the UK, with which it had long had strong economic ties. The two allies forced Argentina, under threat of a complete American trade blockade, to break diplomatic relations with Germany in January 1944. In March 1945 it became a token belligerent, although no Argentinian troops were ever sent to the front. Neutral nations could not remain immune from the general disorientation and dislocation caused by war, although they reacted to the war in different ways.

Switzerland is the quintessential neutral state, whose neutrality, indeed, has not been violated since 1815. Germany actually contemplated invading Switzerland at several points during the war, but the country's mountainous geography and the armed readiness of the Swiss (who were prepared to fight if necessary) would have made such an invasion very costly. The Swiss had foreseen the outbreak of war and were prepared: in 1939 the borders were guarded and the army mobilized. The Swiss army was composed of a trained civilian militia: members kept their rifles in their own homes, and the whole army could be mobilized within 48 hours. In 1939 about half a million were so mobilized, but after the fall of France the numbers were reduced to a quarter million.

Thereafter the Swiss concentrated on watchful neutrality. War materials were supplied to both sides, the press was censored to prevent incitement against any of the main belligerents, and communications were maintained with all. The Swiss, in short, walked a tightrope. But because of their neutrality, their central position and their open communications, the Swiss found that their country played a special role: that of the espionage center of Europe. All the main belligerents had embassies there, and foreigners were, for the most part, allowed freely to walk the streets. The result was a riot of spies, as all the belligerents maintained espionage networks. Some of the resistance organizations did the same: the anti-Nazi German "Black Orchestra" was just one example. Whether the average Swiss man or woman was aware of all this is open to question: for most, the experience was probably one of slightly nerve-wracking, but lucrative, neutrality.

Eire's position during the war was inextricably tied up with its need to distance itself from the UK,

We lived in a world of uncertainty, wondering if we were going to survive from day to day. My husband was away in the RAF as an airgunner, and I'd conditioned myself to the fact that his lifespan was also limited and that our short, happy married life together was over. I lived in a vacuum of loneliness and fright as service in the army, navy, and air force claimed five of our personal friends ... When 1942 came in with the hit-and-run air raids, I began to despair that the war was ever going to end.

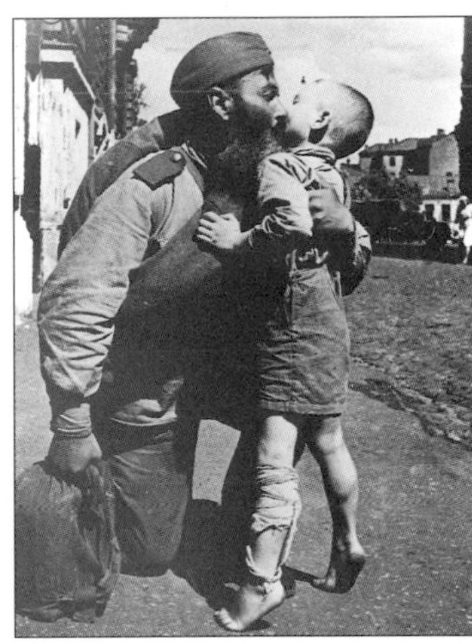

▲ **This Soviet soldier, returning home at the end of the war, may never have seen his son before – or if he had, he may not have expected to see him again, since at least 7 million Soviet citizens died at the hands of the German army. Having defended his country, he would now have to help rebuild it.**

from which it had gained its independence only two decades before: as one historian has noted, Eire "established sovereignty by maintaining neutrality". The strict diplomatic forms of neutrality were always publicly observed, but once the USA joined the war and Eire became less strategically important, its neutrality took on a more pro-British tinge.

From 1942 stranded Allied air crews were sent directly to Northern Ireland, while German air crews were interned. Further, Ireland maintained secret intelligence and strategic liaisons with the USA and UK. Nevertheless, it was clear that an independent foreign policy – independent, that is, from that of the UK – was seen as an absolutely vital factor in buttressing Eire's position as an independent state.

During the war Germany maintained a strong diplomatic and intelligence presence in Ireland, as it did in all of the European neutrals. In this case, however, its intelligence links were pretty fruitless, being confined largely to "inept liaison" with the Irish Republican Army, an organization described in the 1939 Treason Bill as a traitor to the Irish state.

Domestically, the Irish government's response to the war has been characterized as "cautious authoritarianism". It had to cope with internal transport disruption and fuel shortages, and imposed some fuel and energy rationing. It also

Change in a Neutral Country

Argentina is an interesting example of how the war caused internal economic and political change. The closure of its overseas grain markets led to widespread redundancy among rural laborers. Substantial rural–urban migration ensued, providing labor for Argentina's drive to industrialize.

The pressures helped foster a political transformation as well. Argentina was no stranger to strong military influence in politics, but in June 1943 a coup led by the Grupos de Officiales Unidos, high-ranking officers of the German-trained army, overthrew the conservative-led civilian government. Gradually emerging from the shadows was Colonel Juan Perón.

Concerned by the military government's lack of support, he began to cultivate the workers and particularly the trade unions, aided by his mistress (and later wife) Eva. His goal was partly to control the workers by conciliation rather than coercion, partly to divide the opposition to army rule by isolating the middle classes and landowners. Perón's elevation to supreme power in the crisis of 1945–46 was very much facilitated by his mass following of the *descamisados* or "shirtless ones", who responded to his demagogic appeals and wage increases. He remained in power until 1955.

▼ Passersby watch urban farmers harvesting wheat in a square in Zurich, Switzerland. To increase domestic food production, the Swiss (with their legendary efficiency) also cultivated schoolyards, athletic fields and amusement parks. This was not intended to feed a suddenly burgeoning Swiss population: rather, the Swiss fulfilled one of the accepted duties – or privileges – of neutrals, which was to supply food and other materials to both sides in the war.

clamped controls on labor: an order in 1941 removed legal protection for strikers, while the 1941 Trade Union Act restricted rights of collective bargaining. The power and influence of the Catholic Church also ensured that the breakdown in sexual mores was not quite so open as it was in other frontline states.

Sweden's position was even more complicated than that of Eire, particularly after the German invasion of Norway and Denmark had isolated it from the West. It had three discernible elements: fear of Germany, the desire to maintain trading links with the USA and UK, and a strongly protective pro-Nordic bias.

Sweden had refused Germany's prewar offer of a nonaggression pact, and at the outbreak of war reiterated its neutrality while strengthening its defenses. The Germans put heavy pressure on Sweden to allow transit of their troops and armaments, but Sweden held out until the fall of Norway in 1940, only then agreeing to allow German soldiers on leave and certain goods to pass over specified Swedish railways. In June 1941, however, after the outbreak of the second Russo-Finnish war, Sweden was forced to allow the transit of a German armored division from Norway to Finland. Sweden was also forced to supply Germany with iron ore and other materials as the price of being able to import essential goods. By the late summer of 1943, however, the pendulum had swung back, with the increase in Swedish military strength and the deterioration of Germany's position. Sweden then stopped the German traffic on the railways, as well as reducing its exports to Germany.

The decline in German strength also enabled Sweden to renew its trading links with the West. It had negotiated a wide-ranging trade agreement with the UK in the autumn of 1939, which was signed that December, but German strength had made its execution difficult. By late 1943, however, the situation had changed, and Sweden, the USA and the UK entered into a comprehensive war trade agreement.

The third element in Swedish neutrality was its strongly pro-Nordic bias. It gave generous, indeed unneutral, material aid to Finland during its war with the USSR, and the government encouraged Swedish volunteers to join the fight. Sweden felt constrained to refuse transit to Finland for French and British troops, but it also refused transit to German troops to Norway while Norwegian resistance continued. During the war, Sweden was of course a shelter for refugees from the other Scandinavian countries, but it went further: refusing to agree to Germany's attempts to impose its own "new order" on the Swedish government, it also provided shelter for Balts and Jews.

Sweden, then, was probably in the most precarious position of these three European neutral countries. Switzerland sat in the midst of the war, yet its geography protected it. Eire was never in any real danger (although Northern Ireland, as part of the UK, was not neutral, and Belfast was heavily bombed in 1941). Indeed, Eire suffered less from pressure from the UK, the old enemy, than it did from the USA, the protector, since the USA instinctively disliked neutrals. Sweden, however, might well have suffered invasion if the Germans had not found it most useful as a neutral in the early years. After 1943, of course, Sweden too was safe.

Total war, then, affected belligerents and neutrals alike, although the impact varied in form and pressure from country to country. Inevitably, however, those states which were invaded and occupied suffered most. Among the most urgent tasks of the postwar world would be to resurrect an international currency and trading system, to reconstruct a dozen national political systems and to rebuild millions of lives.

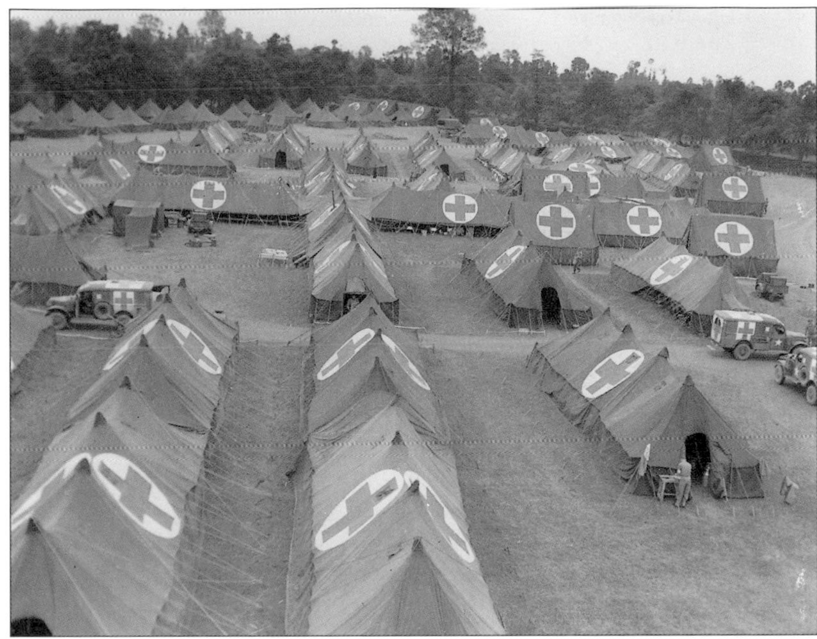

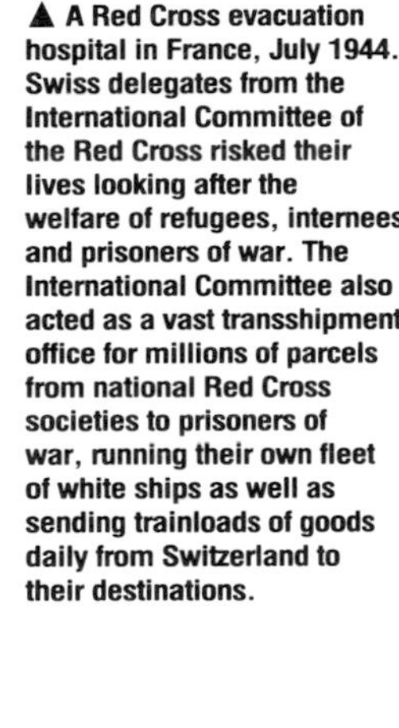

▲ **A Red Cross evacuation hospital in France, July 1944. Swiss delegates from the International Committee of the Red Cross risked their lives looking after the welfare of refugees, internees and prisoners of war. The International Committee also acted as a vast transshipment office for millions of parcels from national Red Cross societies to prisoners of war, running their own fleet of white ships as well as sending trainloads of goods daily from Switzerland to their destinations.**

◀ **General Francisco Franco, the dictator of Spain, led his country from neutrality to pro-German "non-belligerency" and back to neutrality following the vagaries of the war. His main concern was the survival of his regime. Spain greatly profited by auctioning its supplies of tungsten and pyrites, vital for producing munitions, to the highest bidder.**

Datafile

The Japanese economy showed remarkable resilience between 1937 and 1945 but the strains of the enormous war effort in China, the Pacific and Southeast Asia took their toll. The vulnerability of an island economy, heavily dependent upon imported energy and raw materials, was demonstrated starkly during the American onslaught between 1943 and 1945. By the summer of 1945 Japan was crippled.

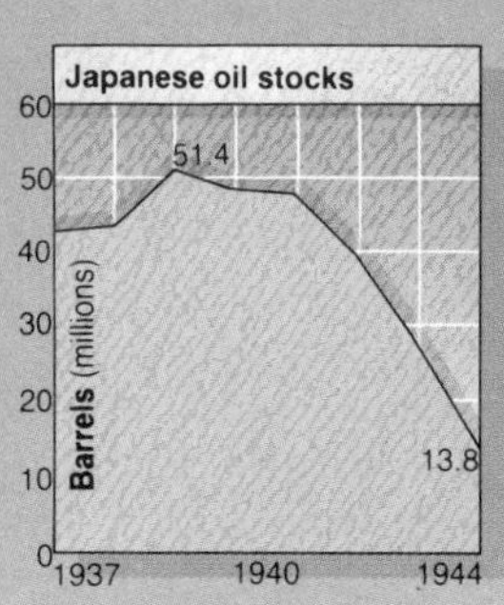

▲ Oil reserves were built up from the mid-1930s so as to sustain the war effort for a limited period. Because Japan was so dependent on imorted oil, the Us oil embargo of July 1941 was a heavy blow. Stocks were rapidly depleted in the middle and latter stages of the war as Japan lost command of the sea and air.

▲ The principal or home islands of Japan. The central island of Honshu has been most important in industrial development, particularly the area extending from Tokyo to Osaka. Hokkaido (right) contains Japan's main coal fields. Much of Japan is so mountainous as to be useless for productive purposes.

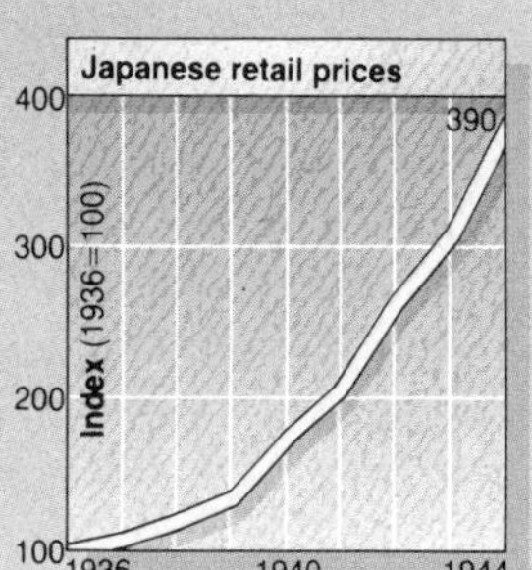

◀ The drive to achieve maximum production stimulated serious inflation. The war in China led to a doubling of retail prices between 1937 and 1941, before the start of war against the USA. Thereafter prices doubled again by 1944, a far worse record than the wartime inflation in the USA, UK and Germany.

► The American bombing campaign and large-scale evacuation temporarily reversed the long-term trend toward urbanization. The capital, Tokyo, suffered the greatest reduction in population, the former imperial capital, Kyoto, the least. Situated on or near the coast, Japan's major cities were easy to reach for US bombers.

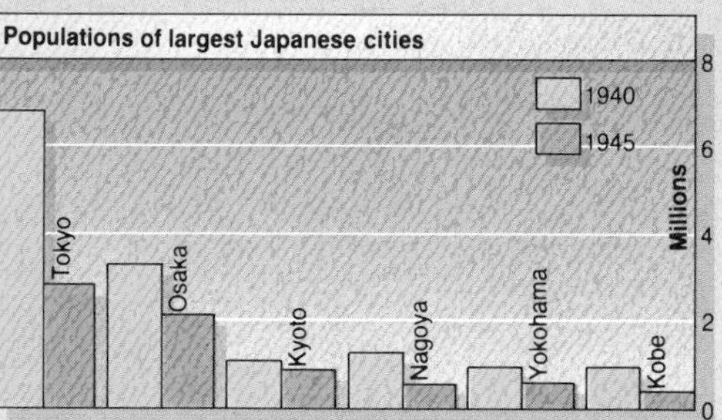

Chronology

1937

July
Japanese invasion of China

1938

March
All employment brought under government control

June
Formation of Patriotic Industry Association

1939

July
National Conscription Ordinance to deploy workers

September
Start of war in Europe

1940

July
Voluntary dissolution of political parties; formation of Imperial Rule Association

September
Neighborhood associations revived

1941

April
Rice-rationing begins

October
Hideki Tojo prime minister

December 7
Japanese attack on Pearl Harbor; USA enters war

December
Invasion of SE Asia begins

1942

April
US "Doolittle" bombing raid; losses of shipping begin

August
US success at Guadalcanal

October
Neighborhood associations take over rationing

1943

November
Munitions ministry created

1944

July
Resignation of Tojo

1945

March
USB-29s begin incendiary campaign against cities

May 8
VE Day; war in Europe ends

July
10 pecent cut in all staple rations imposed

August
Atomic bombs dropped on Hiroshima and Nagasaki; USSR invades Manchuria

August 15
Emperor Hirohito orders surrender: end of war

After the outbreak of the Sino-Japanese war in 1937, the Japanese government became more authoritarian. But, unlike Germany and Italy, the constitution continued to function and elections took place. Governments comprised a mixture of conservative bureaucrats, and active and retired generals and admirals. Their priority was to consolidate foreign and defense policies. Their aims inside Japan were to stimulate morale and unity, through loyalty to the emperor and commitment to the war effort, and by avoiding reliance upon non-Japanese ideologies and cultures. In theory the emperor's power had been restored in 1868 but in reality the government acted in the emperor's name. During the late 19th century State Shintoism was fostered officially and inculcated through the armed forces and the educational system. The emperor's divinity was given strong emphasis. He personified the essence of the nation and had to be obeyed. Failure to do so could lead to disgrace.

Wartime Japanese leaders were consistent in their aims, seeking above all Japanese leadership in Asia, which was expressed in such concepts as the "New Order in East Asia" and the "Greater East Asia Coprosperity Sphere". Prince Konoye, prime minister from 1937 to 1939 and again in 1940–41, advocated the end of party political struggles and the creation of a new spirit of national cohesion. In 1940 political parties formally dissolved themselves and were replaced by the Imperial Rule Assistance Association formed by Konoye. Superficially this resembled European fascist bodies but in reality it proved hollow and accomplished little. Dissenting party politicians paid lip service to the new trends and bided their time until a more sympathetic climate emerged. General Hideki Tojo, prime minister from 1941 to 1944, was the toughest leader of the wartime era and accumulated greater power through holding several different offices simultaneously. But Tojo had no thought of becoming a dictator and regarded himself as the servant, rather than the master, of the system. Limited criticism of the government could be expressed and this became more overt later in the war. The growing setbacks suffered by Japan from 1943 accentuated unease and led to Tojo's resignation in 1944. His successors realized that Japan could not hope to win but they could not consider unconditional surrender, which was demanded by the Allies.

The majority of Japanese gave loyal support to the war, but in the form of stoic endurance rather than fervent enthusiasm. The middle and later stages brought deprivation and hardship which led to the emergence of some discontent. Yet there was no challenge to the government. The trade unions had enjoyed little freedom before the war and were persecuted. Leftwing parties

JAPAN: CHALLENGE AND RESPONSE

The reorganization of politics in wartime Japan

Japan's war aims

The success of the war economy

Reluctance to involve women in war production

Local organizations used for sustaining morale

Repercussions of war for children

had always been weak and were also persecuted. The Japanese police were zealous and brutal in investigating alleged subversion and it was prudent to evade their attention. A watch was kept on rightwing extremists but far more interest was shown in those suspected of socialist inclinations. Ironically the only revolt against the government came from a section of the army which mutinied against the emperor's decision to accept defeat in August 1945.

Expansion and decline in the war economy

From the late 19th century onward Japan's economic development had been remarkable. Industry had developed rapidly, to become predominant within the economy by the 1920s, though it was then hit hard by the depression of the 1930s. The agrarian sector remained large and was regarded as particularly important because rural Japan would preserve essential Japanese values, unlike the modernizing economy which had given rise to the hybrid character of the city. Japan's principal economic weakness was a shortage of vital raw materials.

Japan's invasion of Southeast Asia in 1941–42 gave access to oil, rubber, tin, bauxite, teak, rice

▼▶ Japan prepares for war and moves more closely toward Germany. In a military review, banners waving, Japanese officers express their enthusiasm and dedication in fulfilling the nation's mission (right). This Nazi emblem outside a German club, restaurant and meeting place in central Tokyo illustrates Japan's identification with the Axis powers (below).

▲ Women using machinery in the drive to maximize war production. Pressures for transformation are underlined, as traditional Japan and the kind of roles fulfilled by women were overtaken in the march to Armageddon. Women responded patriotically and wished to contribute more, yet were held back by social conventions linked with male dominance.

and other commodities. Efficient utilization of these resources required integration of dependent territories into the domestic economy, as had occurred with the development of Japanese control in Korea, Manchuria and northern China. In itself this assumed Japan could control sea and air so as to facilitate transport and coordination. In this they failed. From 1943 vast quantities of Japanese merchant shipping were sunk by American submarines. This, plus massive conventional bombing in 1944–45, crippled the economy.

Japan had made a swift recovery from the world economic depression of 1930–34. Production of pig iron almost doubled between 1929 and 1936 while that of steel nearly tripled. Thereafter production was greatly expanded achieving a total of 6.3 million tonnes of pig iron in 1942, nearly 9 million tonnes of ingot steel in 1943, and 6.25 million tonnes of finished steel in 1943. The coal industry boosted production significantly, attaining a total, combined with net imports, of almost 66 million tonnes in 1940. Thereafter production declined to a combined total of nearly 52 million tonnes in 1944 and then to 32 million in 1945. Output in 1941 was one quarter of the UK's total and one ninth of the USA's. Coalfields were inconveniently situated in the northern island of Hokkaido and in the southern island of Kyushu. Production was stimulated in Taiwan (Formosa), Korea, Manchuria and north China–Inner Mongolia with creditable achievements. Overall, however, Japan could not compete adequately in heavy industry with the USA. Heroic efforts were made in shipbuilding but Japan experienced disturbing losses in merchant shipping from the summer of 1942. The percentage of vessels built and sunk during the war reached 42 in 1943 and 45 in 1944. In the course of the Pacific war a total of 38 percent of new shipping was sunk.

Japan's extraordinary attempt to secure full economic mobilization can be seen in statistics for the growth of aircraft, army ordnance, navy ordnance, naval ships, merchant ships and motor vehicles. The peak level was achieved in September 1944, but thereafter a catastrophic fall occurred in consequence of the air offensive, sea blockade and concomitant lack of resources. The price of increased production was a decline in employment in agriculture (from 48 percent in 1930 to 42 percent in 1940) and inflation which reached serious levels. The retail price index reached 175 in 1940 compared with 101 in the United States and 104 in Germany. The government tried to depress it with laws to curb excess profits, and price ceilings, but these were ineffective. Cumulative pressures produced by rising national expenditure and insufficient resources produced a black market, most noticeable in food supplies, especially vegetables, seafood and rice. In 1944 black-market prices were about ten times the official ones. With certain commodities the divergence could be much greater: by the end of 1944 rice had reached nearly 50 times its legal price level. Consumer goods fell rapidly as part of national production, from 40 percent in 1941 to 17 percent by March 1945. Most Japanese used the black market for some of their requirements.

Mobilization of the population developed from 1937, but the increase in civilian employment was small: between 1937 and 1942 it rose from 31 to 33 million. But there were notable changes in employment patterns. Women and girls replaced male farmers who had been called up for the army or other work. The total engaged in commerce fell from almost 5 million in 1937 to 2.5 million in 1944. In manufacturing an additional 2.3 million people were employed with nearly 300,000 more in mining. Men between 16 and 40 and women between 16 and 25 were placed on labor registers but only males were conscripted into the armed forces. About 1.5 million men were conscripted for factories. Later in the war the age of registration was lowered and those employed in essential industries were prevented from switching jobs.

Women in wartime Japan

Before 1945 Japanese women occupied a strictly subordinate role in politics. They did not have the vote and attempts to obtain enfranchisement had made little progress. Within industry women were traditionally employed in textile manufacturing, usually for a short term before marriage. During the war female employment expanded, but only within traditional areas. Many labored in fields and markets, in return for low wages.

Social pressures prevented the transfer of women to jobs with higher status and better pay. Nor would men tolerate the promotion of women to more senior positions. Fewer than 1 million women obtained jobs during the war. Shortages of labor were met by transferring men from less essential areas of the economy to important jobs and by importing Koreans and Chinese as laborers. The employment of students and old men was also encouraged.

General Tojo resembled Hitler in disliking an increase in female employment. He told the diet in October 1943 that Japan should not emulate the UK and USA by compelling women to enter into industry. He said, "the weakening of the family system would be the weakening of the nation ... we are able to do our duties here in the diet only because we have wives and mothers at home". His emphasis reflected the magnitude of the war struggle and pressures from the armed forces for population growth.

The birth rate had to be increased, if possible by as much as almost one half. Propaganda urged earlier marriages and government policy dictated a more generous approach by companies to young fathers. As in Mussolini's Italy and Stalin's USSR, prolific mothers were rewarded. However, the birth rate for 1943–44 showed no increase on the 1940–41 figure of about 2.25 million, and thereafter in fact declined significantly. The attempt by the state to achieve a major increase in the birth rate failed.

◀ The government's bid to enlist civilian support extended to children. Here, girls from Tokyo high schools present aircraft-sound detectors to the war ministry.

▼ Pressures on the economy led to intensified savings campaigns. These were important in stimulating savings and in developing the general theme of war mobilization. This poster shows the savings bond drive supported by the ministry of finance and banks.

Japanese Children and War

▲ Evacuation of children from major cities in Japan resembled evacuation in Europe, as bewildered or stoical children departed for temporary homes. The banner in the upper picture ("Evacuated children let's be good friends...") emphasizes the importance of friendship. The militaristic impact in education emerges in the lower picture where boys are undergoing training for military service.

Japanese children were taught to accept arduous work and to show deep respect for the emperor. Individual desires were subordinated to the requirements of the nation. Militarism was pervasive. All children at school in the 1930s grew up against a national background of conflict and exposure to expansionist rhetoric. Great emphasis was placed on the heroic achievements and self-sacrifice of the imperial forces and these had to be emulated. Boys were taught that there could be no higher calling than the military. The idealized figure of the samurai warrior was used to stimulate interest in war. Japan was opposed by avaricious foes and must defend itself. To accomplish this aim children must accept basic foods rather than relative luxuries and diminished choice in toys. Teachers stressed imperial values and their authority could be reinforced through harsh punishment. Children were influenced by the passion and splendor of militaristic demonstrations amidst the wars in China, the Pacific and Burma. The impact of war communicated itself most forcibly and poignantly with the absence of fathers, elder brothers and other relatives on war service. Many never returned and boys and girls were told that this was inevitable as part of the imperial mission.

While mothers usually bore the brunt of childrearing, circumstances accentuated maternal dominance. Exigencies of war entailed more use of child labor and registration of boys as young as 12 for possible employment. The direct horrors of war were conveyed most savagely through the intensive Allied bombing campaigns in 1944–45. Children were evacuated from large cities and sent to small towns or rural areas. The psychological repercussions of removal from the home environment were profound and caused much distress. Education continued in the cities amidst adverse conditions, sometimes with classes held in underground bomb shelters. Children brought up to believe in the unique progress of Japan leading to the culmination of the imperial mission experienced defeat in August 1945. It was even more bewildering to them than to their parents.

Morale at a local level

To control or influence local opinion, the government attempted to work through local bodies. In September 1940 an edict established community councils and neighborhood associations throughout Japan. They were instructed to achieve moral, social and economic cohesion so as to motivate people correctly. Neighborhood associations were smaller groups, consisting of 10–15 households. In July 1942 a total of 1,323,473 associations existed. In the six principal cities nearly 300,000 associations functioned in April 1942 with around 11 families per association. Associations were instructed to handle local defense, cooperate with councils, prevent crime, support savings movements, and pursue related issues. The most vital single step for councils and associations occurred in October 1942 when the government told them to organize food and clothing rations. Thus the daily lives of all families were affected.

In rural areas leadership came from local elites. In urban areas difficulty was found in securing suitable leaders. Women participated prominently in neighborhood associations and men dominated community councils. Ordinary people were more aware of the work of associations and thought more highly of them than the councils.

A large increase in social legislation also took place, stimulated by the necessity of safeguarding health standards in wartime. In 1938 a welfare ministry was created to supervise medical and social welfare. Measures were introduced to assist mothers and children, orphans and the elderly. Workmen's compensation was improved and hours of work for laborers were defined.

In July 1938 an important health insurance law was introduced. National health insurance associations were created to cater for communities. Families had the choice of joining and premiums were financed by the state and those insured, the latter depending on means. The numbers of insured grew swiftly so that by 1942 almost 16 million people were insured via nearly 4,500 associations. In 1945 these figures had risen to more than 41 million and over 10,000 respectively. Ordinary Japanese benefited appreciably. Public health centers were set up in 1937–38 with the aim of improving general health and sanitary awareness. The impact in rural areas was especially significant. Success was seen in controlling disease including tuberculosis. A health system of basic adequacy existed during the war but younger doctors were usually required for military service.

The positive features of Japan's domestic experience during the war included arduous self-sacrifice, dedication, and acceptance of national objectives. Social cohesion was promoted through enforced cooperation but workers suffered adversely, earning one third less in real terms in 1944 than in 1939 as a consequence of inflation. Absenteeism and poor-quality work in certain spheres showed dissatisfaction with the war. Absenteeism reached 10 percent in war plants in 1943 and was probably higher in 1944. This must be seen in perspective. In total the Japanese people revealed remarkable loyalty to the war-effort, equal to or excelling that seen in the UK or Germany.

▲ The grim reality of war: ambulances convey war injured to hospitals after being landed in ports. The procession of vehicles proceeds through a street market and the citizens betray no surprise at the sight.

▼ Neighborhood associations fostered communal activities and women played an increasingly influential role in them. Some of these children are still helpless infants.

Datafile

The German war machine produced extremely impressive results, supporting the armies in several theaters of war. Slave labor from occupied countries and Germans working long hours helped to compensate for the chronic manpower shortage. Munitions production grew markedly in spite of increasing Allied bombing, with some factories relocated to small towns or rural areas to try to evade the bomber.

Foreign workers

Millions

0 1 2 3 4 5 6

1939 1944

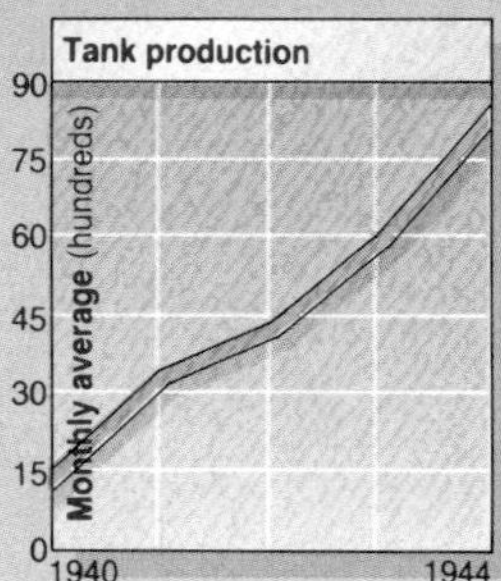

▲ German tank production increased almost six-fold between 1940 and 1944, in spite of a diminishing labor force. This remarkable achievement owed much to huge investment in industrial plant during the war. It was made possible by the expropriation of plant, machinery and, especially, manpower from occupied countries.

▶ With the enlistment of German men, the domestic war-effort depended increasingly on foreign workers. Civilians numbered over 5 million by 1944, POWs almost 2 million. By contrast, the German female labor force remained virtually constant, with fewer than 4 million women industrial workers in 1944, as in 1939.

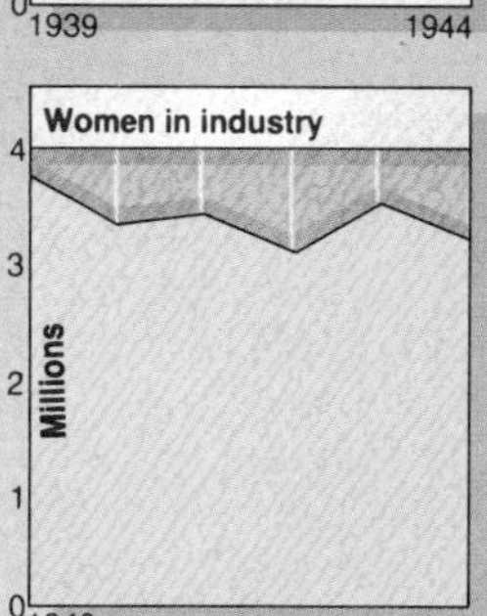

▶ Allied bombing raids became heavier in 1943 as fighters with an increasing operational range were developed. By early 1944 they could defend bombers against German aircraft almost anywhere over the Reich, bringing strategic targets in southern and eastern (as well as northern and western) Germany within range.

Chronology

1939

September 1
Germany invades Poland; start of war in Europe

September
War Economy Regulations introduced

1940

April–June
Germany defeats: Norway, Denmark, Low Countries and France

1941

March
"Women for Victory" voluntary campaign launched

June 22
Invasion of Russia begins

November
German troops halted before Moscow

1942

January
Wannsee Conference plans extermination of Jews

February
Albert Speer becomes armaments minister

April
Central Planning Committee formed: economy to be mobilized for attrition rather than Blitzkrieg; priority given to qualitative weapons' superiority

1943

January
Labor conscription for women aged 17–45

February
German army at Stalingrad surrenders; Goebbels' "Total War" speech at the Berlin *Sportpalast*

1944

June
Allied invasion of Normandy; economy concentrates on quantitative production

July 20
Assassination attempt on Hitler fails

October
Volkssturm created

1945

February
Allied bombing of Dresden

April
Red Army reaches Berlin;

April 30
Hitler commits suicide

May
Soviet army enters Berlin; German army surrenders

May 8
VE Day: war in Europe ends

In Germany, as in all belligerent countries, the home front was vital to the waging of war. Hitler believed this and was determined to maintain domestic support for the war effort. He was obsessed by the need to avoid a recurrence of the "stab in the back", the alleged betrayal of the German army in 1918 by hungry and demoralized civilians, incited by "Jews and socialists". The means were provided by the fruits of Hitler's conquests: occupied Europe was forced to supply foodstuffs, raw materials and labor which cushioned the home front for most of the war.

This strategy was vindicated: throughout the war there was virtually no open defeatism, far less popular revolt. But that did not signify enthusiasm for the war. The memory of World War I remained horrifically vivid. Germany's early victories were popular, but there was clear evidence of war-weariness virtually from the start. Even the victorious campaigns in Poland and western Europe in 1939–40 brought casualties and grief. The Russian campaign (launched in 1941), in particular, caused deep anxiety. Especially after the Germans were defeated at Stalingrad early in 1943, there was increasing despair, but also numbing apathy as reality became too terrible to face. Together with increasing Nazi terror, this ensured that Hitler's regime would not be overthrown from within. Germany would have to be defeated by *foreign* foes.

Political stability in Germany was achieved at the expense of foreigners: prisoners of war and civilian workers imported from occupied countries were used as slave labor; within occupied countries a low living standard was imposed. This accorded with Nazi views about the German "master race" which would dominate "inferior peoples", especially those in eastern Europe. Exploitation of occupied countries and of over 7 million foreign workers in Germany allowed the regime to defer or avoid unpopular policies, like conscripting women into war service, or reducing Germans' rations to, or below, subsistence levels. It also disguised contradictions within a system operated by incompetent, self-interested and feuding policy-makers.

Between 1939 and 1945 more than 11 million German men were conscripted into the armed forces. The German civilian population therefore consisted increasingly of older men, children, youths and – the largest single group – women. They felt the impact of war in differing ways, depending on their family circumstances, where they lived, and whether they were employed. By 1945 few had escaped the effects of the war: to that extent, it was a "total war". But even in 1942 victory was still widely expected, with the prospect of indefinite German rule over a continent-wide helot empire. The long road to defeat in

GERMANY: HITLER'S HOME FRONT

Hitler's social policies for the home front

The structure of the German wartime economy

The response to labor shortages: imported foreign workers

Economic controls and the people

Life under extreme pressure, 1943–45

German resistance to Hitler: the bomb plot of 20 July 1944

1943–45 showed that Germany lacked the resources to hold down its empire at the same time as repelling the forces of the Allies.

The impact of labor shortages

Two major problems afflicted the German war effort: limited materials and shortage of manpower. Resources were already being diverted from domestic consumption to the military before 1939, and further potential for this was limited. While the war necessitated an escalation in armaments production, supplies of food and other essentials had to be maintained to keep German civilians loyal and ready to support the war effort. This was perceived as a short-term problem until plunder from vanquished countries supplied Germany's needs, material and human. By 1939 there was little surplus labor within Germany, since rapid recovery from the Great Depression had absorbed all who wished or needed to work. Labor shortages appeared in critical areas even before the war, while high rates of women's employment in industry, commerce and agriculture indicated that the reserves of willing female labor were already exhausted.

▼ The myth of popular obedience and support for the war-effort was maintained throughout the war. Rallies like this, in the Berlin *Sportpalast* in June 1943, were carefully stage-managed and publicized. Goebbels and Speer used this occasion to stiffen morale among the Party faithful, at a time when the tide of war was turning against Germany.

The shortage of labor on the home front affected businesses large and small. As men were conscripted, factories lost large numbers of workers, including those with special skills. In return, they received substitutes – including foreign workers – who were often unwilling, unfamiliar with the work, and therefore less productive. Almost three million men left industrial work between 1939 and 1944, but the number of women rose only slightly.

In other sectors of employment, including agriculture, trade and commerce, and domestic service, women's numbers actually declined. Even with a doubling of the number of women employed in administration, the total number of employed women in September 1944 was only slightly higher than in May 1939 (14.9 million, compared with 14.6 million). In this "total war" many women, especially from the prosperous middle class, continued their prewar life of relative idleness. The allowance paid to the wife of a serving soldier enabled her to withstand government efforts to attract or shame her into war-related employment. Even after the introduction of labor conscription for women in January

► At first, Hitler's regime relied mainly on propaganda to try to bring nonemployed women into work. This slogan, "You help, too!", was typical, urging women to support their men by working in factory or hospital or on a farm. This convinced few while Germany appeared invincible. Conscripting women into war work was debated virtually from the start. But Hitler's reluctance to impose unpopular measures prevailed until the position began to deteriorate, early in 1943.

1943, many managed to avoid it, to the disgust of those who had no choice but to work.

It is clear that many middle-class women genuinely dreaded the prospect of factory work. Hitler was determined not to antagonize actual or potential supporters by enforcing an unpopular policy. Resentment at home would – as in 1914–18, he believed – be communicated to soldiers at the front, and sap their morale. The "stab-in-the-back" myth thus had its effect, while the availability of foreign labor made the need to recruit women into war work less vital, especially in the early part of the war. Nevertheless, the failure to recruit women equitably for war work was one of the most divisive issues in wartime Germany. Scathing attacks were made on leisured women who had all day to shop for scarce goods, while tired working women faced queues and disappointment at the end of a long day.

Many Germans worked in small concerns, for example, family-run shops or farms. The loss of even one adult male worker – a family member or a hired hand – could have a dramatic effect on a business. Where a wife was left in charge, with the burdens of war economy regulations in addition to customary chores, the family shop might close or the farm go into decline. The employment of a servant at home could release a wife to run a business, which helps to account for there being over a million domestic servants during the war. A small farm might receive assistance from young women enrolled in the Labor Service (a compulsory scheme to indoctrinate the young and provide a pool of cheap labor). But the demands of the Labor Service camp, with its duties, parades and classes, curtailed the time available for work. Farmers much preferred to be allocated a foreign worker.

Supplies and distribution

In 1936 the Four Year Plan gave official priority to rearmament and the drive for self-sufficiency in certain raw materials and foodstuffs over the production and import of consumer goods. While this policy was only partially enforced, the replacement of imported goods by indigenous alternatives was strongly encouraged. Nevertheless, German industry continued to depend on foreign imports, notably ironore from Sweden, which provided 40 percent of Germany's needs in

Foreign Workers

Even before 1939 full employment in Germany provided jobs for 300,000 foreigners. Then came Polish prisoners of war and civilian conscripts and, in 1940, French, Belgian and Dutch POWs. After the invasion of the USSR in 1941, hundreds of thousands of *Ostarbeiter* ("eastern workers") were sent to Germany. In the west, there was "voluntary" migration from now depleted industrial centers to Germany's expanding war industries. But increasingly from 1942 coercion was used in western as well as eastern Europe, to provide substitutes for enlisted Germans. From 3 million in 1941, the number of foreign workers in Germany rose to 7 million in 1944, as 20 percent of the labor force. They were fairly evenly divided between agriculture and industry. The French formed the largest group of skilled workers in industry, while unskilled heavy work fell to the *Ostarbeiter*. Poles were often deployed in agriculture, providing vital assistance on small farms which had lost workers to the armed forces. Later, some were even managing farms where no German male labor remained. Foreign labor, male and female, was invaluable to the war effort.

◄ Foreign workers in a German arms factory, 1943.

◀ Hitler's reluctance to conscript female labor left the task of recruiting women for voluntary work to the Nazi women's organization (*NS Frauenschaft*). But even this attracted few, and in 1940 the average effort put in by each volunteer amounted to one hour per week. The regime depended mostly on small-scale, local enterprises, like this sewing circle in Bayreuth in 1941. With German troops stranded in their first Russian winter, in 1941–42, Nazi Party agencies like the NSF and the Nazi welfare organization (NSV) supervised collections of warm clothes and the knitting of socks and comforters. They also provided emergency aid for air-raid victims and welfare work among evacuees from bombed-out cities. But this was all piecemeal, local and uncoordinated effort. Some women preferred work with the Red Cross or the air-raid protection service, partly because they were not run by Party agencies like the NSF or the NSV.

1940 and still around 25 percent during 1941–44. The conquest in 1940–41 of countries producing essential minerals brought Germany ironore from Lorraine (annexed from France); manganese and oil from the Soviet Union; nickel from Greece and Yugoslavia. Friendly countries supplied other goods, including oil from Romania, wolfram from Spain and Portugal, chrome from Turkey. In addition, Germany's own rich coal deposits enabled it to manufacture synthetic oil and rubber, but these were costly processes which deprived domestic consumers of adequate coal supplies. Once German forces were in retreat (from mid-1943) and resources in the shrinking empire inaccessible, these synthetic goods were vital, if increasingly insufficient. But especially under Albert Speer, as minister for armaments and munitions from 1942 to 1945, Germany's capacity to produce heavy industrial goods was formidable.

Throughout the war, civilians faced rationing and shortages. From the end of August 1939 some foodstuffs were rationed, and at frequent intervals rations were reduced and more goods – for example, soap and clothing – brought under rationing control. For "normal consumers", food rations remained limited but adequate until 1944–45, by which time few goods were unrationed and many commodities completely unavailable. People in special categories, like German workers in heavy industry or pregnant women, received extra rations, while foreign workers received smaller amounts of poorer quality food.

Extra rations might be distributed, on a local basis, in the event of a particularly good crop or to boost morale after a heavy bombing raid. But, otherwise, rations were steadily decreased. Controlling distribution through rationing did not prevent shortages becoming more acute. In winter 1939–40 there was a coal crisis because of transportation problems. Hoarding, bartering and an active black market removed goods from the official economy. Together, rationing, shortages and the black market caused great resentment and a decline in morale.

Other government economic controls, too, were unpopular. Taxation was increased both to raise extra funds and to dampen consumer demand. But the additional "war supplement" tax created problems for small businesses, some of which closed for the duration of the war. Agricultural production was regulated, with regional boards set up to collect and distribute produce such as eggs and milk. This was intended to pace distribution, to avoid periodic shortages, but rural producers greatly resented regulation and

The mood among the working class here is very bad at present! One hears remarks that are very reactionary, provoked by 1) price and wage developments, and 2) the food situation. People from regions endangered by air attacks have been with us for weeks and months now; they can afford the best meals on meatless days and pay cut-throat, profiteering prices for foodstuffs under the counter. It's the working population which has to bear the entire brunt of the economic struggle.

NAZI PARTY OFFICIAL
IN SOUTHERN GERMANY
JULY 1941

▲ **Bombing of German cities damaged not only buildings but also services like gas, electricity and water. People then had to queue for a bucket of water. Bomb damage enhanced opportunities for black marketeers, who had exploited shortages from the start. They hoarded goods and then sold or bartered with them when supplies were low and prices high. They also stole, forged or falsified ration cards.**

tried hard to sabotage it. Arrests, followed by fines or imprisonment, failed to deter others from breaching the war economy regulations.

Increasingly, civilians stood in queues for staple foodstuffs, exchanging grumbles and blaming the government for failing to eradicate the black market. Security service agents mingled with shoppers and reported their grievances. Monitoring the public mood, by means of a plethora of government and Nazi Party agencies, was regarded as vital for avoiding the risk of a "stab in the back" through disaffection. There was particular anxiety about the susceptibility of the industrial working class to this, and sometimes unpopular measures were rescinded. For example, the wage freeze introduced at the start of the war was quickly abandoned because of protests from employers and Nazi Gauleiters about absenteeism and deteriorating work discipline. To try to raise morale, a state housing program and improved social welfare benefits were promised for when the war was over. But this did not meet specific grievances about the way in which the better-off could evade consumption restrictions and war work.

Germany under pressure 1942–45

The absence of men at the front disrupted family life, put family businesses at risk, and meant constant worry among the relatives and friends of servicemen. Conscription and military casualties touched most families. Three million German soldiers, sailors and airmen were killed and many more injured.

In the second half of the war, the major threat to civilians came from aerial bombardment. There had been sporadic daylight air raids in 1940 and 1941. But from spring 1942 Allied bombing of German cities became frequent and systematic. Northern ports like Lübeck, Hamburg and Bremen, and western industrial cities like Cologne, Düsseldorf, Essen and Duisburg, bore the brunt at first. But by summer 1943 the bombers were reaching southern cities like Munich and Stuttgart, where they inflicted heavy damage on civilians as well as on known industrial

complexes. To continue vital production in safety, factories were dismantled and then reassembled in forest clearings, or even underground.

Life under bombing meant not merely terror but also added discomfort. Sleepless nights in air-raid shelters were followed by a day's work and queueing for food. Worse, it could mean the loss of home and possessions, although the government paid compensation. Bombing of north-western towns led to the evacuation of women and children to the rural south. As Allied bombing intensified, evacuees from central and southern German towns and cities, too, were sent to rural areas which were less at risk. Billeting town-dwellers on the rural population often bred hostility between two kinds of lifestyle and expectations.

In the face of these increasing hardships, the government tried to maintain and raise morale through propaganda. After 1933 Germans became inured to constant, repetitive and mind-numbing propaganda, but the war brought a renewed campaign to try to maintain morale and elicit support for the war effort. Although the outbreak of war had not been welcomed, the early victories were hailed as the just settlement of Germany's grievances. New campaigns, however, were unwelcome, especially the invasion of the Soviet Union in June 1941. The failure to win quickly in Russia, after victory had seemed both assured and imminent in late 1941, led to disenchantment and apathy, and to diminishing faith in and respect for Hitler from 1942.

But popular revolt was not on the agenda. Civilians faced the obstacles of everyday life with resentful resignation and harassed anxiety. There were instances of disorder and lawlessness, but these mostly involved crowds of consumers competing for commodities like coal or fruit when a delivery was made; or it might mean people plundering freight trains or stealing agricultural produce from the fields where it grew.

Germans fought for Hitler's cause to the end, especially on the Eastern Front where fear of the advancing Red Army was fueled by Nazi propaganda about "Bolshevik atrocities". There was also fear of being subjected to the same appalling treatment that Germans had inflicted on Soviet soldiers and civilians in 1941–44. In the west and south, the arrival of American forces in spring 1945 was welcomed. Yet SS officers tried to enforce resistance to the last by summarily executing "defeatists" who called for an end to struggle, pointless as it had become by March 1945.

At present, the air war dominates people's minds and does most to undermine their faith in a change of fortunes. The enemy's constant attacks have made a deep impression. Everyone is convinced that worse is yet to come. There is growing anxiety that, once the cities are annihilated, the smaller towns, too, will be destroyed. Low-flying raids cause particularly deep concern. There are frequent complaints about the shortage of air-raid shelter space for the civilian population.

SECURITY SERVICE REPORT
MAY 1944

German Resistance

From 1933 opponents of Hitler's regime either fled Germany, worked there in difficult illicit conditions, or were apprehended by the Gestapo. Socialists and communists, particularly, suffered this fate; but liberals, pacifists and feminists, too, were suppressed. Of the few attempts at a viable coup, only the bomb plot of 20 July 1944 had clear potential. It was the climax of a conspiracy by dissident army officers, former diplomats and politicians, and some socialists and clergymen. Its timing, soon after D Day, has drawn accusations that fear of military defeat was their motive. But many had been involved since 1938; their earlier attempts to kill Hitler had been aborted, often at the last minute. In wartime, opponents faced the additional hazard of being labeled traitors. The burden of resistance to Hitler fell largely on conservatives and soldiers who had earlier collaborated with him, because they had better cover than open opponents. Alarmed by Hitler's war-mongering and sickened by German wartime atrocities, they joined a few former labor leaders and individual Protestant and Catholic clergy in conspiring rather ineffectually against Hitler, more to demonstrate that there was "another Germany" than in the hope of actually succeeding.

◀ On 20 July 1944 a bomb placed close to Hitler in his eastern HQ at Rastenburg by Count Claus von Stauffenberg exploded. The operations room was shattered and injuries sustained, but Hitler emerged almost unscathed. The plotters were quickly apprehended. The refusal of more than a handful of officers to support them (although a great many more knew about the plot) testified to the terror felt by senior soldiers at the prospect of Hitler's vengeance should the plot fail. Their fears were vindicated: around 5,000 people, many merely relatives of the conspirators, were executed in the wake of 20 July. There were no more attempts from within Germany to remove Hitler.

Datafile

UK statistics can deceive the unwary. The "United Kingdom" government was responsible for foreign affairs and defense, but "Northern Ireland" had home rule in domestic matters. Within "Great Britain", Scotland had separate systems of law and education. Foreigners tended to refer to the whole Kingdom as "England" – yet in football Wales, Scotland and Northern Ireland all fielded "national" teams.

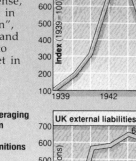

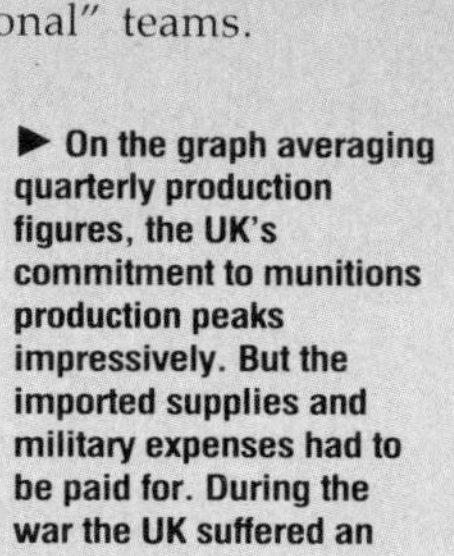

▼ Indictable offences known to the English and Welsh police rose sharply. (The Scottish pattern was similar.) The image of a nation united in self-sacrafice must be heavily qualified: 940,000 working days were lost in strikes in 1940, 3,714,000 in 1944. The latter figure was high by prewar standards.

▶ On the graph averaging quarterly production figures, the UK's commitment to munitions production peaks impressively. But the imported supplies and military expenses had to be paid for. During the war the UK suffered an enormous loss of wealth, of which debts as sterling liabilities were only part.

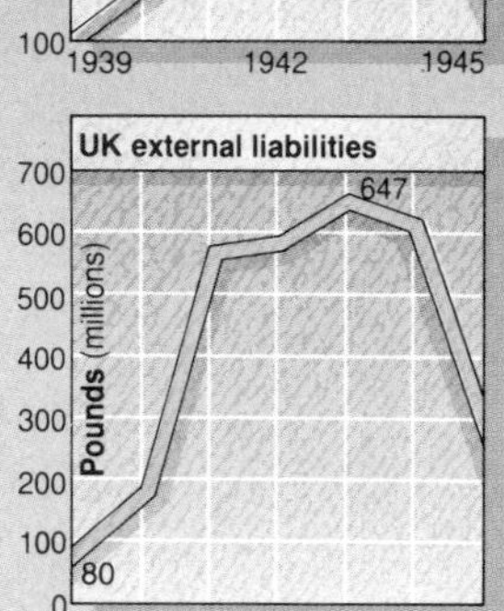

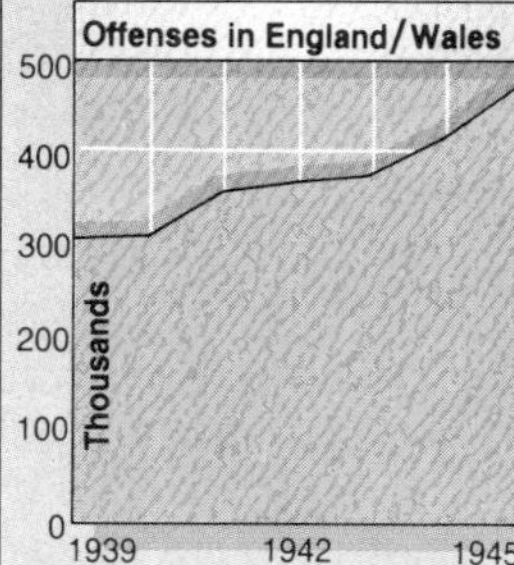

▶ The decanting of child evacuees from likely bombing targets into the countryside was based partly on false assumptions, as the map shows. Sussex (in the southeast) received a lot, but by mid-1940 it was virtually in the front line. Plymouth, later heavily bombed, was a "neutral" area from which evacuation was not supported.

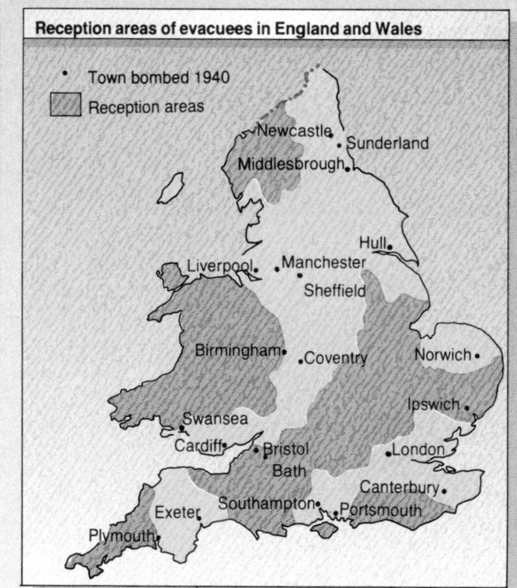

Chronology

1939

March
UK abandons Appeasement and offers guarantees to Poland and other countries

June
Women's Land Army formed

August
Emergency Powers (Defence) Act passed

September 1
Germany invades Poland; start of war in Europe; National Service (Armed Forces) Act passed

September
Internment of aliens begins; first wave of evacuation

1940

January
Rationing begins

April
German invasion of Norway

May
Chamberlain resigns; Churchill becomes prime minister; Germany invades France

July
Evacuation from Dunkirk completed; Local Defence Volunteers reformed as Home Guard; free milk for mothers and small children

August
Battle of Britain begins

September
London Blitz begins

October
Hitler postpones invasion of UK

1941

March
Mobilization of labor by Registration and Essential Work Order; US Lend-Lease Act gives aid to Allies

1942

January
First American GIs arrive

December 1
"Beveridge Plan" published

1943

May
Axis forces surrender in Tunisia; end of war in Africa

1944

June
Allied invasion of Normandy; V1 missiles fall in England

1945

May 8
VE Day; end of war in Europe

July
Churchill resigns; Attlee forms Labour government

September 2
VJ Day; end of war in Pacific

The human story of the UK "people's war" has long been told as one of simple triumph. Never before had ordinary people participated so wholeheartedly in a war and with such a glorious result. According to this view, the war can be narrated as follows.

The UK entered the war on 3 September 1939 under a Conservative prime minister, Neville Chamberlain, who had no will to fight. For months people endured the discomfort of conscription, evacuation, the nightly "blackout" of windows and so on, but very little happened. Then disaster struck in the spring of 1940, first in Norway (see p. 31), then on the Western Front (see p. 32). As the People wished, Chamberlain was overthrown by a revolt of members of parliament from his own party and was replaced by a man of indomitable courage and charismatic oratory. Winston Churchill (known as "Winnie") brought leaders from the Labour and Liberal Parties into his new coalition government. After the fall of France (22 June 1940), heroic effort in the Battle of Britain (Britain's "Finest Hour" according to Churchill), supporting the RAF's brave fighter pilots, thwarted the intended German invasion. First London, then city after city, endured bombing raids, but morale did not falter. The UK "stood alone" – but in 1941 found mighty allies, first in the USSR, then in the USA. After massive effort in industry and agriculture, enduring living conditions which were more and more austere, British civilians saw the Allies triumph. The British People had Saved Democracy.

The UK's unmatched mobilization

Certain statistics support this heroic account. Between 1940 and 1944 The UK mobilized its resources with a thoroughness unmatched by any of its allies, or by Germany. In June 1944, at the time of D Day, the USA had drawn 40 percent of its people into the armed forces or civilian "war work". In Britain 33 percent were in "war work" alone, and the combined total was 55 percent. Employment in "war industries" had been raised at the expense of civilian needs. While labor was directed into engineering, chemicals and other work connected with munitions, employment in consumer goods industries slumped, and much of what remained was devoted to supplying the needs of the armed forces, or producing goods for export which helped pay, in small part, the huge cost of the war effort.

Not even Stalin's USSR went so far in conscripting women. By 1943 it was almost impossible for a woman under 40 to avoid "war work" – unless she had heavy family responsibilities or was looking after a war worker billeted on her. By 1944 women formed 48 percent of the civil service. The proportion of women in

THE UK: A DEMOCRACY AT WAR

The apparent triumph of British society in war

Statistics that support the argument: working women and industrial production

The vogue for planning

Social services in wartime

Two views about wartime Britain

Bevin and trade unionism

engineering and vehicle building rose from 9 to 34 percent. Women in agriculture more than doubled, to 204,000 – and "Land Girls" did farm tasks hitherto performed almost entirely by men. An ex-hairdresser won a horse-plowing competition against a field of men. In July 1943 nearly 40,000 Italian prisoners of war were also at work in the fields. British civilians were exhorted to "Dig for Victory" in their scanty spare time. By 1943 there were 1.4 million allotments, an increase of 75 percent on the prewar figure. Hen-keeping and pig-rearing expanded likewise. The islands were besieged. The British had to feed themselves. U-boats prowled in the Atlantic, and in the worst year, 1943, the UK lost 833 merchant vessels. Bread and potatoes increasingly dominated the diet.

Food rationing, introduced early in the war, intensified. At a peak, in August 1942, one adult could hope for roughly half a kilogram (1lb) of butcher's meat a week: children got less. This was supplemented by 115g (4oz) of bacon or ham, 225g (8oz) of sugar, 225g of fats (of which 115 must be margarine, only 55g (2oz) could be butter) – and so on. Eggs were in especially short supply – the average person might get 30 a year – but the State was careful to ensure that expectant and nursing mothers and children received more (maybe three a week), and a half liter (1 pint) of milk per day. Rationing was perceived by the public as both necessary and fair.

In 1942, when a system of "points" rationing had been extended to canned and dried foods, breakfast cereals, chocolate, biscuits and other groceries, and soap was rationed for the first time, Sir Stafford Cripps, returning from his post as ambassador to the USSR, competed for a while with Churchill himself in popularity. He was a vegetarian and teetotaler and stood for more austerity, not less.

Getting home after long hours of work, perhaps having queued on the way first for food, then for scarce public transport, people went out to "Dig for Victory" or serve in Civil Defence, or train with the Home Guard, affectionately known as "Dad's Army". (It was founded in 1940 to fight

▼ This Land Girl was employed on a drainage project in the summer of 1943. The image suggests that women are doing very tough farmwork. Is it a smile or grimace? The side-view, "Socialist Realist" style permits either interpretation.

▲ **Women, mannishly dressed for industrial work, are manufacturing barrage balloons. These were still doing useful work in 1942, when this picture was taken, though the heaviest air attacks were long over. In that year the Germans carried out a destructive air raid on Canterbury. The addition of a balloon barrage to the town's defenses ensured that a second attack a week later was relatively ineffective. In 1944 balloons ringing London helped ward off V1 flying bombs.**

German invaders had they come.) Elderly people and housewives worked in private homes making aeroplane parts.

Production of aircraft rose from 8,000 in 1939 to 26,000 in 1943 and 1944. Between 1940 and 1942 tank production quadrupled. There was much criticism of graft and inefficiency in "war industry" by workers in it and by outsiders, but this might be attributed as much to a frustrated wish to see Hitler beaten quickly as to the weariness and strains of wartime life. Popular feature films of the day showed people with strong hopes of a better future after victory and a willingness to keep "going to it". Industries and regions which had suffered especially hard in the years of depression between the wars now played their full part. The shipyards of the North were busy again. Scotland got major new industrial works.

The vogue for planning

During the war the prewar regional railway monopolies were run as components of a unified system, prefiguring postwar nationalization. The new ministry of fuel and power (created in 1942) similarly controlled the numerous companies in the coal industry. The idea of "Planning" became immensely fashionable. It was thought that the capacity of the USSR to fight back against the German army (and finally to defeat it) demonstrated the virtues of having a "planned" society.

► **"Digging for victory". Any one, however young, might get open-air pleasure out of growing food. Father and child are spending their Easter Holiday at work together. Potatoes matter far more than sport – a cricket field in North London is being transformed into allotments for people attracted by the "Grow Your Own Food" campaign.**

Books, pamphlets and articles poured out from writers of the left and center full of "plans" for a more just, more efficient postwar country. One series of publications, by distinguished experts, included such typical titles as *Make Fruitful the Land, Start Planning Britain Now, Reconstruction and Peace, End Social Inequality*. In this ambience, the success of the Beveridge "Plan" – published in December 1942, is easy to explain. People queued outside branches of His Majesty's Stationery Office to purchase a dry-sounding, boring-looking, and highly technical official report on *Social Security and Allied Services*. This "Plan" eventually sold 650,000 copies. Within two weeks of publication an opinion poll discovered that 19 people out of 20 had heard of it, and that 9 out of 10 thought that it should be adopted.

Ironically, the chief "Planner" of Britain's wartime civilian life was a wholly uncharismatic Scot, a former civil servant, Sir John Anderson. His lord president's committee had central control of home and economic policy from 1941. In the following year, after much public clamor, a ministry of production was created to oversee supply to the armed forces. Despite the unpopularity of controls with businessmen, and the irritation felt by the general public over civil service "Red Tape", it was generally accepted that Total War demanded governmental and official dominance of the lives of all citizens.

War and Social Services

That many people in the UK lived in dire poverty was dramatized by evacuation. One and a half million school pupils and mothers with small children were suddenly decanted, in September 1939, from cities likely to be bombed into rural and suburban "safe" areas, where verminous slum-bred children shocked genteel householders. Concern for "social welfare" in general was heightened in some quarters.

In early 1941 the Trades Union Congress lobbied the minister of health about inadequate existing provision for health insurance. The government set up a committee of civil servants, chaired by the famous social scientist Sir William Beveridge, to survey all social insurance. The resulting "Plan" took a range of existing schemes and gathered them into one overall scheme to provide for the citizen "from cradle to grave". It was based, Beveridge wrote, on three assumptions: (1) that family allowances would be given for all children; (2) that a National Health Service would be provided; (3) that mass unemployment could be avoided. Before the war ended family allowances were introduced.

▲ During the war milk consumption increased considerably. The decision taken in July 1940 that mothers and small children should get free or cheap milk came to symbolize the beneficial impact of war on social services. This poster's designer, James Fitton, had been well-known before the war as a leftwing cartoonist.

Politicians and the people

The central institutions of British public life came out of the war with their prestige enhanced. The House of Commons was physically destroyed in the greatest air raid on London, on 10 May 1941, but parliament continued to meet, a beacon for democrats all over Europe. Buckingham Palace was also bombed, though less severely, but this served to join King and People in a common tribulation. George VI, Queen Elizabeth and their daughters provided a symbol of decent family life in austere and difficult days.

The political parties all had cause for satisfaction. The Conservatives, many of them hostile to Churchill in May 1940 when he ousted Chamberlain, came to delight in possessing the allegiance and leadership of the great man. The Liberals were overjoyed when Beveridge joined their ranks late in the war. But Labour was to be the electoral beneficiary of the Beveridge Plan. The party's leaders served loyally and effectively in the war cabinet, gaining the party a wholly novel reputation for solidity and common sense. At the same time its backbench members of parliament served it well when nearly 100 of them voted against the government in February 1942 over its less than total acceptance of the "Beveridge Plan". The Labour Party was identified with the planned future which so many people wanted, and swept to a landslide victory in 1945 in a general election held before the war with Japan had ended.

The British historian Henry Pelling has commented that "Parliament, the political parties, the Civil Service, local government, the press, the law, the trade unions – all emerged from the war with slightly different surface features but basically unaltered. There had not been much of that 'inspection effect' which is supposed to be one of the by-products of war; or, if there had, it had found most institutions not unsatisfactory and so served to reinforce the view which so many people in Britain still retained: that somehow or other, things in their own country were arranged much better than elsewhere in the world...".

The UK had escaped invasion, which would have tested all its institutions to the limit. It had endured air raids on a far smaller scale than those which the German economy survived in 1944. Its moral reputation stood, or seemed to Britons to stand, very high in 1945. St George had slain the dragon of Nazism.

▲ The "Dad's Army" image of the Home Guard is typified here by the grizzled veteran closest to the camera. But by 1943 the average age of the 1.7 million part-time soldiers in this service was under 30 – it was used to train boys of 17 and 18 prior to their call-up into the army.

Doubts about wartime Britain

Yet within a quarter of a century of World War II the UK became a troubled, manifestly unsuccessful, country. Not surprisingly, the triumphalist account of the war outlined above came under fire, not only from "New Left" Marxists, but also from the so-called "Radical Right". It can be held that the UK's very success in wartime helped to preserve antiquated institutions and incompetent industrial practices, while confirming its imminent demise as a great power.

There is no contest over the evidence. From 1940 the UK depended on American friendship. "Lend-Lease" from 1941 to 1945 provided vitally important food and armaments but represented client status. On 14 August 1945, the day Japan surrendered, the British treasury advised the cabinet that the UK faced a "Financial Dunkirk". External disinvestment for the sake of the war had reached 4,000 million pounds. Shipping, a source of invisible exports, had been reduced by 30 percent. Visible exports were running at no more than four tenths of their prewar level.

From the right, the British historian, Corelli Barnett, has argued that the establishment of a welfare state, which was incumbent politically on any postwar government, whatever its political coloration, was an economic absurdity in these circumstances. But the apparently successful organization of the war economy had convinced politicians of diverse hues that anything was possible, given Planning.

Barnett argues – with much justice – that the British war economy was in fact bedeviled by longstanding inefficiency ("uncompetitiveness" in postwar terms). Small-minded, small-scale industrialists snarled at trade unionists obsessed with traditional practices. The UK preened itself on its wartime technological feats. Yet it had long lagged behind Germany in applied science and technology. The Germans developed a better jet fighter far more quickly, while British production figures were being boosted artificially by the manufacture of obsolescent planes: the Whitley bomber, almost useless for its task when the war started, nevertheless continued in production till 1943. An *official* British historian has remarked that of the Whitleys produced – six times as many as originally planned – "many, perhaps most ... scarcely left the Aircraft Storage Units". In productivity, anyway, the US aircraft industry far outstripped the British. The UK's machine tool industry was so inefficient that dependence on America was inevitable. The UK's tanks were so botched in design that, by the summer of 1942, American Grants and Stuarts far outnumbered British Crusaders in the armored divisions of the British 8th Army.

But the view that the British people were deluded by their own propagandists, who were themselves deluded, perhaps points to unusual cohesion and unity within the society. It is said that a high-risk rescue worker, when praised by King George VI for getting a young girl out of a wrecked house after she had been buried for four and a half days, replied, "It's all in the day's work, sir, we all get the same pay." He was wrong; skilled rescue men in fact got more than the basic Civil Defence wage. But such expressions of solidarity abound in writings from the war years, as do memories of them. An aircraft-factory worker recorded: "There was very little absenteeism caused by the raids ... we knew that if we didn't turn up our mates would be worrying. You would see men staggering at their work from lack of sleep, snatching a ten minutes' doze in the canteen over their food, and still, when knocking-off time came, going off with a

▼ November 1942, King and Queen inspect pies in the canteen of a Manchester aircraft factory. Such shots of royalty mixing with workers emphasized that this was a "People's War". Industrial canteens were amongst Ernest Bevin's favorite causes. In 1943 he pushed through an act fixing catering wages against the votes of over 100 Conservative MPs.

▲ May 1940, a newly completed housing estate has been converted into a temporary internment camp for "aliens". Some 27,000 people born in enemy countries were rounded up that summer. A few of them were indeed anti-British. The vast majority were not, including many Jewish refugees as well as distinguished German and Italian anti-fascists. Some 11,000 were deported to Canada and Australia; and many of these lost their lives when their ship was torpedoed. Camps on the Isle of Man filled up with the rest. Conditions were fairly comfortable. Gradually most "aliens" were released. Despite their treatment, many joined the services or did other war work.

cheerful, 'See yer in the morning, boys'."

The evidence is contradictory. A coherent society is a tolerant one. The UK interned fascists early in the war, but communists who called for peace until Hitler invaded the USSR were allowed freedom of speech, though their newspaper, the *Daily Worker*, was suppressed. Conscientious objectors to the war were not hounded as they had been in World War I. By-elections continued to be fought, though the coalition partners refrained from opposing each other. From 1942 to 1945 Independents and candidates for the new Common Wealth Party (which was socialist) – eight in all – defeated government nominees and entered parliament. They did not oppose the war; they argued that it should be conducted on a more efficient, usually socialist, basis.

But the hysteria of the summer of 1940, when rumors of "fifth column" activity made neighbors suspicious of each other, and thousands of European refugees from Hitler's Europe were interned as "enemy aliens", does not suggest a self-confidently harmonious UK. It can be argued that resentment between classes intensified in the 1940s. Certainly it was present in the strike wave of 1943–44. There was a rise in working-class wages at a time when people had little to spend them on. But the earnings of farmers rose fastest of all: were these men heroes justly rewarded or profiteers?

Huge crowds cheered Churchill and the King in London on Victory in Europe Day. The last word must rest with them. They believed that they had made history and triumphed.

Ernest Bevin and Trade Unionism

The Conservative rebels who voted against their prime minister in May 1940 wanted a coalition, and so Chamberlain fell because Labour would not agree to join one led by him. The party which commanded the loyalty of the most active trade unionists held a key position in a war which would be won by industrial effort. Even so, it was imaginative of Churchill to bring Ernest Bevin, from outside parliament, into office as minister of labour. Bevin was general secretary of the Transport and General Workers Union. He duly became the most powerful Labour man in the war cabinet.

Self-educated, he had had his first involvement with "Transport" as a drayman. He had not joined a trade union until he was nearly 30, but between the wars had become a leading union figure. Strongly anti-communist, he was also anti-Nazi.

The Essential Work Order of March 1941 and the registration of the entire labor force during that year gave Bevin enormous power over every civilian man and boy, girl and woman. He used it to direct labor to where it was considered most essential. But he also employed it in the promotion of his long-time aims: collective bargaining, recognition for trade unions, and improved conditions.

The Trades Union movement was vastly strengthened by wartime full employment and the propitious work of Bevin. Between 1938 and 1943 the number of unionized workers increased by over a third, to 8,174,000. Bevin's own TGWU became the first union to top a million members. Paradoxically, Bevin's old enemies the communists, when they switched to support the war effort after Hitler's invasion of the Soviet Union, were able to recruit scores of thousands of new party members. Still more paradoxically, they wielded their new influence against strike action, without much obvious success.

▲ Bevin (right) starts a loom to open an exhibition.

Datafile

For every five persons killed in the course of World War II, two were Soviet citizens. The toll in blood was horrendous, though the exact tally will probably never be known. Civilian losses reached 7 million due to death, deportation and wartime exertions, to which must be added the wounded, the mutilated and the orphaned. The trail of destruction left a third of Soviet industry in ruins.

Steel smelting
Tonnes (millions)
17.9
8.1
1940 1945

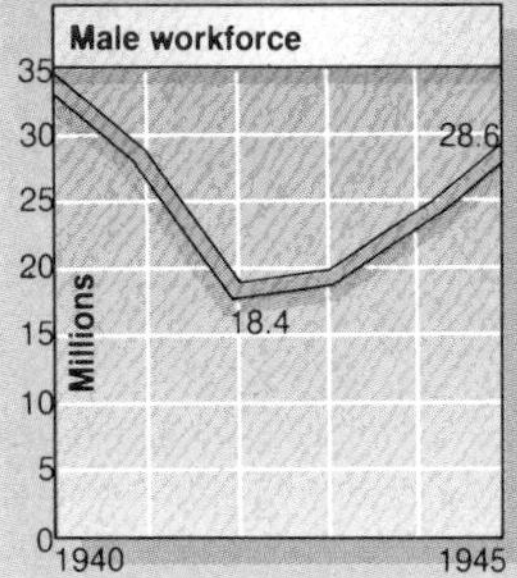

▲ German occupation of large areas of the western USSR made huge inroads into the industrial labor force available to the Soviet authorities, losses steadily made good by the vast increase in the number of women and young persons (16–25) employed in factories, and pensioners over 50 recalled to their work-benches.

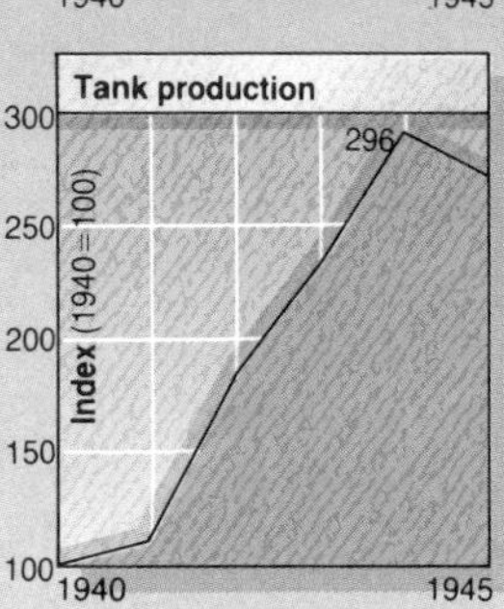

► Even before the German attack of 1941, Stalin prophesied that the next war would be won in the machine-shops, the factories and the foundries. The USSR won the industrial war outproducing Germany in spite of a vast forced industrial migration eastward and enemy occupation of major industrial regions.

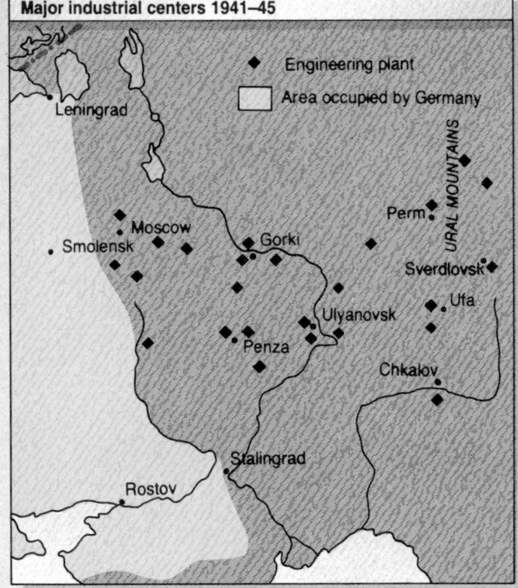

► Between July and November 1941, in the first stage of an unprecedented industrial evacuation, 1,360 major military plants were shifted into the eastern interior, turning the Urals, western and eastern Siberia and Central Asia into a great arsenal. The map shows new or expanded engineering plants east of the Front.

Chronology

1939

August 23
Soviet-German Nonaggression Pact signed

September 1
Germans invade Poland; start of war in western Europe

September
End of Soviet–Japanese border conflict (since July 1938); USSR occupies Eastern Poland

November 30
Soviet-Finnish War begins (till March 1940)

1940

June
Working day increased

1941

June 22
Germany invades USSR

July
Stalin makes first statement since invasion, calling for loyalty to motherland; ration-cards introduced in Leningrad and Moscow

August
Emergency war plan for Soviet economy

September
Siege of Leningrad begins (lifted January 1944)

November
Evacuation of industrial plant completed

1942

March
Soviet economy rallies

April
Nonagricultural labor conscripted for harvest

May
Central partisan staff set up to organize guerrilla warfare

1943

January
State defense committee takes action against abuses in distribution of commodities; Germans surrender at Stalingrad

1944

August
Red Army invades Romania

1945

April
Red Army reaches Vienna and Berlin

May 9
VE Day; end of war in Europe

August
USSR declares war on Japan and invades Manchuria

September 2
VJ Day; end of war in Pacific

It would be difficult to argue with the proposition that even more than the Bolshevik Revolution of 1917, the Soviet–German war, known as "The Great Patriotic War of the Soviet Union 1941–45", shaped the USSR as a world power, cemented its political system and colored its society.

By virtue of the magnitude of the Soviet achievement in defeating Nazi Germany and the scale of Soviet losses this was a war that was militarily and morally different, if not unique. Not only was it the greatest land campaign in history, but also a war of enslavement and extermination on the part of Nazi Germany and one of almost unprecedented barbarism.

During the 1,418 days of the war the Red Army fought along a front varying in length from 3,000 to 6,200 km (1,860mi–3,900mi), launched 9 campaigns and 210 operations by groups of armies; seven of the campaigns and 160 operations were offensive in nature. Each hour of the war cost 587 lives, each day 14,000 died; only 3 percent of the younger generation (17–21) survived. Society suffered the loss of the equivalent of two five-year plans; 30 percent of Soviet industry was left in ruins.

Authority and organization

Certain characteristics of Soviet society stood the state in good stead in time of war. To move to a war footing required surprisingly little administrative adjustment given the complex overlapping levels of Soviet administration and authority. Stalin, after recovering from his stupor during the first days of the German attack in June 1941, imposed his personal authority in a system of "super-centralization", directing military affairs through the *Stavka* (the equivalent of a general headquarters, for which the general staff worked directly) and the state defense committee (the GKO, with Stalin at its head and an entirely civilian membership), the latter body all-powerful and operating either by decree or direct personal intervention.

Thus Stalin had eyes and ears everywhere. The system worked on naked personal authority and personal control, subjecting governmental, military and administrative organs to the communist party and police primacy. Stalin ran the war much as he ruled in what passed for peace, with an all-powerful, rigidly centralized directorate headed by himself as party general secretary, as defense commissar and supreme commander and as chairman of the state defense committee, overriding "legal norms" and with Lavrenty Beria and the NKVD (the People's Commissariat of Internal Affairs) at his back. The system had strengths but was cumbersome; lack of direction from "the center" caused it to seize up.

THE SOVIET GREAT PATRIOTIC WAR

The unprecedented scale of the war in Russia

Stalin and the structures of internal authority

The industrial achievement: a second industrial revolution

The impact of war on the countryside

German barbarism and Soviet resistance

The restoration of Russian culture and the Orthodox Church

In the western regions a wartime regime was introduced by the hastiest of decrees. The initial military mobilization went smoothly, bringing 5 million men to the colors within little more than a week. But all too soon, as the German army drove deep, the lack of prewar contingency planning for industrial evacuation and strategic dispersion produced a grave crisis. Before the war Stalin had ordered the military authorities to plan for achieving rapid victory through offensive action "with little loss of blood". The Red Army was unprepared for a protracted war. In the German invasion of 1941 it suffered enormous battlefield losses; the invaders overran grain lands, mines and factories and engulfed millions of Soviet citizens, Stalin's labor force. Improvised industrial evacuation under fire accomplished astounding feats on the orders of the GKO, though the real heroes were Soviet railroad workers, who transported 2.5 million troops to the front in August–September 1941 and 1,360 heavy plants to the interior, to the Urals, western Siberia and the Volga region. Once plants were in the interior there came the mammoth task of building new factories, but machines were set down and set to work on the bare earth. Workers lived in grim barracks or in scooped-out dugouts – in a fierce winter.

The USSR survived the rending crisis of the late autumn of 1941. Total collapse was staved off by the fact that the Japanese did not attack Soviet Asia, by communications which did work (unlike those of World War I), by an extraordinary capacity for improvisation and by the ability to operate under conditions of wholesale disorder, by enduring grievous losses, by eliciting effective responses when "the authorities" also did their jobs properly and by the slow but horrifying popular realization what German rule meant: the enslavement and extermination of the "Slav rabble".

Resistance and economic mobilization

The emergency mobilization bore horrendously on the civilian population, called upon to man the militia, dig tank-traps and fortifications, undergo reserve training, patrol as "home guards" and, above all, increase war production. In Leningrad the population stood fast against German besiegers, even though starvation soon began to pierce. It was the start of almost 900 days of siege (8 September 1941–27 January 1944). Moscow endured martial law and siege conditions. Leningrad lived on its civic pride, for there was little else. In mid-November 1941 the GKO accepted revised production plans for 1942: 22–25,000 aircraft and 22,000 tanks, all with a labor force now fallen from 27 million to 19 million. Women and youths were drafted into the work places while in the interior, desperately housed and far from adequately fed, workers in the newly transshipped factories started production as walls went up around them. The Kharkov Tank Factory turned out its first 25 T-34 tanks 10 weeks after it had been reassembled.

Full scale military-industrial and economic mobilization effected what amounted to a second industrial revolution in the USSR. Soviet workers won what Stalin called "the battle of machines", turning out 78,000 tanks, 16,000 self-propelled guns, 98,000 artillery pieces. New production methods speeded up the great flow of tanks. The cost was heavy: 55 percent of Soviet national income now went for war purposes, as opposed to 15 percent in 1940. This was all-out mobilization, demanding everything from the Soviet people.

Under the slogan "All for the front!" the country was ravaged and scavenged, in a hunt for its brain and brawn. Food and shelter, never

▼ The shocking cost of the combat was not borne only by adults. The war bore terribly upon children, through death and the chaos of evacuation.

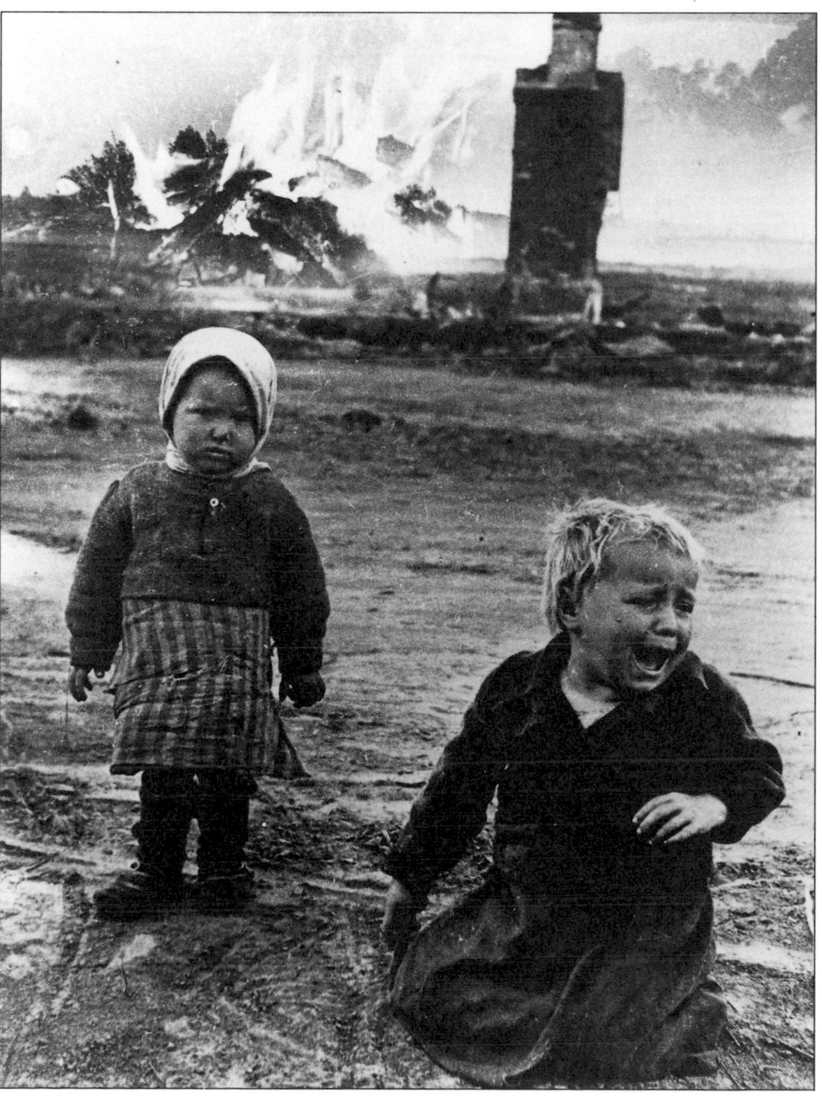

abundant in the USSR, presented daily problems. A "differentiated" rationing system for bread was introduced at an early stage, with increased rations for miners and the apprentice training schools: by 1943 three million were receiving augmented rations. But the breakdown in transport in Moscow in the winter of 1941–42 meant that only 30 percent of the meat ration and 34 percent of the sugar ration was "honored". At the end of 1941 in Zlatoust (a town to the east of the Urals) factory canteens were so flooded with evacuees that they failed to function. As Soviet war expenditures rose four times, civilian consumption fell by as much as 40 percent, while consumer items – sugar, jams, meat, matches, knitwear – were steadily vanishing. In January 1943 the GKO acted against abuses in the supply of food and goods, enlisting the trade unions to control factory stores and canteens and to combat "speculators". Communal feeding eased the situation, with 25 million being so fed by 1944.

The housing situation, already a chronic problem, now became desperate. In the Volga, Urals and Siberian regions, 8,500 workers survived in dugouts, more than 12,000 workers used derelict barracks. Funds were allocated for housing in the east but the problem remained acute, compounded in 1943–44 by the terrible devastation discovered in the west in the wake of the retreating Germans – millions of dwellings, thousands of hospitals and sanatoria ruined, large-scale depopulation. In addition, a depleted medical service had to deal not only with general public health but with widespread infectious diseases in the liberated areas. Provision had also to be made for the thousands of war orphans as well as the war wounded. Enormous burdens fell from the outset on Soviet women, young and old: women and girls were recruited for industry, the mines and railroads or for front-line service as nurses, snipers and aircrew in a unique bomber squadron. Women formed 26–35 percent of the labor force in mining, 48 percent in the oil industry; in all, a 56 percent increase in the number of women employed throughout industry.

War and agriculture

The Soviet peasantry – work-horse and whipping-boy of Stalin's prewar policies of social reorganization – shouldered an immense burden, supplying 60 percent of the Red Army's wartime manpower, responsible also for feeding the Red Army and the general population in the desperate "battle for bread". During the war the agricultural labor force was more than halved by the beginning of 1943 – (down from 16.9 million men and 18.6 million women to 4 million men and 11 million women) – but the travail began much earlier. Soviet farming, divided into collective farms (*kolkhozy*) and state farms (*sovkhozy*), did not escape the crisis of the early months of the war. It lost most of its mechanized equipment to the Red Army; lorries and tractors were used as gun-tows. Women and juveniles, even the old, had to substitute, literally harnessed to plows. Under the wartime labor regime the compulsory minimum of unpaid work or "labor days" (a labor unit equivalent to one and a half days in peacetime)

was set at 150 in cotton-growing areas, 100–120 in the grain and livestock regions. Severe penalties were enforced by the NKVD and the communist party for violations.

German occupation in the early months deprived the USSR of 47 percent of its grain-bearing lands, 87 percent of its pigs. During the early part of the war, in 1941–42, no tractors were supplied to the farms, whether collective or state, nor were spare parts or fuel readily available. In an attempt to compensate for the lost resources, fresh land was developed in the Urals, in Siberia, Kazakhstan and Soviet Asia but again shortage of labor, constantly drained off for the army and industry, and less fertile land hampered this effort. The low point was reached in 1943 when the harvest produced only 63 percent of its 1940 figure. Yet it cannot be said that the Soviet peasantry failed in the "battle for bread". With astounding resilience, though gripped by extreme hardship and suffering dire poverty, the Red Army was fed and the population at large did not actually starve. Little money was paid to the collective farmers, only "payment in kind" – minuscule amounts of grain and potatoes, if they materialized at all. As for exploiting the "private plots" (peasant's allotments), these too were subject to compulsory state deliveries with a *fivefold* swingeing tax increase on any income derived from them or from livestock. The peasant could hardly be described as a war profiteer on these terms.

◄◄ The war launched the USSR on a second industrial revolution, with a huge arsenal building up in the deep rear, in Siberia, far from roving German bombers. In this tank factory, workers were reminded of the front: "The Red Army needs *Your* tanks, it needs *Your* aircraft – Stalin", "The Red Army needs modern combat aircraft like it needs bread...".

◄ Peasants are exhorted to year-round work to produce food. "The Battle for Bread" was vital to the survival of the USSR. Its most difficult year was 1943, when the harvest yield fell to 20.6 million tons, but this crisis was overcome. This war differed from World War I in that the population and the armed forces were fed, not luxuriously but enough to keep body and soul together.

◄ Women, children and old men dig antitank ditches outside Moscow. Soviet women, young and old, bore the full brunt of the war. In a land of brutal shortages and hardship, women were called upon to work as air-raid wardens, to dig the huge antitank ditches, and to labor in the munitions factories.

German barbarism and Soviet resistance

Not only protracted hardship and constant sacrifices but also terrible savageries came to dominate this war. Behind the German wire an inhuman fate awaited Soviet prisoners of war. They were deliberately left to die, used in death-camp experiments or simply murdered, though from their massed ranks the Germans plucked Soviet non-Russian ethnic minorities for forced service under German arms. From the ranks of Soviet prisoners General Vlasov, a Soviet army commander captured in 1942, attempted to form an anti-Stalinist movement and a "Russian Liberation Army", only to run foul of Nazi political and racial fanaticism. In search of slave labor the Germans deported some 5 million Soviet citizens to operate their factories or run their farms. Stalin, out of revenge for real or suspected "collaboration", brutally deported ethnic-German Soviet citizens, the Crimean Tartars and ethnic minorities from the north Caucasus to Siberia and Central Asia.

The partisan movement in occupied Soviet territory, which grew from an organization dominated by party members and the NKVD into a geniune mass movement, generated more atrocity and unbridled savagery as the Germans intensified their war of reprisal and retribution, employing ferocious anti-partisan units in the

Winter had already come when Sverdlovsk received Comrade Stalin's order to erect two buildings for the plant evacuated from the south. ... It was then that the people of the Urals came to this spot with shovels, bars and pickaxes: students, typists, accountants, shop assistants, housewives, artists, teachers. The earth was like stone, frozen hard by our fierce Siberian frost. Axes and pickaxes could not break the stony soil. In the light of arc-lamps people hacked at the earth all night. ... Their feet and hands were swollen with frostbite, but they did not leave work. ... On the twelfth day, into the new buildings with their glass roofs the machinery began to arrive. ... And two days later, the war factory began production.

"PRAVDA"
18 SEPTEMBER 1942

▲ **Peasants donate money for the defense fund.**

► **"Russia's sons are on the march! Young and old together. Strike down the insolent invader, take your revenge! Let the tyrant tremble...his hour has struck. In each fighting man there stares out a hero." In an evocative poster the emphasis is on Russia: Red Army troops, "ancient Russia" in the background. Stalin himself called the sons of Russia to war in a speech of November 1941, invoking the great names of Russia's past. This was a confession that the people might not fight for communism but they would struggle for "Mother Russia". The victory of patriotism was celebrated with this great parade in Red Square, Moscow, on 9 May 1945.**

Bandenkrieg (guerilla war). Soviet authorities steadily expanded their control, the partisan movement acquired its own central staff supervised by Stalin who shrewdly insisted on sending partisans printing paper and duplicating machines, as much a weapon as machine guns for re-establishing a Soviet presence. With 88 million people under enemy occupation the reintroduction of Soviet authority, however, indirect, assumed great importance, discouraging collaboration and encouraging resistance.

Once the USSR was in control of the former occupied areas, "re-Sovietization" followed. Programs frequently included harsh measures – mass propaganda, the closing of churches, the reimposition of collectivized agriculture, the struggle against "nationalist" (ie separatist) movements, notably in the Ukraine and in the Baltic states. Among the younger generation, deprived by the Germans of elementary education, Russian literacy had to be established.

The victory and the cost

Eventually the victory was won. The Soviet system, for all its imperfections, had withstood a test to destruction. Total breakdown, confidently predicted by Hitler, had been averted. Total mobilization, going into high gear in 1942–43, ensured victory. The communist party, initially discredited and displaced, recovered ground as it aligned itself, as Stalin did himself, with a "patriotic culture", though in 1944 the ideological screws were again tightened. The *Komsomol*, the Young Communists League, made an enormous effort, marching to the battlefields and into the factories and suffering a slaughter of the innocents. The younger generation suffered hideously.

Within a population so often "brazenly coerced or rudely abandoned", popular attitudes varied widely, but morale and motivation held up. To some the war actually seemed to be a boon; to others the generation of a "front-line morality", demanding but satisfying, brought relief from the cynicism and parasitism of prewar days. The realization of the dread nature of German occupation policies exercised a powerful, all-pervasive influence: a captured Soviet official put it bluntly: "... in the long term the people will choose from two tyrants the one who speaks their own language. Therefore, we will win the war". The resurgence of Soviet power did not fail to have an impact, though recognition of the cost induced a certain war-weariness in 1943–44, offset in turn by the "promise of better things to come", "a party in our street" (a promise not fulfilled). Communal resilience and personal fortitude, coupled with the encouragement of patriotism, encouraged and deepened popular responses.

The alliances with the USA and UK, whatever the strains, and even the late opening of the Second Front in Western Europe, bolstered feelings that Russia was not alone. To their amazement but deep satisfaction and self-esteem, Soviet people discovered that leadership, discipline and patriotism brought results, engendering devotion to duty and steadfastness in danger.

May 9 was an unforgettable day in Moscow. The sponaneous joy of the two or three million people who thronged the Red Square that evening – and the Moscow River embankments, and Gorki Street, all the way up to the Belorussian Station – was of a quality and a depth I had never yet seen in Moscow before. They danced and sang in the streets; every soldier and officer was hugged and kissed ... Nothing like this had ever happened in Moscow before.

ALEXANDER WERTH

Church, State and People

On the day of the German invasion of the USSR, Sunday 22 June 1941, Stalin maintained a sinister public silence. It was eventually broken in early July, not by Stalin but by Metropolitan Sergius of the Moscow Orthodox Patriarchate. He denounced "the Fascist brigands", vowed that the Church would not forsake "its people" and blessed the "forthcoming heroic exploit" of these same people. This was the first of many unofficial but strikingly effective Church announcements, urging in the name of Russia and the cause of Christianity all-out resistance to Hitler and his pagan hordes.

Far-reaching consequences for Church, State, party and populace flowed from wartime tribulations. If party purists sneered that it was only the "weak-minded" and the "ideologically unstable" who turned to the Church and to religion for individual or collective solace, comfort or hope, this confirmed that ideology alone could not sustain flagging spirits or inspire complete confidence in the Soviet cause.

Stalin himself recognized this, reduced the ideological rasp and embraced Russia's heroic past which he acknowledged as encompassing the Orthodox Church as a staunch defender of Russia. In return the Church not only encouraged morale but made collections of money from the "weak-minded".

On 4 September 1943 Stalin met with the three patriarchs of the Russian Orthodox Church and later concluded a compact or concordat. Though very far from ushering in religious freedom, it at least emplaced the Orthodox Church as a recognized, legitimized institution within the Soviet system.

Ecclesiastical government (the holy synod) was re-established, a limited number of seminaries were opened to meet the dire shortage of priests, the number of bishops was increased and churches long closed reopened. Thanks to the war the Orthodox Church won for itself a legal position which it has never entirely lost: it was a gain of historic proportions.

РОССИИ ДВИНУЛИСЬ СЫНЫ
СТРАШИСЬ, О РАТЬ ИНОПЛЕМЕННЫХ!
РОССИИ ДВИНУЛИСЬ СЫНЫ;
ВОССТАЛ И СТАР И МЛАД; ЛЕТЯТ НА ДЕРЗНОВЕННЫХ,
СЕРДЦА ИХ МЩЕНЬЕМ ВОЗЖЕНЫ.
ВОСТРЕПЕЩИ, ТИРАН! УЖ БЛИЗОК ЧАС ПАДЕНЬЯ.
ТЫ В КАЖДОМ РАТНИКЕ УЗРИШЬ БОГАТЫРЯ"...

900 DAYS: THE SIEGE OF LENINGRAD

Not since the siege of Paris in 1870–71 had the world witnessed the harrowing spectacle of a modern city the size of Leningrad (with over 3 million inhabitants) enduring the nightmarish privations of the besieged. But any comparison stops there. What made the siege of Leningrad wholly unique was not only its duration, 30 terrible months, but also the total involvement of the civilian population of all ages in the city's defense and in the desperate struggle for survival in a huge city trapped between the German and Finnish armies. On 8 September, 1941, German troops cut Leningrad's last land link with the outside world. Hitler, however, set his face against an all-out assault on Leningrad itself and decided instead to starve the city out.

Already, on 2 September 1941, the bread ration had been cut to conserve stocks, the harbinger of dreadful things to come. Starvation loomed up; the first of many to die began dropping at their machines or their desks. Substitute "food" appeared, with cotton-seed oil cake, soup stock from wheat flour and water, and sheep gut for meat "jellies". In the factories still working employees were fed on salted water flavored with a cabbage leaf. Bread, compounded with sawdust, came only as a wafer-thin ration. Medieval horrors reappeared: animals – dogs, cats, or mice – disappeared, all eaten. The mutilation or dismemberment of corpses gave rise to suspicions of cannibalism.

In January–February 1942 between 12,000 and 20,000 (possibly more) individuals died daily: 73,000 died from hunger in February alone. Later in 1942 the decision was taken to evacuate more than half a million people. The population fell to below 1 million by August 1942 and dropped to 637,000 by May 1943. A partial lifting of the blockade eased the grimmest of the privations but salvation came only in January 1944, when the Red Army smashed the German ring.

Leningrad had held out against all the odds and at the cost of stupendous losses and gruesome hardship. "Leningraders" held their heads high, refusing to let the city die.

► A Soviet antiaircraft gun and its crew. In September 1941 Leningrad was woefully ill-prepared to face a prolonged siege. The city's food stocks were only sufficient for one month. Fuel supplies for power stations and factories met only peacetime needs. During the fierce winter of 1941–42 the death toll rose steeply, with 73,000 civilians dying from hunger in one month, February 1942, alone. The sub-zero temperatures, however, spared Leningrad the horrors of an epidemic. Throughout 1942–43 the city struggled to survive and it only barely managed to do so.

► After an air raid. German pressure on the besieged city included constant bombing and shelling. In July 1941 Hitler had decided to abandon a frontal armored assault on the city, insisting rather that Leningrad (like Moscow) would be bombed and shelled into rubble. In the course of the siege, German long-range guns killed 65,000 inhabitants of the city.

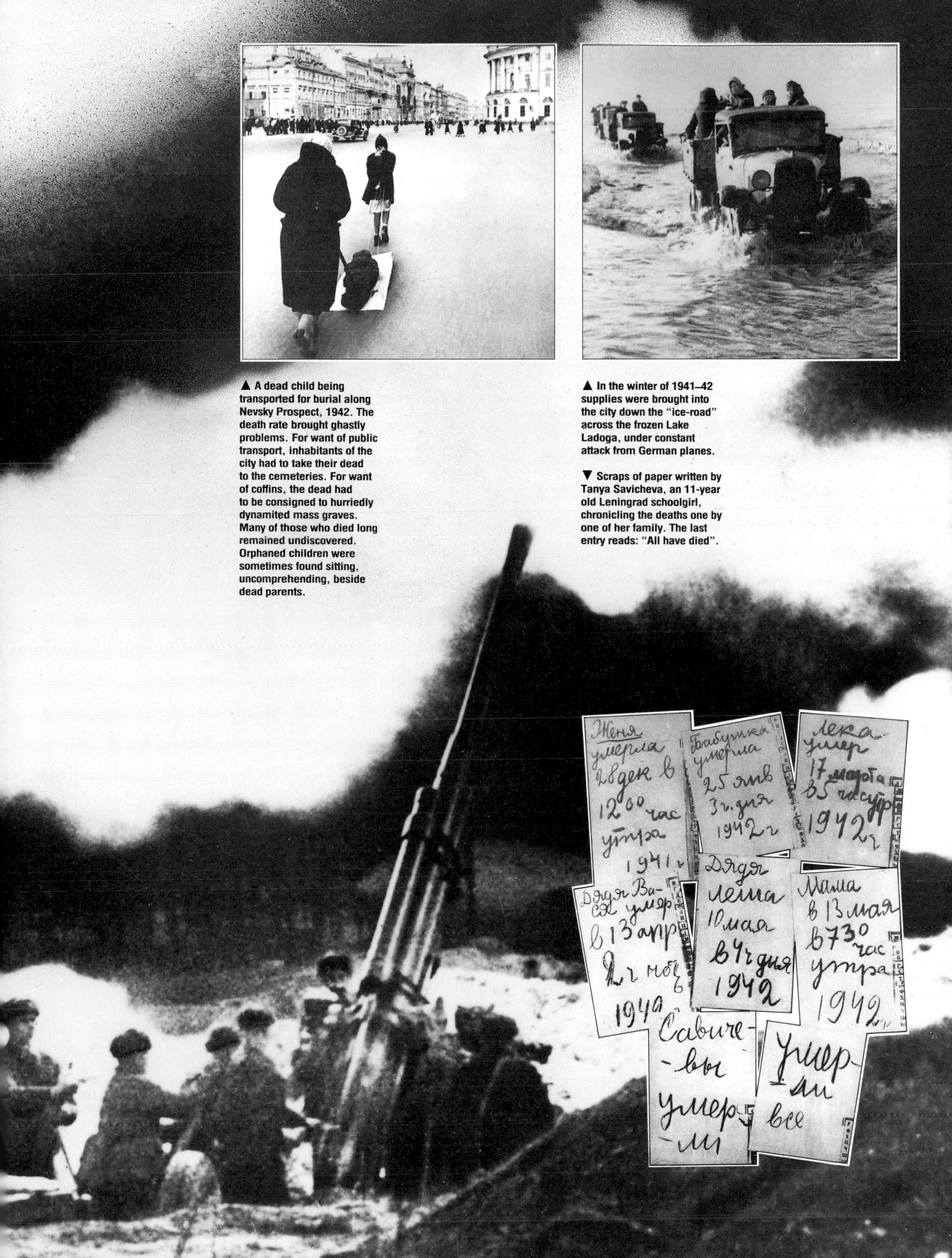

▲ A dead child being transported for burial along Nevsky Prospect, 1942. The death rate brought ghastly problems. For want of public transport, inhabitants of the city had to take their dead to the cemeteries. For want of coffins, the dead had to be consigned to hurriedly dynamited mass graves. Many of those who died long remained undiscovered. Orphaned children were sometimes found sitting, uncomprehending, beside dead parents.

▲ In the winter of 1941–42 supplies were brought into the city down the "ice-road" across the frozen Lake Ladoga, under constant attack from German planes.

▼ Scraps of paper written by Tanya Savicheva, an 11-year old Leningrad schoolgirl, chronicling the deaths one by one of her family. The last entry reads: "All have died".

Datafile

Although far from the fighting, the American home front experienced great and transforming changes: substantial internal migration resulted in permanent shifts in the population, the rapid increase and improvement in health care led to a healthier and longer-living citizenry, and the growth in industrial production fueled by the war consolidated the USA's position as the world's leading industrial power. Less amenable to counting were changes in attitudes: the birth of the teenager, the change in the expectations of many women, and the tendency for the electorate to ask not what they could do for their country, but what their country could do for them.

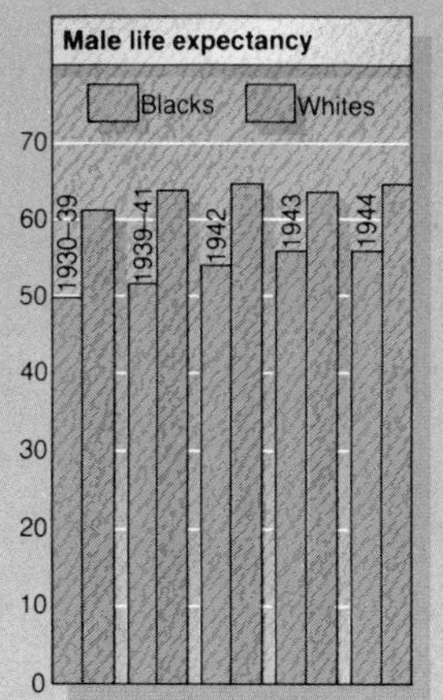

▲ The development of health care and the increase in well-paid jobs both contributed to the rapid increase in life expectation of black and white males – although the black male had considerably further to go.

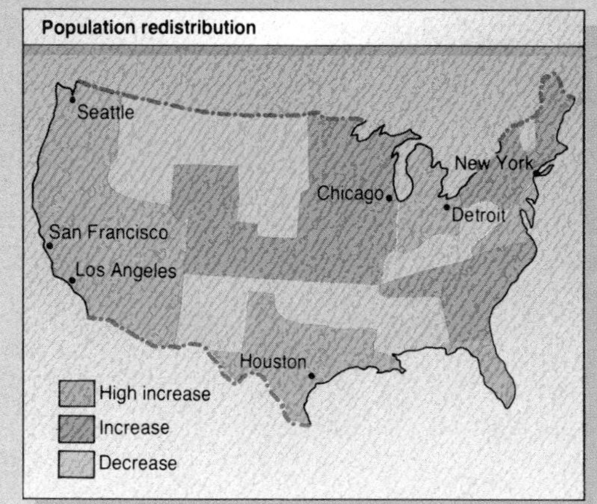

◀ One reason for the improvement in the lot of the black population was migration from sharecropping and scrub farming in the South to better jobs in the cities of the South and North. The white farm population also joined the rush.

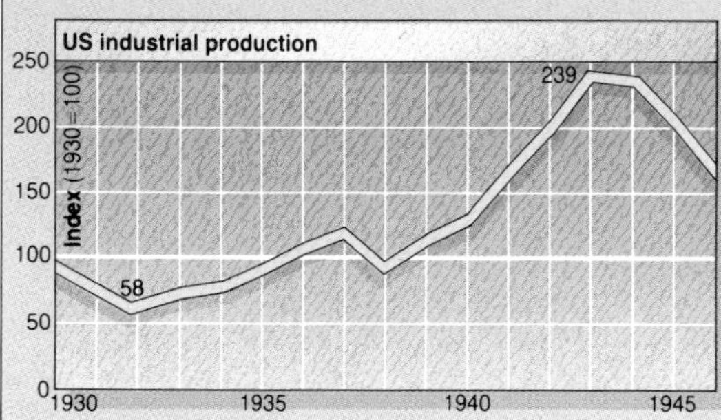

◀ The chart well demonstrates the role played by the war in pulling the USA out of the Depression. In 1932 the US government's Index of Industrial Production stood at 52; by 1940 it was 125 and in 1943 239. Production then began to fall as the USA began if anything to overproduce: by 1946 it was back to 170.

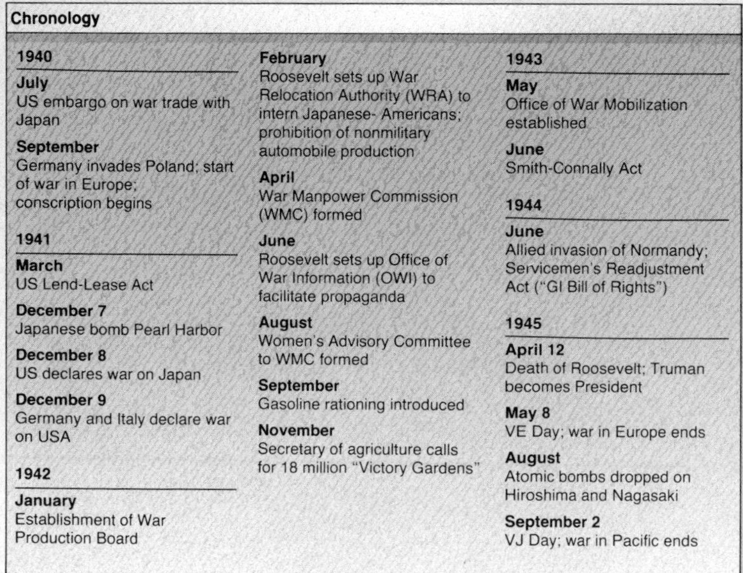

Chronology

1940

July
US embargo on war trade with Japan

September
Germany invades Poland; start of war in Europe; conscription begins

1941

March
US Lend-Lease Act

December 7
Japanese bomb Pearl Harbor

December 8
US declares war on Japan

December 9
Germany and Italy declare war on USA

1942

January
Establishment of War Production Board

February
Roosevelt sets up War Relocation Authority (WRA) to intern Japanese- Americans; prohibition of nonmilitary automobile production

April
War Manpower Commission (WMC) formed

June
Roosevelt sets up Office of War Information (OWI) to facilitate propaganda

August
Women's Advisory Committee to WMC formed

September
Gasoline rationing introduced

November
Secretary of agriculture calls for 18 million "Victory Gardens"

1943

May
Office of War Mobilization established

June
Smith-Connally Act

1944

June
Allied invasion of Normandy; Servicemen's Readjustment Act ("GI Bill of Rights")

1945

April 12
Death of Roosevelt; Truman becomes President

May 8
VE Day; war in Europe ends

August
Atomic bombs dropped on Hiroshima and Nagasaki

September 2
VJ Day; war in Pacific ends

The American home front was turbulent and disorganized, but productive. Mobilization, of both the economy and the labor force, developed over a long period and intentions were never completely implemented. It produced social tensions, between races, between capital and labor, between army camps (and munitions-making boom towns) and local populations, and between the generations. Civil liberties were an early casualty. But order eventually emerged out of chaos, and by good luck, an abundance of resources and planning, the USA developed into the Arsenal of Democracy, producing a substantial proportion of the arms, munitions and ships used by the Allies. The USA proved to be a good example of the truism that war is an engine of economic, social and political change. During World War II the average standard of living vastly improved, and the relationship of the citizen with the government was transformed. Seen in a long perspective, World War II produced social developments that were more notable than the economic achievement.

Business and mobilization

Because the USA contained a very vocal isolationist and antiwar element, and had remained neutral during the early days of Japanese and German expansion, there had been little preparation for war. At first, mobilization of the economy was difficult to effect. American business had suffered profoundly during the Depression (1929–36), and once the economy began to pick up in the first years of the war (1941–43), it was loath to forgo production for the profitable domestic market in order to concentrate on war-related contracts. The automobile industry, for example, anticipated selling 4 million cars in 1941: consequently it fought against converting factories for the production of military aircraft instead. While the country was neutral, the political costs were too high for President Roosevelt to attempt to force business to act against its will. Once war was declared (8 December 1941), new rules came into play. But throughout the war big business successfully insisted that war-related contracts be drawn up with provisions which ensured very healthy profits.

The economy was never fully mobilized. During 1940–41 reluctance to convert was a key reason; thereafter the reluctance of major industries (such as steel, electric power, automobile manufacture and railroads) to expand capacity constituted a drag. Businessmen had acute fears that if they did so expand, postwar markets would be flooded, and yet another depression would ensue. Consequently, after wartime expansion, production was hardly greater than output during 1921–25. Unemployment did not dwindle until mid–1943.

Major characteristics of American wartime mobilization

The reluctance of business and industry to convert for war production

Wartime's social problems: prostitution and migration

The internment of Japanese-Americans

Capital and labor in wartime

Citizens and the state: the growth of "big government"

The slack in the economy can be illustrated by looking at the mobilization of the female population. From 1940 British women were subject to the draft, and 70 percent were engaged in full time work. In the USA the figure was 29 percent: in short, 71 percent of the female population over 18 stayed at home. This was partly due to male insistence that they were fighting for the right of women to remain in their traditional sphere; as one soldier protested, if, for example, women joined the armed services, American men "would throw away our own self-respect – our right to pledge in earnestness to 'Love, Honor and Protect' the girls we want to marry when we get back." This sentimental servitude, however, could not have been imposed if the economy and the armed forces had seriously needed their services. As it was, at the peak of war production in November 1943 the country was turning out 6 ships a day and 6 billion US dollars' worth of war-related goods a month; a year later many such factories were being closed, as output exceeded requirements.

Wartime social problems

Mobilizing the economy to such an extent caused great upheavals and social tensions. New factories were built and old ones enlarged, with the result that many areas of the country were transformed. The mushrooming of munitions plants produced boom towns to which thousands flocked for work. Migrants suffered a lack of adequate housing, health provision and schools, while inhabitants saw their home towns change from prudish small places into rambunctious and immoral boom cities.

▼ **Employees in Chattanooga form a "V" for victory, to celebrate their work on defense orders.**

One result was the widespread phenomenon of teenage prostitution. Many were in fact as young as 12 years old; they were called Victory Girls, Patriotutes, Cuddle Bunnies or Roundheels. When arrested by the police, they frequently claimed to be performing a patriotic service by maintaining military morale. However, it was a service sometimes performed for free: federal housing projects spawned what were called "wolf packs" – clubs for teenagers where girls paid their dues by copulating with every male member. One historian has noted that "one determined young initiate joined a club with ninety members."

These developments were amongst the saddest but least surprising results of mobilization, since by mid-1942 there were hundreds of thousands of "eight-hour orphans" wandering the streets of the cities and boom towns, their latchkeys frequently hanging around their necks. The fact that it was "teenage" prostitution which was decried is notable. Before the war there were two age categories for the population, adults and children; during the war an adolescent subculture developed. By 1945 teenagers as such attracted a considerable amount of parental, sociological and advertising interest; several magazines devoted to their interests had sprung up, such as *Seventeen*.

Mobilization also encouraged a substantial internal migration. Sixteen million men and women left home to join the armed services, and another 16 million moved around the country, either wives and sweethearts following the servicemen, or workers moving to other jobs. A million blacks moved from rural areas in the South, some to southern urban areas, but a substantial proportion to urban areas in the North and West. The West, and particularly the Pacific Coast, was the main gainer in the great wartime population shifts. Shipbuilding and munitions-building centers drew workers in particular to California, which rejoiced in a 37-percent gain in population during the war. Rural blacks were not alone in leaving the land for the bright lights: the farm population as a whole fell by 20 percent.

▼ The USA as the Arsenal of Democracy. President Roosevelt called for the USA to assume this role on 29 December 1940. His purpose was to encourage the public to buy Victory Bonds. The V-for-Victory campaign, launched by the BBC to encourage people in the occupied countries, swept the USA in 1941.

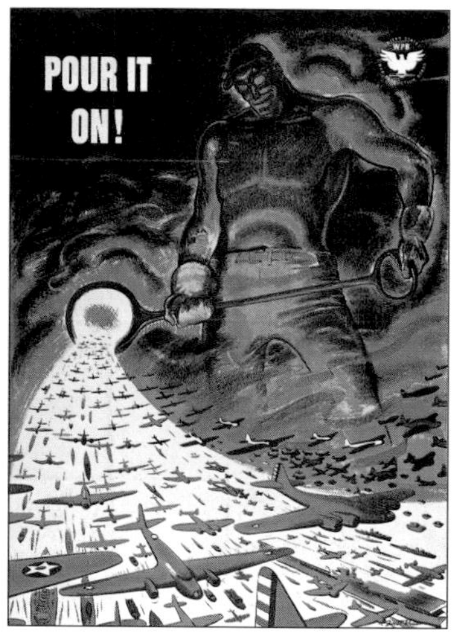

▼ This man is buying seeds for his Victory Garden. Such gardens were an important source of food by late 1942, given the flight from the farms.

Indeed, the stampede of workers from the land to the towns might well have caused a serious food crisis, but for the Victory Gardens. At first they sprang up spontaneously, but the farm security administration then launched "An Acre for a Soldier" in rural areas. The idea was that the cash proceeds from the crop grown on that acre would be devoted to such things as soldiers' canteens. City dwellers pitched in as well and by April 1942 more than 6 million people were digging "for vitamins and victory". At first the federal department of agriculture was pretty scathing about the movement, but as the number of military conscripts increased and others moved to well-paid jobs in the munitions factories, ordinary agriculture faced a serious crisis in manpower as early as 1942. In fall 1942 the secretary of agriculture called for 18 million Victory Gardens. The gardens were now seen as a resource that might help to offset the threatened loss of millions of tonnes of food.

The economic results of the migration from the land to urban areas, then, were beneficial: production was increased and overpopulation of the land was eased. But the consequence was social problems, particularly as a result of the shift in the black population. Racial tensions erupted in areas of the country where they had previously been unknown. Conflict followed, sometimes violence.

Over one million blacks joined the armed forces, but segregation into separate fighting units, and the relegation of blacks to the most menial duties, ensured deep-burning resentment. Similarly, there was an increase of 800,000 blacks working in manufacturing, but there too their positions were overwhelmingly low-paid and menial. What this did was to underline for all to see the gap between American protestations of equality and democracy, and the reality for those of a different color or race. (Japanese-American citizens suffered even more acutely, those on the West Coast being rounded up and herded into concentration camps.) The reaction of blacks to their treatment during the war may have been the basis for the civil rights protests of the 1950s and 1960s.

Living standards and the state

In short, minority groups suffered, but so did those in the majority, if they were perceived by those in power to be working against the war effort. One such group was labor, both organized and unorganized. Attempts by businesses to ensure that they did not suffer losses, but indeed profited, during the war were seen as perfectly natural; attempts by workers to ensure that their wages kept up with prices and that they retained the freedom to move jobs were seen as attacking the war effort. One result of the New Deal recovery program during the 1930s had been to bring labor into the center of the political process. It had been acknowledged that the political wishes of the workers were as important as those of business interests. During the war this position came under attack, from business and from a Congress dominated by the Republican Party. They combined in an attempt to turn back the clock. In 1943 the Smith-Connally Act was passed,

which forbade unions from contributing to candidates for office in national elections. The unions' response was to set up the Political Action Committee to canvas and get out the vote for favored candidates.

Indeed, labor managed largely to sustain its recently won political position because of its crucial role in the war effort. Workers' patriotism was frequently much more obvious than that of business: for example, inmates of San Quentin prison in California staged a near-riot to support their demand that they be allowed to produce war materiel. Skilled labor was relatively scarce, and employers found it necessary greatly to improve working conditions in order to keep their work forces. Certainly the economic position of labor was far better in 1945 than it had been in the late 1930s.

Overall, in spite of the many problems which erupted during the war, by 1945 the USA was in

◀ The proportion of women over 18 who worked outside the home – only 29 percent – was much lower in the USA than in any other belligerent country save Germany. Of those who worked in industry, the proportion who were black increased enormously. At the same time, the numbers of black women who worked as farm laborers and domestic servants dropped sharply.

▼ This Dodge plant near Detroit produced army truck bodies, but the lack of a fast-moving production line and the relaxed attitude of the workers produced a scene different from that of today. At least some of these workers had either been unemployed or working in lower-paid jobs before the war.

Japanese Internment

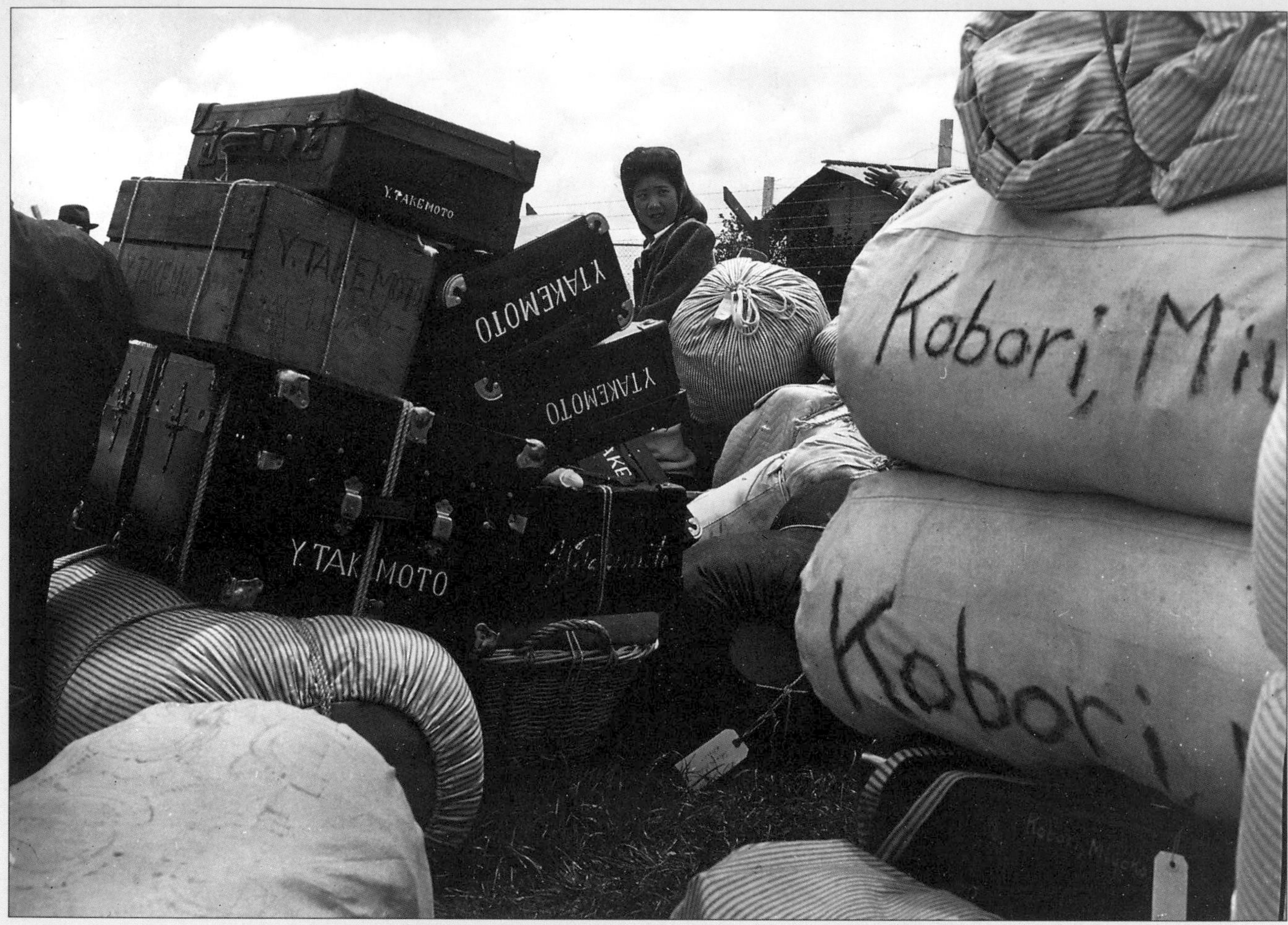

▲ A young evacuee of Japanese ancestry identifies her baggage at the assembly center in Salinas, California, prior to being sent to a War Relocation Authority Center. She may well have ended up in one of the ten concentration camps set up to incarcerate 110,000 persons of Japanese ancestry, the majority of whom were American citizens. It took the American government a generation to admit that a great injustice had been done to a group of citizens against whom no charge of attempted subversion or sabotage was ever made.

One consequence of the bombing of Pearl Harbor (7 December 1941) was the American government's decision that all those of Japanese ancestry living on the West Coast, whether immigrants or native-born American citizens, constituted a threat to national security and should be removed from their homes and incarcerated in camps. The US army argued that no military necessity required such a step, and the FBI had in any case already rounded up those Japanese aliens whom it deemed dangerous. But the idea of a security threat was pushed by Henry Stimson, the secretary of war, and by his assistant secretary, John J. McCloy, with the support of various politicians and pressure groups. President Roosevelt supported the civilians' assessment of danger and in February 1942 signed the order authorizing the establishment of a War Relocation Authority (WRA) to move the Japanese-Americans away from the Pacific Coast.

On the ground that speed was of the essence, no hearings were granted to the "evacuees", and within weeks 110,000 residents of Japanese ancestry, 65 percent of whom had been born in the USA, were resettled in ten WRA camps. The WRA insisted that there was no coercion, but the reality was camps surrounded by barbed wire and watchtowers, forced loyalty oaths, attempted dispersal of the internees all over the USA to "Americanize" them and in general, their treatment virtually as enemy aliens.

There is a good deal of evidence that the wartime fear of subversion was driven by racism, since there was no general round-up of German-Americans or Italian-Americans purely on the basis of their nationality. A statement by the governor of California, Culbert L. Olson, in February 1942, is enlightening: "You know, when I look out at a group of Americans of German or Italian descent, I can tell whether they're loyal or not. I can tell how they think and even perhaps what they are thinking. But it is impossible for me to do this with the inscrutable orientals, and particularly the Japanese."

The task of the WRA was aided by the cooperation of the leadership of the Japanese American Citizens League and of the national leadership of the American Civil Liberties Union. It was left to the Northern California branch of the ACLU, by arguing the cases of some of the internees, to make clear publicly the illegality of the WRA's procedures.

On 17 December 1944 the war department announced the revocation of the order excluding Japanese Americans from the West Coast, and on the following day the Supreme Court ruled that the WRA had no legal authority to detain a "concededly loyal" US citizen and no legal authority to impose conditions for his release.

economic and social terms a much fairer place than before the war. It is clear that civilians in the aggregate did not suffer serious inroads into their material standard of living: on the contrary, total personal consumption rose in real terms during the war. Expenditure per head on food, for example, rose from 560 US dollars in 1941 to 700 in 1945. Overall, real disposable personal income rose 25 percent during the same period. What is equally important is that the economic benefits were spread across the board. The economy expanded substantially, and from 1941 every group benefited. By 1945 the average annual income was almost 3,000 dollars, more than double what it had been in 1939.

One of the most important developments during the war was the very great increase in medical provision. Medical care had been wholly inadequate before the war: although the population had increased by nine million between 1929 and 1939, the number of doctors and dentists had remained static. Added to this were hunger and pollution, and limited public-health facilities. Consequently, the physical condition of a substantial proportion of the recruits for the army and navy was appalling: the overall rejection rate was 50 percent, rising to 70 percent in some areas. One benefit of the 1939–41 defense boom was the building of thousands of new health clinics and community hospitals. But in many cases, no sooner were they built than they lost their doctors, drafted into the armed services. By 1943 there was one doctor for every 100 servicemen and one doctor for every 3,500 civilians. This was ironic, because the civilian death rate of over 10 per thousand in 1941 was higher than that of the armed services, at only 5 per thousand in 1942–45.

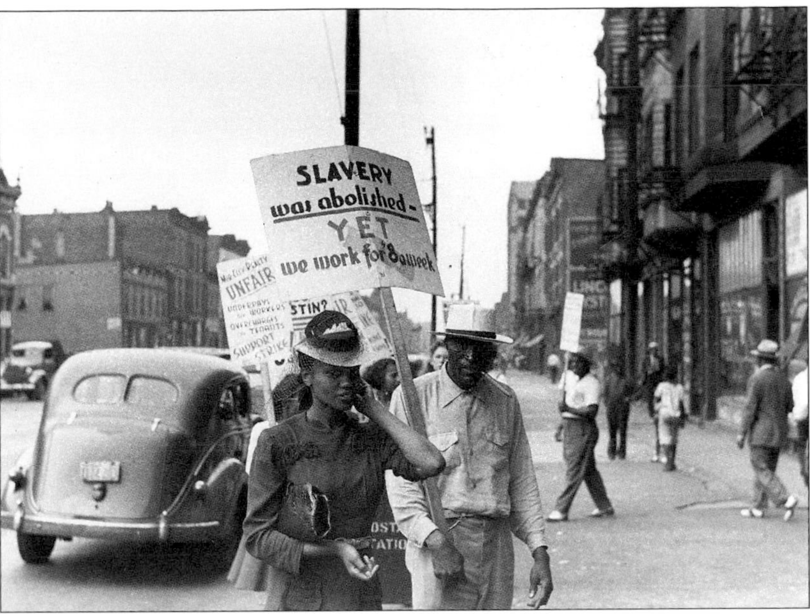

▲ Employers and white trade unions combined to freeze blacks out of defense industries, or to restrict them to low-paid jobs. Black leaders threatened a march of 50,000 to 100,000 blacks on Washington in July 1941 to demonstrate that American democracy was a mockery. Eleanor Roosevelt, the president's wife, convinced them to call it off, but other strikes, such as this one in south Chicago, continued.

Improving health care was given a much higher priority than it had received in peacetime. By 1944 the numbers of doctors and dentists graduating were double the highest numbers in any prewar year, while the number of hospital beds per thousand people increased by 50 percent. Before the war the American Medical Association had objected (solely from greed) to such improvements as health insurance and group practices. Now they were swept away by court order, and by the end of the war the amount of money spent on accident and health insurance was 100 percent greater than in 1939. The results were astonishing: life expectancy rose by three years in the period 1939–45, and for blacks by five years, while the death rate for infants dropped by more than a third.

These wartime gains were consolidated by the Servicemen's Readjustment Act of June 1944, popularly known as the "GI Bill of Rights". This provided a wide range of benefits for the 16 million veterans and their families, who by 1950 constituted about one-third of the population. Veterans were entitled to government help to attend high school or college, to buy a house or take up farming. Millions availed themselves of the provision, particularly to get an education. The results of this aid – welfare wholly without stigma – was to enhance "a powerful wartime trend to the creation of a genuine middle-class democracy".

The GI Bill of Rights was symbolic of another permanent change in American life: the transformation of the relationship between the state and the citizen. Before the war, contact between the average citizen and the government hardly occurred. But contact was regularized during the war, with the draft, social security and the office of price administration. Furthermore, the growth in wage and salary levels and the lowering of the tax floor ensured that many came into contact with the internal revenue service for the first time. The availability of veterans' benefits extended this trend. In short, the war saw the growth of "big government", a permanent development of American life.

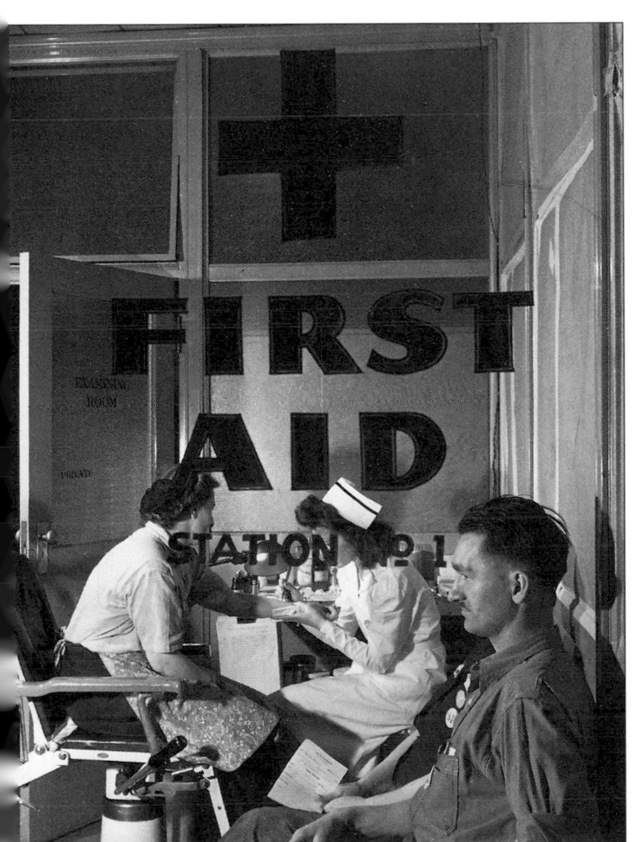

◀ The health crisis was probably most acute in the war industries: its workers had frequently been rejected by the draft for health reasons, the machinery was often dangerous, and conditions could be squalid. Notable exceptions were those run by Henry J. Kaiser. He provided a modern clinic or hospital for each factory or shipyard, and pioneered a health insurance plan for all of his workers and their families. In other war plants, the presence of women workers benefited all, in that sanitary and safety conditions frequently improved dramatically: hot meals might be provided, and operations requiring heavy manual labor were often mechanized. This aircraft factory had first-aid stations dotted throughout the plant.

PART 4

THE FRONTLINE CIVILIANS

Datafile

Bombing campaigns against civilian populations began in World War I. Raids made then caused little damage but had considerable psychological impact. The experience led to two contradictory strategic positions: first, a belief in the power of bombing to devastate enemy civilian populations, destroy morale, and bring about defeat; second, a fear of bombing, which stimulated preparations against bombing raids. In World War II the initial Japanese and German successful use of tactical bombing in China and western Europe seemed to confirm the efficacy of bombing. But the larger and sustained bombing campaigns of the Allies failed to destroy morale.

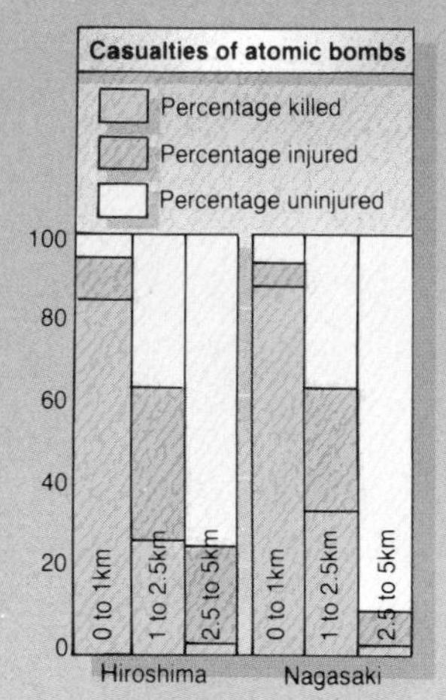

▲ **The atomic bomb, first used in 1945, raised the power of bombing from disruption to obliteration. Chemical bombs produced local explosions. An atomic "explosion" created a huge yield of energy. It almost totally destroyed the square kilometer around the blast and sent out waves of intense heat and radiation up to 5km (3.1mi). Two bombs were dropped before Japan surrendered. They had slightly different impacts – Hiroshima was flat, Nagasaki hilly – but both took a ghastly toll of human life.**

► **The impact of bombing was determined more by the existence or absence of protection than by the number of bombs dropped. The short US bombing campaign against Japan caused far more casualties than the Allies' sustained bombing of Germany. The Japanese urban population was concentrated in lightly built cities, where it was exposed to enemy planes for want of defenses.**

Bombing campaigns

10% 4% 85%

Tonnage dropped on countries
Total 1,591,000 tonnes

7% 35% 58%

Civilians killed by bombing
Total 860,500

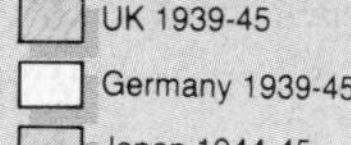

► **The bombing campaign against Germany in 1943–45 was mounted by RAF Bomber Command and the 8th and 15th USAAF. Bomber Command replicated the German campaign against the UK of 1940–41 with its "area bombing" of German cities. The USAAF concentrated on specific industrial and military targets.**

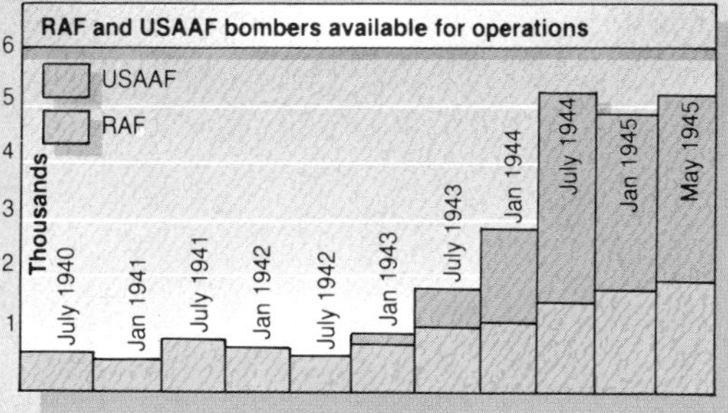

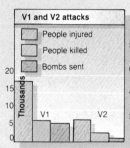

V1 and V2 attacks

People injured
People killed
Bombs sent

Thousands
20
15
10
5
0
V1
V2

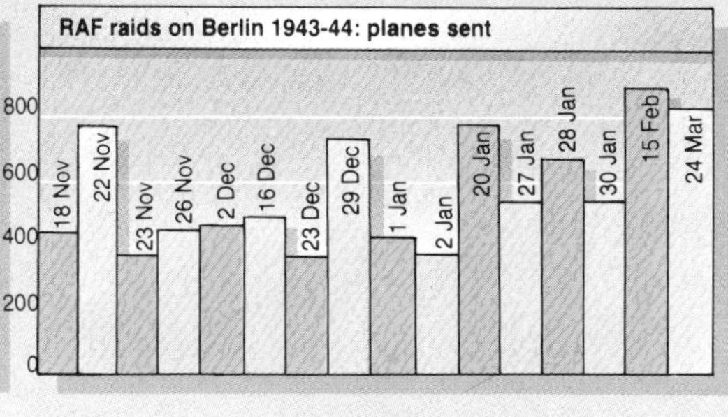

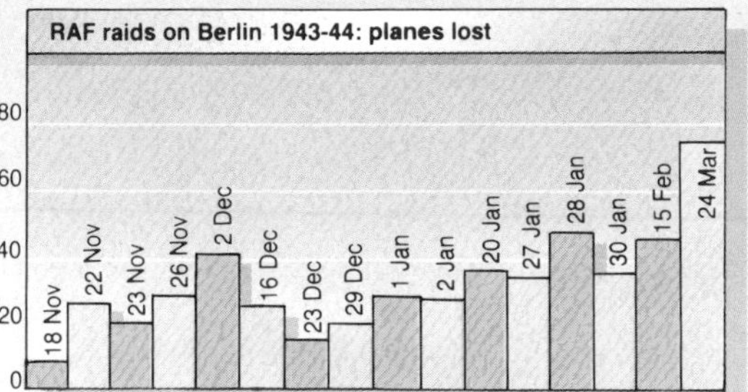

►▲ **In November 1943 the RAF began a major assault on Berlin, the German capital. Increasing numbers of planes were sent, but losses mounted and soon the "Battle" had to be called off. Hitler's "secret weapons", the V1 flying bomb and the V2 rocket, also created initial shock but low casualties overall.**

In January 1942, when the Japanese were advancing toward Singapore, some 600 civilians were killed and over 1,500 injured by bombs. A report in the Melbourne *Herald* of 19 January stated: "Europeans here have given up talking about 'white' and 'coloured' races... The European ARP workers, who braved death alongside Singapore's splendid body of Asiatic wardens, roof-spotters and fire-fighters, have learnt things not easily forgotten. European women shielding children in the same shelter with Chinese mothers, who have exchanged smiles of relief as a stick of bombs passed a few hundreds of yards away, have discovered many things which will not vanish when Singapore's ordeal has passed."

The rhetoric here may seem eerily familiar. Substitute "middle-class" and "working-class" for "white" and "coloured" and this would read exactly like many reports filed by anti-Nazi US pressmen in London in the fall of 1940. The Australian reporter's optimism was wholly ill-founded. Not only did Singapore fall, but the "Asiatics" whom the Europeans called "coolies" tended to desert their work when air raids were threatened in Japanese broadcasts and leaflets: little had been done to provide shelter for them.

The effect of high explosives on human flesh is horrible. Is every use of it against civilians morally evil? What if these civilians, in "total war" are proud to be "frontline" fighters like their soldiers, as was said of Londoners during the Blitz? Such questions require debate. Bombing in World War II was used in several different ways: with precision against nonhuman targets of military significance; like artillery against enemy troops; in cooperation with land forces; or in deliberately indiscriminate assault on enemy populations.

Aerial bombardment was less than 30 years old in 1939. It was first used by Italians against Turks in Tripoli in 1911-12. In the Balkan wars which followed, a Bulgarian aeroplane dropped 30 bombs on Adrianople in one day, causing only six casualties. The technique appeared somewhat futile. But in January 1915, zeppelin airships of the German navy opened the first "strategic" air offensive. By the end of the war there had been 52 raids by these slow and vulnerable dirigibles, which killed 498 civilians and 58 military and injured 1,357 people. In May 1917, 21 Gotha bomber planes appeared over Kent and killed or injured nearly 300 people in 10 minutes. Twenty-six more Gotha raids in the next 12 months brought the tally of casualties to over 4,000 civilians.

Douhet and theories of bombing

As air forces expanded after the war, extrapolation from such figures suggested that the bombing of cities could cause huge of casualties and

THE BOMBING OF CITIES

The origins of strategic bombing

Douhet's theories and their impact

The story of the Blitz

Allied bombing of Germany: force and effect

The bombing of Japan

The first use of the atomic bomb: bombing brings results

◀▼ Some scenes of war are almost interchangeable. This fire, left, happens to be in London (1940–41). Moscow citizens, below, took shelter in the underground railroad system. The photograph insists that there was good order in Stalin's Metro.

that defense against it was difficult if not impossible. In 1930 an Italian, General Giulio Douhet, crystallized thinking with his claim that armies and navies were now best used for defense while air forces first destroyed opposing fleets in the air, then killed, maimed and demoralized the enemy populations till their clamor forced governments to sue for peace.

Events in the 1930s gave credence to Douhet's predictions. The bombing of Shanghai by the Japanese in 1932 caused Stanley Baldwin, the British Conservative leader, to conclude that "the bomber will always get through. The only defence is offence, which means that you have to kill more women and children more quickly than the enemy if you want to save yourselves." The British Royal Air Force developed long-range bombers. However, the attack which seems most to have focused fears of aerial bombardment was conducted by German pilots cooperating with an army offensive by General Franco's troops in Spain (and also carrying out an experiment in the effect of their current weapons). On 26 April, 1937, a market day, the center of Guernica, the Basque capital, was destroyed, at a human cost of 2,500 casualties.

In Germany Hermann Goering had been impressed by Douhet's ideas, but his air force, in 1940, did not comply with their requirements. It was designed for close cooperation with fast-moving German armies, in Blitzkrieg, "lightning war", and lacked long-range bombers. However, the Dutch capitulated swiftly when their capital and Rotterdam were bombed.

Meanwhile, in May 1940, after Churchill became prime minister of the UK, the long-range RAF bombers had been unleashed; the UK had committed itself to a policy of "strategic" night bombing of Germany. That the British raids on the Ruhr in May and June 1940 were wholly ineffectual should not obscure the fact that they happened.

The UK and the Blitz

Fortunately for the UK its fighter command was just able to withstand the assault of the short-range German Messerschmitts. Confusingly the British date the start of their "Blitz" from the point where German Blitzkrieg faltered. Goering should have continued to concentrate the German air force's attack on fighter aerodromes. But on 7 September, London was the target. Huge fires blazed in the city's East End. The officer in charge of an inferno in the Surrey Docks involving 100ha (250 acres) of timber sent a message to his superiors: "Send all the bloody pumps you've got; the whole bloody world's on fire." There were over 2,000 casualties on the first night of

▲ A street in Clydebank, near Glasgow in Scotland, after the Blitz of 13–14 March 1941. The photographer's skill has given the shot a hint of optimism. No other British town suffered such extensive damage. But this was Scotland's only really severe experience of bombing.

▲ A British rescue worker carries a raid victim. Many of those employed in rescue teams had been building workers in peacetime, intimately acquainted with the construction of dwellings – and with hard physical toil.

the Blitz, 430 of them dead, and thousands of East Enders were made homeless. For 76 consecutive nights (barring November 2) London was raided, usually heavily.

Again, had the German air force concentrated on bombing the East End, with its poor and ethnically mixed population, working-class morale might have been decisively affected. Some government ministers feared that it was anyway. But the sheer range of places in the vast city hit by accident or design encouraged the conviction that all Londoners were in it together. As it was, the war in the air turned in fighter command's favor in mid-September. Hitler decided to postpone invasion of Britain and turned his attention to the USSR. London was given ample chance to prove that sustained bombing of a nation's capital was not enough in itself to bring victory.

The Blitz on Britain falls into four main phases: (1) from 7 September to the end of the month, when invasion was still possible; (2) October and November, when Londoners' morale was under siege until ... (3) The heavy raid on Coventry, an important center of war industry, on 14 November, which marked the beginning of a phase where the capital was only one target among many; (4) January to May 1941, when London was still much visited, but when bombers tended to concentrate on the western ports which were termini for British supplies from America.

Many quite small places, especially on the east coast, were raided heavily relative to their size. From Orkney to Cornwall, few areas of Britain were really remote from the bombing. The Blitz became a national experience. But London – most would have thought this appropriate – took the first heavy blows and set examples of civilian behavior and morale for the rest to follow.

Why did London's morale remain high, and attract admiration from the whole English-speaking world, including, importantly, the USA, where Roosevelt, standing for his third term, had to convince Americans that his anti-German posture was justified? In the first place, during the "Battle of Britain", many parts of London had been bombed as the German air force searched for fighter-command aerodromes. Awesome though the September Blitz was, people had seen bomb damage and heard of casualties before. Civil defense and post-raid preparations, though much criticized, proved just adequate. Furthermore, the earliest phase of the Blitz coincided with victory for fighter command. When Churchill broadcast to the people on 11 September, he announced that present events compared in importance with the crises in which England had seen off the Spanish Armada in 1588, and Britain had stood firm against Napoleon. One of his younger

Early last evening the noise was terrible. My husband and Mr P were trying to play chess in the kitchen. I was playing draughts with Kenneth in the cupboard... Presently I heard a stifled voice "Mummy! I don't know what's become of my glasses." "I should think they are tied up in my wool." My knitting had disappeared and wool seemed to be everywhere! We heard a whistle, a bang which shook the house, and an explosion... Well, we straightened out, decided draughts and chess were no use under the circumstances, and waited for a lull so we could have a pot of tea.

LONDON, NOV. 1940

female subjects had already committed to paper feelings which many shared. A couple of nights before, frightened sitting indoors by the thump of bombs, she had gone out to survey London from Hampstead. "It was a beautiful summer night, so warm it was incredible, and made more beautiful than ever by the red glow from the East, where the docks were burning. We stood and stared for a minute, and I tried to fix the scene in my mind, because one day this will be history, and I shall be one of those who actually saw it. I wasn't frightened any more..." In this context, even the behavior of those people, mostly poor, who sought illicit shelter in the deep underground stations of the London Tube, mysteriously became acceptable. The 177,000 Londoners who were sleeping in the Tube in late September, simply by being in the city, were Making History, as they calmly arranged their bedding on station platforms. Special trains were soon taking better-off people to Chislehurst Caves, in Kent, which had been developed as a paying private enterprise. They had their own church, electricity, barber and entertainment.

But it is impossible to say whether London "morale" would have stood up under any circumstances. The only safe conclusion is that it was volatile. The flow of adrenalin might enable individuals and groups to endure, even perversely enjoy, flames, wrecked buildings, corpses in the open, but depression might well follow. In this connection it was crucial that the German raids were not only relatively light by standards achieved by the RAF itself over Germany later in the war, but widely dispersed over a very large city.

By contrast, Coventry was a smallish city out of which, on 14 November, the German air force seemed to have ripped the heart – to express which German propaganda coined a new verb, *coventrieren*, "to coventrate". Casualties were over 1,400; 40ha (100 acres) of the city center were destroyed. Several hundred shops were put out of action. As refugees – "trekkers" – poured from the city, those who stayed had to eat from emergency mobile canteens.

Coventry set a pattern for the Blitz outside London at its most severe, in two respects. Firstly, morale was severely affected by the hammering of a city's center, especially if, as in the case of Plymouth in March and April 1941, the raiders came back again and again. Even in the huge Merseyside conurbation which included Liverpool, eight successive nights of bombing – the "May Week" of 1941 – produced torrents of trekkers and wild rumors of huge casualties and social collapse. But, secondly, as in Coventry, "war production" was little affected.

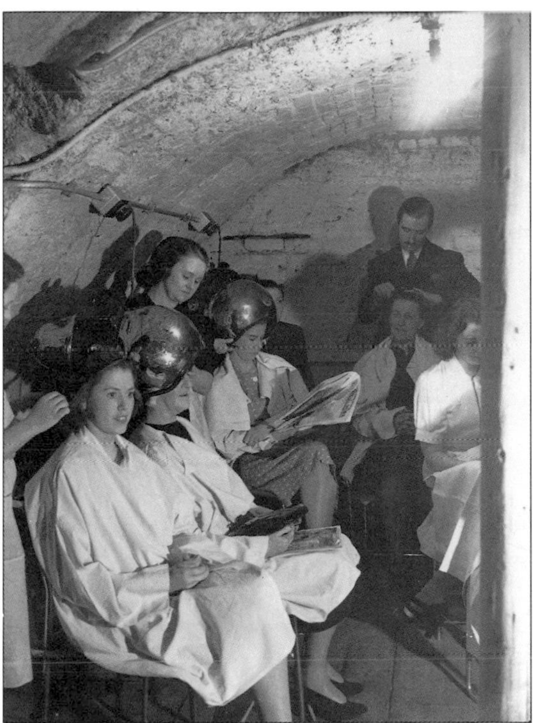

►▲ Work continues after the air-raid warning has sounded, in this shelter located under a hairdressing establishment in London's fashionable West End. Carefully posed pictures such as this helped sustain the conviction that normal services must and would continue. However, a barber whose small shop had disappeared might well have been angered by the contrast with his own lot.

▼ Britain's "Queen's Messenger" convoys of mobile canteens – 18 fleets of 12 vehicles each, ready to rush to any bombed town were almost all provided by American generosity. They went into action in 1941, but this one is helping out after a V1 has struck, in July 1944.

Events in the small shipbuilding town of Clydebank, near Glasgow, exemplify why this was so. Before two successive nights of raiding hit Clydebank in March 1941, labor relations in the important John Brown's shipyard had been conspicuously bad. Now all but seven of the town's 12,000 houses were damaged; 35,000 out of 47,000 people were homeless; the night population dropped to 2,000 as the inhabitants took to the moors. Yet "a good proportion" of John Brown's men turned up for work on the morning after the Blitz, and the vast majority came back after an average absence of a fortnight.

The Allied onslaught on Germany

If London could take it, Berlin could take it. If Coventry could survive heavy bombing, so could Hamburg. The British learnt only after the war that the policy of "area" bombing by night, which was favored by Churchill and his favorite scientist, Frederick Lindemann, who argued that flattening German working-class houses would break morale and bring victory, was ill-conceived – and immensely wasteful of the lives of brave bomber crews.

In April 1942 RAF bombers raided Rostock, a beautiful old town. They destroyed a lot of buildings, but killed just 204 people. Rostock was back to nearly a hundred percent production after only two days. On 30 May of that year, 1,043 bombers flew for Cologne. That this was a "Thousand

The Bombing of Dresden

The raid on Dresden by British planes on 13–14 February 1945 became at once a fiercely controversial episode. Harris, head of bomber command, believed that terror raids on German cities could win the war. The USSR wanted sharp raids to assist its westward advance. Churchill, at this stage, was actively in favor. In two successive attacks 805 British bombers poured 2,600 tons of bombs into Dresden's center. The second attack was carried out in excellent conditions and achieved exceptional accuracy. The Americans followed this up with daylight raiding which pinpointed the railway yards.

So far, so rational. But even before the full scale of the raid's results was appreciated, Churchill began to recoil from it, referring to it in a minute as "a serious query against the conduct of Allied bombing." Why was this?

First, though Dresden was a major center of communications through which German troops might have passed to get to the Russian front, it was not obviously an important target. Secondly, its population was swollen, not only with wounded troops and refugees who had fled before the Soviet advance or from other bombed cities, but with Allied prisoners of war. Thirdly, the former capital of Saxony was a very beautiful town, hitherto unbombed, a historic center for art and music. Fourthly, while Harris liked attacking such old German towns – "built", as he had said of Lübeck, which his men had destroyed in 1942, "more like a fire-lighter than a human habitation", Portal, chief of air staff, wanted to attack German oil supplies, and so prevent their planes and tanks from moving. Fifthly, the firestorm in Dresden was horrific. It is impossible to establish exactly how many died: 50,000 seems a reasonable estimate – a much greater toll relative to population than that of the Hamburg firestorm.

► The devastated center of postwar Dresden.

◄ May 1945, Berliners flee their burning houses as the Soviet army pours millions of shells and rockets into the small area still defended, and uses flamethrowers to incinerate soldiers and civilians alike. Land forces could still match the havoc caused by bomber planes.

Bomber Raid" mattered to the British public, who were thoroughly restive over military reverses in the Far East and North Africa. Of the air crews involved, 898 actually claimed to have reached the city: they dropped nearly 1,500 tons of bombs. When Goering heard this phenomenal figure, he refused to believe it. Yet in return for the loss of 291 British airmen, only 474 Germans were killed. There was great destruction, but the official British historians conclude that "Within two weeks the life of the city was functioning almost normally."

However, Churchill believed that British raids were having "a devastating effect", and told Roosevelt so. In January 1943 a joint Anglo-American bomber offensive was agreed. While the USAAF would attack targets of key military importance with precision, by day, the RAF would keep up "area bombing" by night. So from March 1943 to March 1944 the "strategic" employment of air power independently of other arms reached its European wartime peak. Assault by bombers seemed the only way to hit Germans at home so as to give support to the USSR. Initially, British night raids concentrated on the Ruhr.

My mother and I went into an open field. I remember lying between the furrows watching the Christmas trees – the target markers dropped by the pathfinders – falling slowly down... like huge, very bright, silvery fir trees hanging in the sky, one in each corner of the square, which was the target, Halberstadt.

PETER SAABOR
APRIL 1945

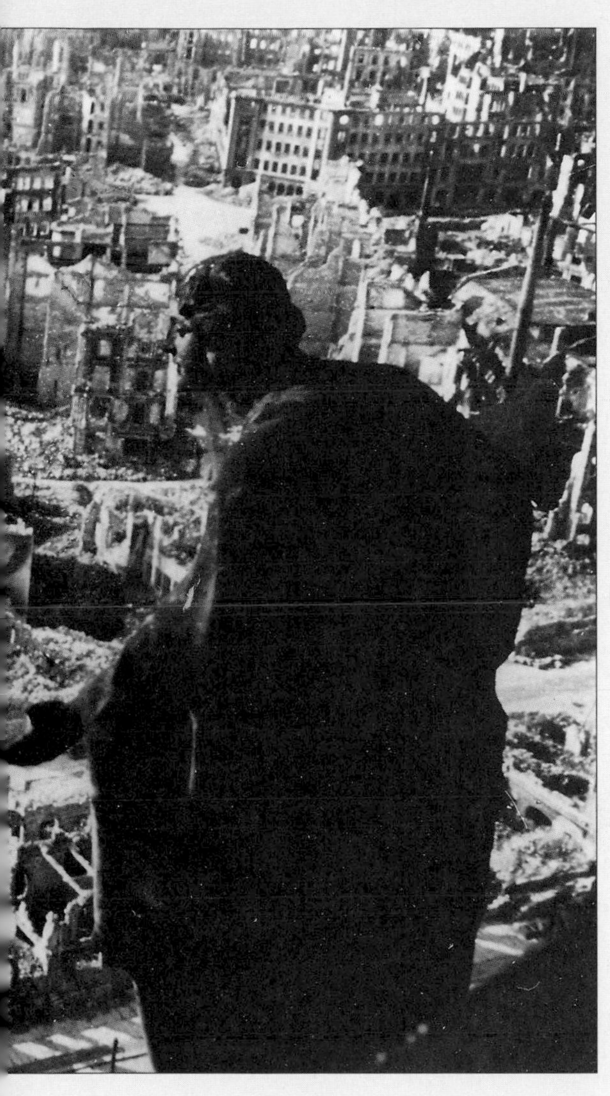

Then, in late July and early August came the Battle of Hamburg. Bombers wrought vast damage and created the first "firestorm", an effect produced on July 27-28 by an attack with incendiaries on a residential area. Twenty-two sq km (8.5sq mi) became a "lake of fire" with air temperature up to 1,000 degrees centigrade. Huge suction was created. Trees were uprooted. People were flung alive into the flames by winds of over 240km/hr (150mph). While high-explosive bombs forced people to shelter, once inside they suffocated, and were reduced to ashes as if in a crematorium. Calculation of the casualty numbers was impossible. Refugees with apocalyptic stories poured from Hamburg into nearby provinces. But most of the workers soon returned. Meanwhile, the US Eighth Air Force, which had made two daylight raids on the city, had lost 43 bombers, about a sixth of its effective force.

Sir Arthur Harris, head of British Bomber Command, believed that he could win the war by devastating Berlin. Impressive damage was caused: Wilhelmstrasse and Unter Den Linden battered, the Chancellery and many other famous buildings destroyed. After the November raids, more than 400,000 people were homeless for over a month. By 17 December, a quarter of Berlin's total living accommodation was unusable. But German fighter planes were excellent, German air defense in general subtle. The RAF grew increasingly ineffectual.

What about the effects on German arms production? It tripled between 1942 and 1944. Tank production, 760 a month at the start of 1943 when the Allies agreed on their air offensive, had reached 1669 by July 1944.

One hope of patriotic Germans was that Hitler had secret weapons up his sleeve. He did: unmanned aircraft. In summer 1944 the VI "Flying Bombs" were unleashed and sent puttering and growling across the channel to southern England. The British disliked them intensely, but neither these nor the later V2 rocket bombs had significant effect except on individuals whom they killed or maimed and their families.

In Europe Douhet's theories seemed to be disproved. Neither steady bombing over a long period, nor sudden devastating attack on a key city, nor even the use of shock secret weapons had made decisive impact.

Bombing finally brings results

The Japanese experience might suggest that Douhet's theories received a new lease of life. In fact it completed their supersession. Japan, by August 1945, had suffered the most traumatic exposure to air power yet seen – its air defense was so weak that it had trained suicide pilots to ram the huge bomber planes which were destroying the country. Yet it was by no means clear that all, or even most, of its people were ready to surrender. The weapon which brought surrender was indeed a bomb, but one of a type which made obsolete in a second previous notions about international conflict.

Bar a gallant little raid by B-25s taking off from an aircraft carrier in April 1942, which had landed 16 tons of bombs on Tokyo and shocked the Japanese at the peak of their fortunes, the US air offensive began on 29 June 1944, from western China. The B-29 Superfortresses which made a few long journeys from China to Japan were the

▼ After a raid on Mannheim in 1944, when the town experienced frequent "light" raids by RAF Mosquitoes. For the victim of a "light" raid the experience might be fully as traumatic as seeing a huge city crashing around one. Young children might suffer nightmares well into adulthood.

▼ USAAF precision bombing – a German aircraft factory smashed on 22 February 1945. It was revisited several times before VE Day.

▲ **Bombs fall toward Kobe, Japan. The USAAF bombing program attacked 65 Japanese cities. The number of planes involved rose steadily – 45 in May, 600 in June, 800 in August. Millions of leaflets warned Japanese in advance as to which cities were next in line. By the end, Allied battleships, too, were able to sail close to shore and bombard ports and industrial centers with impunity.**

biggest bombers of the war, and altogether remarkable. They could fly for 6,500km (4,000mi) at 10,700m (35,000ft) carrying 7.5 tons of bombs at the same speed as a Battle of Britain fighter. Once they got bases which were easier to supply, on the Mariana Islands, within 2,400km (1,500mi) of Japanese cities, they were in a position to achieve spectacular results.

Japan was exceptionally vulnerable. Its air force had been destroyed. Its crowded cities were largely built of matchwood and paper. Attacks by day began on 24 November, 1944. In March 1945, night attacks using incendiaries commenced. The first destroyed about a quarter of Tokyo in a firestorm. In this case 14 US bombers were shot down – thereafter losses were negligible. Similar raids on Yokohama, Osaka, Nagoya and Kobe followed. By the summer Japan's industries had been laid to waste. A quarter of its dwellings had been destroyed. Rural people who were by 1945 eating only four-fifths as much rice as in 1941, now had to cope with 10 million refugees from the bombed cities. Famine was imminent. On 25 March 1945 the government had ordered the mobilization of all people over 13 and under 60 (except pregnant women and the sick) in a "People's Volunteer Corps." If invasion came, they were to fight with bamboo spears for want of other weapons. It was still feared that this remarkable people might fight to the last, taking perhaps half a million Allied soldiers with them on their way to self-destruction. This was the logical basis for Truman's decision to use the atomic bomb, though there were also other motives. Many have since doubted whether this experiment would have been inflicted on a "white" population.

This outrage against humanity calls for the intervention of the civilised world. This is not war, not even murder, it is pure nihilism. But it will not in any way alter the Japanese war aims... but merely strengthen the will to fight to the last.

JAPANESE RADIO, SINGAPORE
MID AUGUST 1945

Hiroshima

On 6 August 1945 one bomb, 3m (10ft) long and 0.7m (2ft) in diameter, exploded 500m (1,600ft) above the city of Hiroshima. The temperature at the hypocenter of the explosion was three times that needed to melt an iron bar. Uranium 235 was the agent of destruction.

People suffered skin burns 3.5 km (2mi) away. Of 76,000 buildings near the hypocenter, 70,000 were destroyed beyond use. After the initial slaughter, more people died of burns, blast injuries, and radiation effects. The city's population was swollen with Korean workers and military personnel, to perhaps 420,000. The exact death toll will never be known, but at a high estimate, 140,000 people had died from direct exposure to the blast by the end of 1945. Effects of radiation thereafter may have brought deaths to 200,000 by 1950.

On 25 March 1945 the Japanese government had ordered a general mobilization of all people over 13 and under 60. Schoolgirls in Hiroshima were detailed to do simple physical tasks connected with civil defense in the city center. Shigeko was one who survived, terribly burnt. She was still heavily scarred when she told her story 40 years later.

"I was thirteen years old; I was ordered into the center of the city to clean up the streets. We were working about 1.6 meters from the hypocenter. I heard the aeroplane; I looked up at the sky – the plane had a pretty white tail, it was a sunny day, the sky was blue. This had happened many times before, so I didn't feel scared. Then I saw something drop – white I think – and pow! – a big explosion knocked me down. Then I was unconscious – I don't know for how long. Then I was conscious but I couldn't see anything, it was all black and red. Then I called my friend, Toshiko; then the fog goes away but I can't find her. I never see her again. Then I see people moving away and I just follow them. It is not light like it was before, it is more like evening. I look around; houses are all flat! I could see straight, clear, all a long

distance. I follow the people to the river. I couldn't hear anything, my ears are blocked up – or maybe my consciousness is blocked out. I am thinking – a bomb has dropped!

I heard a baby screaming – that woke me up and I could hear things, but nothing sounded loudly. Then I heard a man say, "let's go to the hills" and I thought maybe I should go back to school. People started to push; I was afraid I would fall into the water, and I wanted to leave with the others. There was an old woman on the ground and people were stepping on her. I couldn't help her up! I didn't know my hands were burned, nor my face. Very very difficult...all these years my regret...why couldn't I help her? It still hurts me.

Someone gave me oil for my hands and face. It hurt; my face had a swollen feeling, and I couldn't move my neck. My eyes were swollen and felt closed up. I got to my school yard and sat down, put my head back against a wall and – unconscious again."

Shigeko's parents had survived and found her at the school after four days searching.

For over a year after the Hiroshima bomb fell, few Westerners had more than a vague impression of the horrors it had caused. Then on 31 August 1946 the *New Yorker* abandoned satire, cartoons and fiction and devoted an entire issue to a reconstruction by John Hersey of the experiences of six people who had survived. It did so in the conviction that few of us have yet comprehended the all but incredible destructive power of this weapon, and that everyone might well take time to consider the terrible implications of its use.

One of Hersey's interlocutors was the Reverend Tanimoto, pastor of the Hiroshima Methodist Church. He quoted a letter from Tanimoto, reporting the unshaken patriotic faith shown by citizens after the explosion. Thirteen-year-old schoolgirls sang *Kimi ga yo*, the national anthem, as they died under a heavy fence which had fallen on them. A Methodist professor was temporarily buried under this burning house with his student son. The boy said, "Father, we can do nothing except make our mind up to consecrate our lives for the country. Let us give *Banzai* to our emperor." Then the father followed after his son, *"Tenno-heika, Banzai, Banzai!"* As a result, Dr Hiraiwa said, "Strange to say, I felt calm and bright and peaceful spirit in my heart..." The pastor concluded, "Yes, people of Hiroshima died manly in the atomic bombing, believing that it was for the emperor's sake."

▲◀ Hiroshima, seen (above) after the bombing, had stood on a plain. Nagasaki's site was hilly, its population smaller and less concentrated, and this led to different effects when the second atomic bomb hit it three days later. Hiroshima's bomb used Uranium 235, producing explosive force equal to 13 kilotons of TNT. Plutonium 239 gave Nagasaki's bomb the equivalent of 22 kilotons, but produced fewer casualties. These victims (left) may have been among up to 70,000 who died at Nagasaki by the end of 1945. The city had just before suffered two "conventional" raids, without, in either case, the prior warning normally given by the USAAF.

Datafile

In both Asia and Europe the occupying Axis powers made intensive attempts to utilize the economic and manpower resources of defeated enemies to fuel their own war machines. Their approach was overtly exploitative, and in every instance it was pursued against a complex background of collaboration or resistance by small minorities and unwilling participation by the bulk of the subject population.

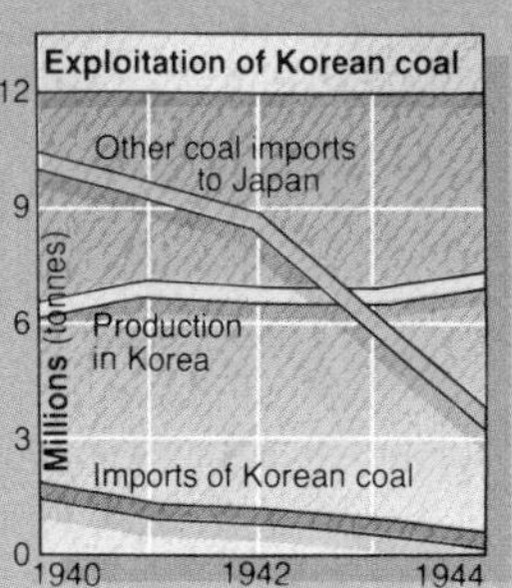

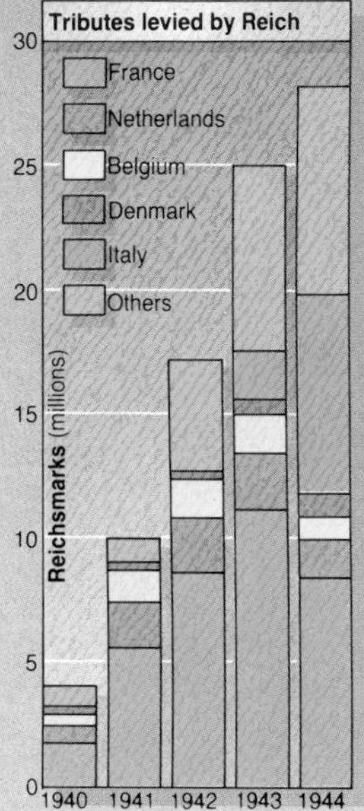

◄ Germany levied direct contributions on the European countries which fell under its sway. France, the largest and most prosperous of its client states, was the most prominent contributor, but the burden fell heavily on every area in which the Germans were able to erect the machinery of exploitation.

▲ Japan strove hard to extract essential raw materials from its subject peoples to fuel its resource-poor domestic industrial economy. Japanese control over the sources of supply was total and uncompromising, but supply difficulties (eg sinking of merchant ships) placed severe constraints on the economic returns.

► One product of military occupation was the disruption of normal patterns of family life and procreation. In occupied Guernsey, the birth rate fell steeply to approximately half the prewar norm, while the percentage of illegitimate births rose even more abruptly – from 5.5 percent in 1938–39 to 21.8 percent in 1944.

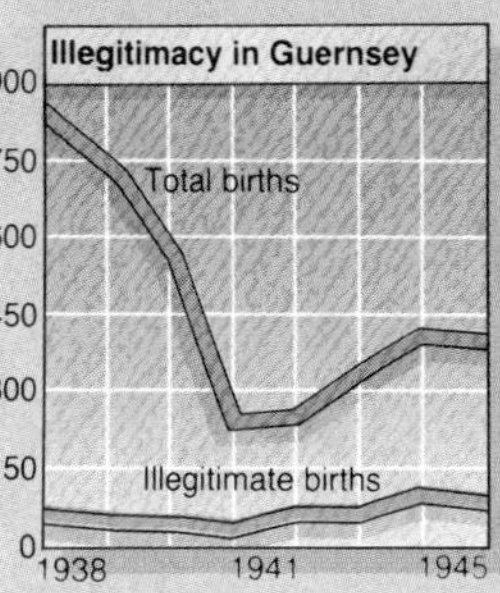

▼► At the peak of Axis fortunes in 1942, the zones of occupation covered immense areas of the globe: all but a tiny portion of mainland Europe and Scandinavia and the western part of the USSR were in German hands, while approximately half the Pacific islands, most of Southeast Asia and the East Indies, and the northeastern quarter of China had fallen to the Japanese. In many areas occupation lasted for years and was underpinned by an extensive administrative and police systems.

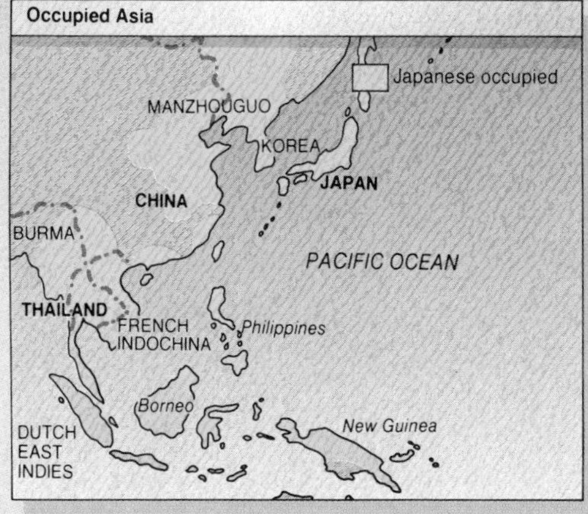

During World War II the Japanese ruled two areas of China: first, Manchuria in the north and northeast; second, a very large but continually fluctuating zone along the coasts and rivers of the south. The Chinese experience of Japanese occupation during World War II was characterized by a level of brutality, cynical exploitation and social, political and economic dislocation that was exceeded only by the horrific ordeal of European Jewry.

Japanese treatment of Chinese territories

Manchuria had been occupied in 1931. It served as the testing ground for the machinery of domination and exploitation which was later to be employed throughout Japanese-conquered territories.

The Japanese established a nominally independent state, Manzhouguo, under the Chinese-led regime of the restored emperor, Pu Yi. The occupiers sponsored a massive inflow of settlers and capital, which resulted in new cities and industries, but they treated Manchuria as a zone of economic exploitation. Resources were ruthlessly stripped away to meet the needs of the Japanese war machine.

The regime of Pu Yi attempted to ape the glories of the former Manchu dynasty, but it was an empty sham. Government officials up to the highest level were pawns in the hands of their Japanese advisors, and real power rested with the local Japanese army, the Guandong army, which contained most of Japan's radical imperialists. The trappings of Manchurian independence were only a thin veil for the intolerance of any effort to better the lot of the populace or any resistance to the demands of the occupiers.

The Japanese occupation of China itself was even more overt in its disregard of local needs or sentiments. The government of Weng Jing-wei, made up largely of nationalist rivals of Jiang Jieshi, marked only a belated realization that full conquest was best pursued through the mobilization of local political elements hostile to both Jiang and his communist enemies. No real power was transferred to collaborators and they were soon given every reason to rue the misguided nationalism which had placed them in Japanese hands.

The brutal apparatus of military occupation operated unchecked throughout the war, with the daily record of routine persecution being occasionally interrupted by outbursts of the utmost savagery. The efforts of the Japanese to realize the economic potential of the large cities and of their rural and riverine hinterlands were often hampered by their very rapacity, by the consequent flight of much of the population and by Chinese sabotage.

OCCUPATION AND RESISTANCE

Notwithstanding the territorial gains of the Japanese in 1937-41, much of China remained a fluid war zone throughout the conflict. At its best, Japanese military treatment of Chinese civilians in contested areas was characterized by indifference to safety and casual brutality. At its worst, it degenerated into murderous savagery. When Nanjing fell in December 1937, the conquerors exposed the prostrate city to a month-long reign of terror which left a third of it in ashes and hundreds of thousands of inhabitants dead. Later in the war, when attempts at creating an alternative collaborationist focus for civilian loyalties failed to stem the spread of communist guerrilla activity, the Japanese instituted the infamous "Three All" policy: take all, burn all, kill all. They laid waste entire rural areas and exterminated all who did not flee.

The immediate effect of Japanese atrocities was often a form of shocked compliance. In the long run, however, the occupiers only succeeded in catalyzing the hitherto muted nationalistic sentiments of the population. Nor did the massive structural damage inflicted on Chinese society advance the cause of Japanese hegemony. Mass destruction and death, and the even more massive internal migrations these forced, did heavy damage to the old traditions of village life. The administrative, intellectual and economic centers in the great cities were similarly battered out of shape by military onslaught and then strangled by exploitation.

The end result, again, was only to render Chinese civilization more receptive to 20th-century ideas of nationhood and populist political action. It would be simplistic to view the postwar history of Chinese communism as the direct result of Japanese occupation, but there can be little doubt that the heavy hand of the invader helped set loose the social and political forces that would bring Mao Zedong and his colleagues to power within four years of the war's end.

Japan's imperial dream

The dazzling series of victories won by Japanese armies in 1941-42 left a huge area, stretching from Burma through Malaya over the Dutch East Indies and up to the Philippines and Hong Kong, to be administered by the conquerors. Japanese objectives in establishing their rule were confused and to a certain extent contradictory, but they did revolve around three basic principles: first, the eradication of European and American influence, be it cultural, economic or political; secondly, the exploitation of mineral, agricultural and manpower resources for the benefit of the Japanese war effort; thirdly, the establishment of Japan as the leader of Asia in all walks of life with the new territories grouped around it as satellites, subject to varying degrees of control from the center. In a sense Japan was attempting to assume the role once played by Imperial China on the wide Asian stage.

Japanese propaganda stressed the pan-Asiatic nature of the war effort from the start. The war itself was called *Dai Toa Senso* (the Greater East Asia War) and its objective the establishment *Dai Toa Kyoeiken* (the Greater East Asia Coprosperity Sphere). Peace and stability were to be restored to the area, which Japanese arms would protect from Anglo-American exploitation, and all of the constituent peoples would cooperate to fashion a self-sufficient economic zone. The major proponent of the scheme was the semidictatorial prime minister, General Hideki Tojo, who overcame entrenched foreign-office resistance to create a new Greater East Asia Ministry in November 1942. This body's policy role tended to be minimal, the naval and military commands retaining a firm grip on major political and administrative decisions, and while it did provide some manpower to replace ousted colonial officials, its prime mission was cultural. Huge efforts were made to replace the various European tongues with Japanese as the second language of the captured territories, and a series of conferences, group study tours and goodwill missions were staged in an attempt to prove that Japanese

▼ **False humanity.** A Japanese propaganda picture of a Chinese boy receiving medical attention from the occupying forces. The reality was far different: apart from cynical attempts to promote collaboration through some measure of correct treatment, the Japanese treated Chinese civilians either with callous indifference or murderous savagery, the latter being particularly common when any form of resistance was encountered.

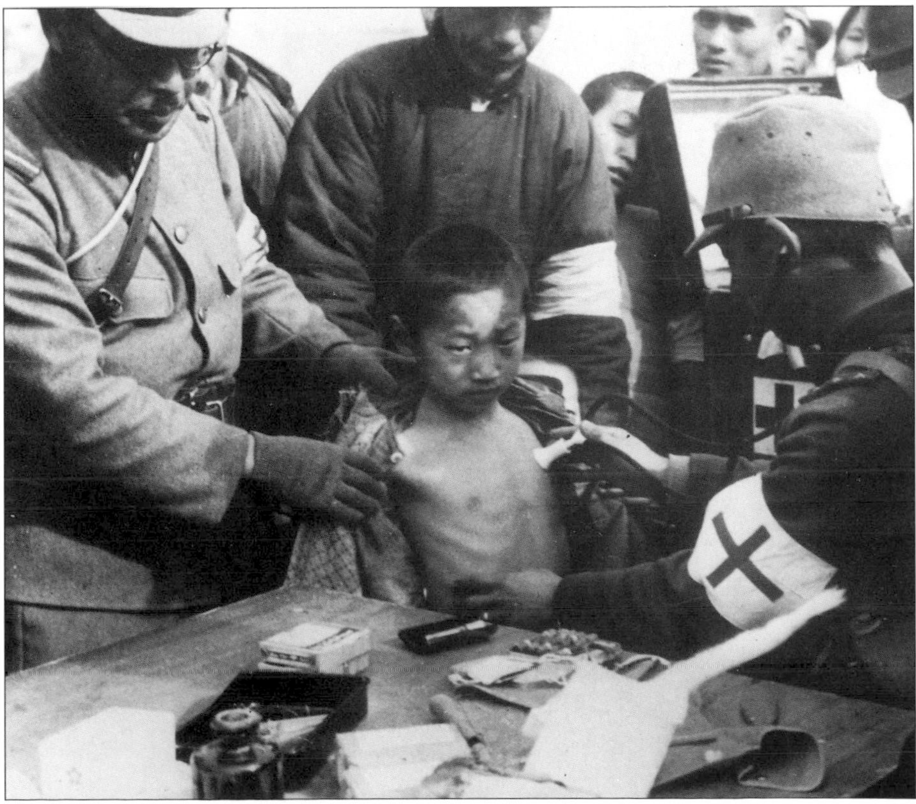

culture was at once more natural and more dynamic than that of the decadent West. Japan's attitude to and treatment of the heterogeneous territories under its sway varied according to past history and strategic importance. By the time of the Greater East Asia Conference in Tokyo in the autumn of 1943, five countries had been recognized as independent allies of Japan: Manzhouguo, China, Thailand, Burma and the Philippines. In the first two cases, the areas in question had been under Japanese control for years. In the third and fourth cases the Japanese were able to establish collaborationist governments in the existing political framework of countries with long histories of independent existence. In the Philippines another puppet government was eventually set up, but the occupying forces never really broke through the barriers of a heavily Europeanized culture and an active resistance movement.

Elsewhere Japanese policy followed a different line for reasons which were primarily economic. The conquest of Malaya and the Dutch East Indies gave Japan direct access to the tin, rubber and oil resources it required for self-sufficiency. The occupying forces were very slow to move away from direct naval or military rule, or to encourage any sort of nationalist movement. Initially at least, promises of greater control of their own efforts were made only to limit obstruction to and encourage participation in Japan's exploitative economic policies. As the war progressed, and as the possibility of Japanese defeat became more and more real, the occupation forces began to foster nationalism in a more wholehearted fashion, hoping thereby to create some form of united Asian front against advancing Anglo-American forces.

The realities of Japanese occupation

All over the conquered territories Japanese rule was brutally exploitative. However elaborate the pretence of independence or self-government, the military used coercion to meet its ends and responded violently to any form of resistance.

European civilians were herded up and confined in conditions little better than those prevailing in Japan's notorious prisoner-of-war camps. Eurasians were treated with harsh suspicion as polluters of indigenous ethnic purity, and Chinese communities were singled out for systematic oppression. Regular occupation troops were responsible for much of the almost routine violence of the system, but any overt or suspected sign of recalcitrance or resistance was dealt with by the *Kempeitai* (Japanese military police). Their cruelty was paralleled only by Hitler's Gestapo. The military police were particularly fond of pulling out their prisoners' fingernails, which soon gave rise to the grim joke: "Do your nails need a manicure?"

The labor demands of Japan's overextended war effort were met in part by the mass conscription of native labor. Despite the fact that they were often citizens of countries in national alliance with Japan, these laborers were forced to work in conditions of great deprivation and brutality. Thirty thousand Burmese civilians

◀ **Foreign manpower for the Japanese war effort. Japanese policy was characterized by the massive use of impressed labor from occupied zones. These Korean youths are under training in a Japanese agricultural school – other races received no such tutelage but were simply herded straight to their task.**

◀▼ **The Sino-Japanese conflict had a peculiar ideological dimension. The Japanese made periodic attempts to convert disgruntled nationalists to the spurious doctrines of coprosperity, while the Guomindang and their communist rivals devoted as much time to propaganda attacks on each other as to countering Japanese efforts.**

died while helping build the Burma–Thailand railway, the worst but by no means the only example of forced human sacrifice inflicted on the Asian population. Economic policy was equally rapacious. Trade between Japan and its new dependencies was almost entirely one way – a flood of raw materials being sent to the home islands without any adequate reverse flow of manufactured goods (commodities which most areas had formerly received from prewar European trading partners). The Japanese also debased local currencies, causing runaway inflation through their habit of printing money as required for their own needs. Whatever their reaction to the initial Japanese arrival, few Asians were able to sustain ideas of liberation or Japanese friendship for very long. Even the behavior of the departed European overlords seemed preferable to the new "coprosperity".

The motives of those East Asians who collaborated with the occupying forces were generally mixed. A minority may well have swallowed the initial Japanese message of liberation, friendship and prosperity, but few were paying it any more than lip-service after even brief exposure to the realities of occupation. Some collaborated in the hope of shielding their countrymen from the worst excesses of Japanese rule, and in one case at least there is some evidence to suggest that the puppet regime was giving covert support to the guerrilla resistance. While there was widespread approval of the defeat of the white colonial powers, the response to the forced importation of Japanese culture was rarely positive. By the last years of the war, all but the blindest were collaborating only in the hope of Japan's defeat bringing true independence rather than a return to colonial rule or European domination.

The Dutch East Indies

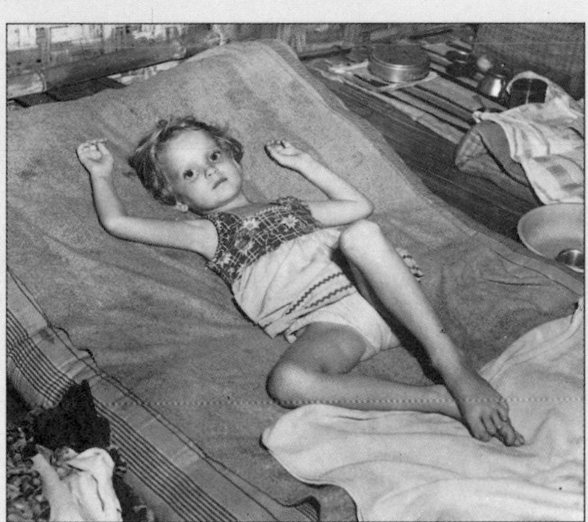

◀ **The savagery of neglect: a girl of seven in a recently liberated internment camp in Batavia. The Japanese interned the resident European populations in the colonial countries which fell under their sway.**

▼ **Indonesian sailors of the Dutch East Indies squadron before the successful Japanese invasion of 1942. The Japanese occupation heightened nationalist sentiment, and when the Allies attempted to reimpose Dutch rule in 1945, they were confronted by armed nationalist resistance.**

The Japanese occupation of the Dutch East Indies provides an interesting study in contrasts and paradox. Rich in natural assets, particularly oil, on which the whole Japanese economy depended, the area was viewed foremost by the Japanese as a strategic asset. After the Japanese conquest the islands were kept under firm military or naval control.

The Dutch East Indies had no tradition of national unity save that imposed by the colonial power. These ties were violently severed by the Japanese, by the defeat of Allied forces, by the removal of all colonial administrators from their posts and by the imprisonment and deliberate humiliation of the resident white minorities. The long-term Japanese goal was the creation of new, notionally independent states under their own hegemony. The military and naval commands, however, were loath to surrender control over such important resource areas, and in the short term the islands were subject to rigid Japanese administration. But imported bureaucrats proved unequal to the task and effective power devolved on their native subordinates. This, and the necessity of using Indonesian as the official language, fostered nationalism in spite of Japanese conservatism.

Two factors combined to accelerate the move toward Indonesian statehood. The first was the collapse of the plantation economy under the strains of the failing Japanese war effort. The native population became united in opposition to the occupiers. The second was the series of defeats suffered by Japanese forces in 1943–44 which made the occupiers more prepared to make concessions to nationalist leaders like Sukarno, in the hope of rallying the Asian population.

On 17 August 1945 Sukarno and his followers proclaimed the independence of Indonesia. While this was premature according to the Japanese schedule, the occupying forces made no move to intervene. In violation of the surrender terms, which required them to retain control until the arrival of Allied forces, the Japanese allowed the nationalists to take over administrative functions (including the guarding of interned Dutch nationals) and actually supplied the new republic with arms.

▲ German victories were accompanied by conspicuous triumphalism, like this daily military parade down the Champs Elysées in Paris. This high profile, with a German military presence in provincial as well as metropolitan areas, was calculated: it made resistance activity both more difficult and more hazardous. But, as defeat loomed and resistance forces gained in confidence, this committed German troops to policing large areas – a costly exercise for a conqueror with severe manpower problems.

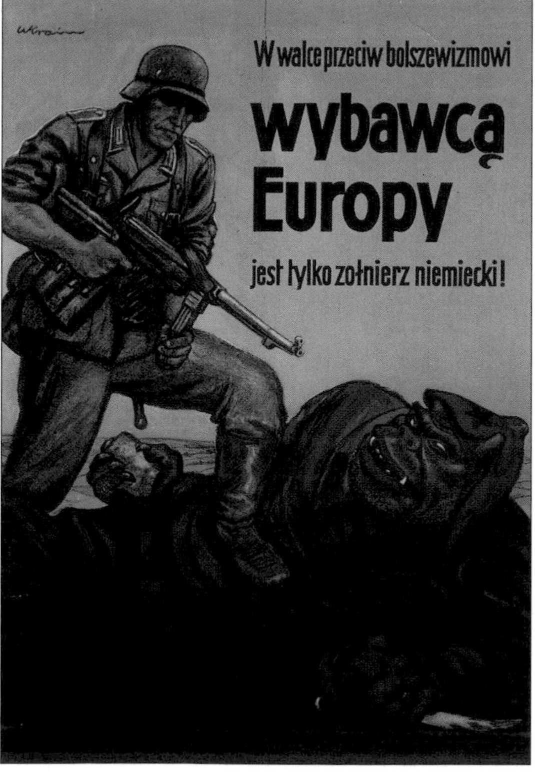

W walce przeciw bolszewizmowi
wybawcą
Europy
jest tylko zołnierz niemiecki!

► Even in countries where there was little attempt to conciliate the native population, Nazi propaganda tried to elicit compliance by stressing the possibility of a worse fate. With the Red Army's counterattack in 1943, Poles were warned that "only the German soldier can save Europe in the struggle against Bolshevism".

Occupied Europe: Hitler's New Order

Hitler went to war in September 1939 not merely to recoup the losses imposed on Germany by the Treaty of Versailles in 1919 but to create a "New Order" in which Germany would dominate continental Europe, take over Russian territory, and destroy its twin enemies: Jews throughout Europe and Russian "Bolsheviks". Nazi ideology propagated the idea of a "New Europe" in which European nation-states would accept German hegemony as being in their own best interests, with Germany leading a European crusade against communism. In most countries which came under German rule during the war, this vision had its enthusiasts.

To create his "New Order" Hitler was determined to secure "living space" (*Lebensraum*) for the German "master race", including non-German "Nordic" or "Aryan" peoples like Norwegians and Dutch. Non-"Aryans", in both western and eastern Europe, would have the task of serving the "master race" within a large-scale economic bloc (*Grossraumwirtschaft*). Germany's needs could thus be given priority; above all, its food supply would be guaranteed against any future blockade, as a vital part of Hitler's overall strategy. (See p.146.) In order to acquire "living space", German forces would have to take military

and political control of virtually all of Europe, either directly or else through puppet surrogates.

Through prewar annexations and wartime conquests Germany secured mastery over most of Europe. From fall 1939 to fall 1944 German soldiers fought on fronts outside the borders of the Reich, extending and defending an empire which at its zenith in 1942–43 had some 180 million inhabitants. This "Greater German Reich" stretched from the Atlantic to the Caucasus, from Scandinavia to the Mediterranean. Only Switzerland, Sweden, Eire, Turkey, Spain and Portugal were neutral; the last two were clearly pro-German and, like Turkey and Sweden, they supplied Germany with vital raw materials for most of the war. Switzerland exported machinery and armaments to Germany; the USSR, too, provided Hitler's war machine with essential supplies, including fuel oil, until the German invasion in June 1941. Finland was nominally neutral and independent, but inclined to favor Germany once the invasion of the USSR had begun; this was a case of "my enemy's enemy is my friend", after the bitter Russo-Finnish War of 1939–40. (See p.55.) Romania, Bulgaria, Hungary and Slovakia were allied to Hitler, nominally sovereign but entirely subordinate to him.

Mussolini's Italy was initially an ally which shared (belatedly) in the French campaign and then in the invasion of the USSR. But Italy's disastrous campaigns in Greece and North Africa made it a liability, and the price of a German rescue in both these areas was Italian acceptance of German domination. Italy's defection in July 1943, after Mussolini's fall, brought German invasion and occupation. Thereafter, Italians, in the north of their country especially, were as much at the mercy of German brutality as were citizens of other countries.

The redeployment of German peoples

The *Lebensraum* Hitler sought was more than simply extra territory: it was assigned the status of a "promised land" of prime territory, stretching across Europe and inhabited by "Nordic" peoples, along with enough non-"Nordic" workers to service their requirements. The most fertile, mineral-rich, industrialized and climatically benign areas would form the "Greater German Reich", with most Slavs and other non-"Nordic" peoples banished to the wastes of Siberia or other inhospitable areas. The permanent need for vigilance along the borders between the two kinds of territory would, Himmler believed, keep Germans alert and strong.

These plans required a vast movement of peoples from one part of Europe to another. While the Soviet-German pact was in force, from August 1939 to June 1941, hundreds of thousands of "ethnic Germans" left eastern European countries like the Baltic States, Romania or the USSR itself, to settle either in Germany or in land annexed to the Reich from Poland. Citizens of the Reich, too, were encouraged either to settle in the former Polish territories or to undertake a tour of duty there to assist new settlers. At the same time, many Poles and other non-German inhabitants of this area were deported to the "General Government", the residue of Poland after German and Soviet annexations. Once German forces had conquered most of European Russia, in 1942, Himmler dreamed of settling Germans there, as frontiersmen. And from 1941 there were mass deportations of Jews from all over Europe "to the east", that is, to extermination camps (see p.212).

These migrations occurred over and above complex military operations and, like them, required manpower, transport and fuel. It is a measure of the primacy of the ideological goal of

▶ A Ukrainian woman reading a Nazi poster. In the USSR the native population was mostly treated as hostile. But in the Ukraine Germany's struggle against *Russian* Bolshevism was portrayed as a common cause. This was contradicted by German policy, which alienated Ukrainians who had at first welcomed German soldiers as liberators from Stalinism. But German refusal to rehabilitate Christianity and to dissolve the collective farms was crucial in forfeiting the support of the major disaffected nationality in the USSR.

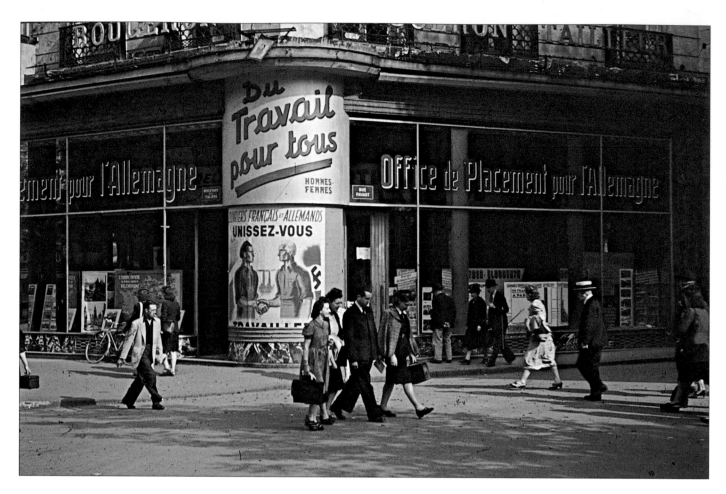

Lebensraum that Germany's already scarce resources were depleted by the demands of Nazi programs of settlement and genocide. Even in 1944, with the Allies closing in on Germany, precious fuel and transport were utilized to ensure that Hungarian Jews were conveyed to the death-camps at Auschwitz.

The pattern of occupation

All countries under German rule suffered expropriation and exploitation of resources, both human and material. Their peoples suffered the humiliation of being at best second-class citizens in their own land, and many were forced to assist the continuing German war-effort. Their material resources were worked mercilessly for the same purpose. In some cases, industrial plant was dismantled and removed to Germany, followed by the "voluntary" migration of workers from, for example, France or Belgium seeking employment. Poles had been forcibly drafted into work in Germany from late 1939, often as farm labor, and, especially from mid-1942, forced labor from the USSR was put to work in Germany. As the war turned against Germany, from 1942, compulsion was increasingly used to draft foreigners into Germany, where they worked mostly either in factories, fueling the German war-effort, or in agriculture (see p.148). France was the principal source of skilled labor, while the unskilled came chiefly from the USSR.

For no two countries was the experience of German occupation identical, and in each country German rule was in a state of flux throughout the period of occupation. Structures of government were constantly being adapted to meet changing circumstances, such as the initial conquest, levels of collaboration and resistance, and ultimate military defeat. But for any country German occupation was a deeply degrading experience. Western European countries, it is true, were mostly spared the barbaric brutality meted out in eastern Europe. There, Nazi racial prejudices branded Slavs as "inferior", fit only to serve "Aryan" Germans and to live, consequently, at a much lower standard than the "master race". This mirrored the treatment of Polish and Soviet forced labor in Germany itself. Above all, most of Europe's 9 million Jews were to be found in the east, chiefly in Poland and the USSR, with fewer

▲ Before forced labor conscription began in 1942, the Germans encouraged western European men and women to migrate to work in Germany. This recruitment office in Paris gave the impression that French and Germans were equal partners in the venture. Nothing could be further from the truth.

▼ In 1940 France was divided into two: the German-occupied area in the north and along the western seaboard, and the southern Vichy zone. French people could move between the two areas only if their identity papers satisfied German border police.

The Netherlands

In May 1940 the queen and cabinet escaped to Britain and formed a government-in-exile. A German civilian Reich commissioner was appointed – Arthur Seyss-Inquart, who was personally responsible to Hitler. Early German rule was conciliatory: the occupiers believed that the Dutch, as a "Nordic" people, would collaborate in their racial war. At first Dutch laws remained in force and Dutch civil servants continued to administer government. They issued identity cards, photographed and fingerprinted everyone, distributed ration cards, and issued a racial questionnaire. Thus German authority and Nazi ideology were gradually imposed, although Dutch officials tried to mitigate their worst effects.

The chief collaborators were in the Dutch national socialist movement (NSB). Its leader, Anton Mussert, was a useful pro-German propagandist, portraying the Dutch as the "second strongest branch of the German tree", but he lacked real power. He helped to organize the Netherlands Legion, which fought on the Eastern Front. There were also collaborators, who, often unknown to their friends and neighbors, spied and informed. But, in greater measure, there was opposition to German rule virtually from the start, from communists, the Church, the universities, among many others. Symbolic items signifying loyalty to the royal family – white carnations, the color orange – were widely displayed. London-based "Radio Orange" encouraged subversion and a boycott of all German or pro-German events, from radio programs and concerts to football matches.

Antisemitic measures, however, provoked both open and clandestine resistance. Jews were identified and segregated, and sometimes brutally attacked in the streets by the NSB. Revulsion at one such incident in Amsterdam in 1941 led to a 48-hour strike and demonstrations there and elsewhere. Many Dutch families concealed Jews to prevent their deportation. But some 104,000 out of 140,000 Dutch Jews perished, with Seyss-Inquart a zealous executor of Nazi racial policies. He also brutally tried to eradicate resistance. Nevertheless, it flourished: underground newspapers spread information and boosted morale; German offices and vehicles were blown up; several leading NSB members were assassinated. Five million Dutch workers were sent to Germany, but 300,000 fugitives were concealed and given forged ration cards by resistance groups.

Winter 1944–45 brought unspeakable suffering. To assist the Allied advance into Holland, the Dutch government-in-exile called for a railroad strike to cut German supply lines. The response was massive, as were the German reprisals. Some 16,000 died of cold and hunger as the occupier refused to allow food supplies to towns in western Holland. Liberation by the Allies was completed only in early May 1945.

▲ Nazi propaganda attempted to persuade other "Nordic" peoples actively to embrace Germany's struggle. This poster calls on "volunteers from Germany, Netherlands, Flanders, Denmark, Norway to apply for the Waffen–SS", under the slogan: "Peoples of similar blood fight together against the common enemy".

than a million in the west. German conquests made Jews throughout Europe an easy prey: almost 70 percent of all European Jews perished in the Holocaust, including 90 percent of Polish Jews.

Apart from their implacable attitude toward Jews, the German occupiers in western Europe were more conciliatory, especially while Germany was still victorious. They exhorted French, Dutch and other western Europeans to recognize their alleged common cause, and, especially after the German invasion of the Soviet Union in June 1941, propaganda urged western Europeans to support Germany's campaign to eradicate "Bolshevism". With Hitler's initial insistence that "none but Germans shall bear arms", it was their duty to provide munitions and other supplies for the German army by hard work in factories, whether in their own country or in Germany. But the insatiable demands of the German army led to appeals for western European volunteers for the "struggle against communism", and many joined units of the Waffen-SS.

Occupied peoples and their rulers

Nazi racial theorists regarded Norwegians, Danes and Dutch, especially, as "Aryan", and therefore worthy of a favored place in the "Greater German Reich". Propaganda was directed at "educating" them to accept their destiny. But most western Europeans continued to prize national independence above a mirage of loosely defined superiority under Nazi control. In every western occupied country and some eastern ones, however, there were collaborators prepared to accept German hegemony and German plans for their future. There were also in every occupied country – in east and west – groups of resistance fighters, who would not accept German rule, and who engaged in activities from sabotage to helping Allied agents in a plethora of small local counter-offensives against the German occupiers.

But collaborators and resisters were at the extremes of reaction to German occupation: in both cases, they were very much in a minority. The vast majority of citizens in occupied countries simply tried to survive, accommodating themselves to German demands where refusal to do so could bring severe retribution. Especially in eastern Europe, the occupiers were quick to take reprisals, usually in response to acts of resistance. The total destruction of the Czech village of Lidice in 1942 was a particularly barbaric example. But there were also martyrs in western countries like France, where the massacre of women and children at Oradour-sur-Glane in 1944 was perhaps the most notorious example. Major atrocities were perpetrated elsewhere, notably in both Italy and Serbia (the German-occupied area of Yugoslavia).

On the whole, western civilians tried to maintain as normal a way of life as possible. This was not the same as collaboration: it was merely pragmatism, given a seemingly invincible German army. Until 1943 it looked as if German occupation might last for 50 or 100 years – for a lifetime. Accommodation to the "New Order" was the obvious option in western Europe, after German forces had quickly destroyed countries' actual and

We now experience the grotesque spectacle that after the tremendous starvation of prisoners of war, millions of foreign laborers must be recruited to fill the gaps which have appeared in Germany. With the usual unlimited abuse of Slav people, "recruiting" methods were used which can only be compared with the blackest periods of the slave trade. (They are allowed only the most limited education, and can be given no welfare services.) We are interested in feeding them only insofar as they are still capable (and they are given to understand that in every aspect we regard them as inferior).

DR OTTO BRÄUTIGAM
REICH MINISTRY OF THE EASTERN TERRITORIES
25 OCTOBER 1942

Denmark and Norway

▲▲ The Danish resistance had very advanced radio technology. Here Danish resistance workers in Odense listen to a radio broadcast from London in 1945. BBC transmissions were good for morale in occupied countries. Listening to them was a clear act of defiance, and a capital offence. They brought an alternative view of the news (contrasting markedly with the German one) which became increasingly cheering as the war progressed. They could also bring coded instructions for resistance workers' activities.

▲ Vidkun Quisling was leader of the tiny Norwegian fascist movement (*Nasjonal Samling*), a fringe political party in the 1930s. He naively believed that "Nordic" Norway could prosper under the "New Order". From 1940 to 1945 he served as a puppet for the occupation forces. He was executed in 1945.

The Nazis expected collaboration from "Nordic" countries like Denmark and Norway, but they remained overwhelmingly anti-German. In Denmark, uniquely in the Nazi empire, national institutions continued to function. The king and his cabinet remained, whereas Norway's monarch fled to London and formed a government-in-exile. Norwegian political parties (except Quisling's) were dissolved and power vested in a German civilian Reich commissioner, Josef Terboven. In Denmark, by contrast, parliament continued to function, within limits imposed by the Germans; this included appointing the collaborator Erik Scavenius first as foreign and then, in 1942, as prime minister. But in 1943 free elections were permitted and gave the social democrats 90 percent of the votes. Resistance forces increased sabotage and a wave of strikes ensued. The Germans imposed martial law and dissolved Danish institutions, prompting heightened resistance activity. The freedom council, founded in September 1943, worked closely with the British SOE, as well as helping Jews to escape. Nazi antisemitic policies, which Quisling enthusiastically embraced, were generally deplored by Norwegians and Danes alike, and many Norwegian Jews were conveyed to safety in Sweden. Through its government-in-exile, the Norwegian resistance also had SOE support and provided invaluable logistical intelligence for SOE. Their agents' destruction of the Vemork heavy water plant in 1943 hindered German atomic research. The exiled Norwegian navy ferried agents across the North Sea.

In both Norway and Denmark, resistance activity obstructed the German war-effort to the very end. The Danish freedom council called for a general strike on D Day and in early 1945 sabotaged transport conveying German troops to the front. Norway, with its long coastline, was of paramount strategic importance to the Germans, which committed Hitler to maintaining over 300,000 soldiers there.

potential military strength in 1940. Thereafter, the likelihood of imminent liberation seemed negligible. The UK was on the defensive after Dunkirk, and the USA was at first neutral and then at war in Asia.

Accommodation to circumstances made life under German authority in occupied western Europe less harsh but still far from pleasant. For a start, the economies of these countries were taken over and made subservient to Germany's military needs. The vanquished had to bear the costs of occupation, at an inflated price dictated by the Germans. The surplus from this was used to acquire raw materials, foodstuffs and manufactures for the German war-effort and to meet consumer demand among German civilians. Other than in the Netherlands, banks were taken over, giving Germany greatly enhanced reserves of gold and foreign currency. In some countries, for example France and Belgium, occupation policies led to roaring inflation. The exchange rate between the Mark and other currencies was set at an artificially advantageous level for Germany. All external

► General Charles de Gaulle escaped to Britain in June 1940. The Allies eventually recognized him as leader of the "Free French", although his arrogance repelled many, especially Roosevelt. But his charismatic image inspired loyalty in many including resistance leader Jean Moulin, in 1942–43 his agent in France. Moulin brought together noncommunist resistance activists.

▼ Pétain, here visiting a small town, was head of the Vichy French State, a rejection of the French republican tradition, which seemed to promise stability in 1940, after defeat; but it dissipated goodwill by increasingly slavish accession to German demands.

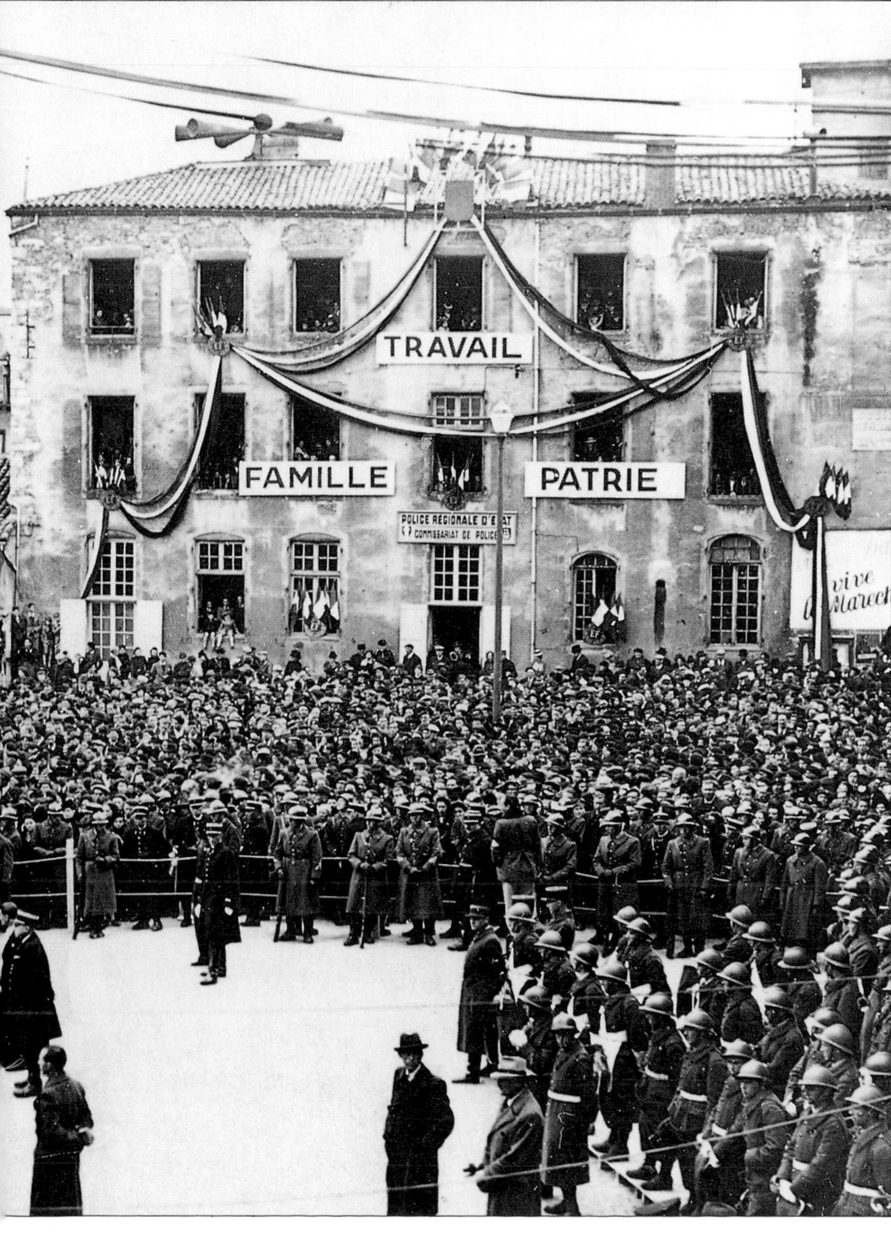

trade was controlled by Germany. Firms were seized, raw materials and fuel channeled into German projects, and Jewish businesses confiscated. Altogether, the raw materials and industrial capacity of France and Belgium, especially, were irresistible and invaluable.

In addition, there were daily reminders of German mastery. Occupation forces patrolled the streets, relaxed in bars and cafes, and conspicuously dispossessed the natives of anything of value from art treasures to girlfriends. Indigenous fascist groups, to be found in most countries, had expected to come into their inheritance after Germany's military victories. But, even in "racially desirable" countries like Norway or the Netherlands, Hitler was reluctant to permit willing collaborators to rule.

He wished to leave his options open and hoped to be able to incorporate "racially desirable" peoples and their lands into the "Greater German Reich" after the war, as he had done with Luxembourg and Alsace-Lorraine in 1940. Therefore he at first bypassed the small indigenous fascist parties in Norway and the Netherlands and installed a German commissioner, personally responsible to himself, as the chief civil authority. In Norway, Josef Terboven, and in Holland, Arthur Seyss-Inquart, assumed this office. The Dutch fascist, Mussert, was recognized as "Leader", but was never assigned real power. The appointment in 1942 of the Norwegian fascist leader, Vidkun Quisling, as head of a puppet government, merely ensured that he was an easy scapegoat in 1945.

There was no uniform pattern of occupation control in western Europe. In Belgium, where King Leopold remained throughout the war, the German military commander took over the institutions of civil government. By contrast, the Dutch civil service departments were charged with administering the Netherlands under the German commiss and his four deputies. Queen Wilhelmina, like King Haakon of Norway, fled to London. Perhaps the most genuinely autonomous western country was Denmark, where the king and his cabinet continued to rule until 1943, when the Germans declared martial law and introduced direct military rule to combat sabotage and strikes.

Anomalously, non-"Nordic" France was permitted its own native government in the southern Vichy area, under Marshal Pétain. While Vichy had nominal autonomy from 1940 to 1942, thereafter it was no better than a German puppet state. Vichy's "National Revolution" asserted the values of the old right wing in France, in opposition to the republican tradition from 1789 onward. Its motto "Work, Family, Fatherland" summed up a ragbag of prejudices which were similar to many of those harbored by Nazis in Germany. Independent workers' organizations were destroyed, official youth organizations were founded, and a monopoly political party, the Legion of Ex-Servicemen, authorized. Antisemitism, xenophobia and clericalism became official policy. Vichy's chief achievement was to ensure that traditional conservatism was utterly discredited in France by the end of the war.

▲ In the USSR armed partisans used local conditions to telling effect in harassing the invader and sabotaging communications.

▲ The requirement that Soviet citizens constantly and conspicuously wear an identification number emphasized their relegation by the German occupiers to the status of subhumans. Place-names and road signs appeared only in German; the victors required information, the vanquished victims merely awaited commands.

▶ In every occupied country, the Gestapo rapaciously sought out opponents. Resistance leaders soon realized that well-informed agents could, under torture, reveal enough to destroy an entire network. To avoid this, the information given to each agent was kept to an absolute minimum, so that there was very little for him or her – involuntarily – to reveal.

German oppression

The level of German oppression in western Europe clearly depended on Germany's fortunes during the war. Until 1942, with the prospect of German victory seemingly both assured and imminent, the demands of the occupiers were relatively mild. For example, the voluntary migration of labor to Germany sufficed not least because forced labor was conscripted from Poland and, from 1942, the USSR. But the failure to win quickly in Russia in 1941-42 strained existing resources. The need for increased numbers of workers for German industry led to the compulsory recruitment of hundreds of thousands of French, Dutch and Belgians for work in German industrial cities. At the same time, rations in the occupied countries declined to a level of around three-quarters of the calorie intake of Germans, whose rations were decreasing. Much of Germany's food supply came from France, whose contribution was as great as that of all the eastern occupied territories together.

In the east, as in the west, resources were plundered for the benefit of Germany. If yields were lower in the east, it was because resources were poorer and the methods of exploitation, while no less ruthless, were often mutually contradictory. The brutality and inhumanity which characterized German treatment of Poland's land and people from autumn 1939 clearly indicated what other Slav populations could expect at Germany's hands. The major difference between occupation policy in Poland and that in western Europe was the prominent role played by the SS in the east. Special SS "task forces" (*Einsatzgruppen*) were attached to German army units, charged with eliminating actual and potential Polish leaders, including the intelligentsia, army officers, the clergy and the nobility. In the areas absorbed into Germany, Poles were to form a docile class of manual laborers for unskilled work. Young Poles were to receive only the rudiments of education necessary for the menial tasks assigned to them. On the other hand, Polish children of "good blood" were to be educated in Germany, with or without their parents' consent. In the General Government, conditions were only slightly better. Any sign of dissidence among Poles was countered with conspicuous brutality.

If German conduct in Poland was barbaric, it plumbed new depths in dealing with the peoples of the USSR, who became, in German parlance, *Untermenschen* (sub-humans). Here the SS "task forces" were used on a large scale. Unleashed to massacre Jews and Communist Party officials, they also indiscriminately murdered thousands of ordinary Russians as well. This was not some unintentional byproduct of expropriation and exploitation: it was deliberate brutality and mass murder. It often actually ran counter to German interests: at a time when labor was scarce on the German home front, huge numbers of Soviet prisoners of war were rounded up and left to starve or freeze in primitive camps. The zeal with which Red Army soldiers (all classed as "Bolsheviks") were liquidated in 1941-42 as racial and political enemies was that of the fanatic. But the failure to win a quick victory in Russia gave the initiative to pragmatists who saw in Russia's millions the kind of basic manual laborers who could work for the German war-effort.

At the start of the Russian campaign, German forces had a strong potential advantage: the many subject nationalities within the Soviet Union were implacably hostile to Russians generally and to Stalin and the Communist Party in particular. But this opportunity was thrown away in an orgy of killing and destruction. The Ukrainians who welcomed the German "liberators" in 1941 received a cool response. Here, however, there was an exception to the Germans' normally hostile attitude to eastern Europeans: an appeal was made to Ukrainians, as to western Europeans, to serve the anti-"Bolshevist" war-effort by working in German armaments factories. Even so, the saddest comment to be made about German occupation policy in the USSR is that, for ideological reasons, it was in practice no better, and often even worse, than the horrors recently experienced by Soviet citizens at Stalin's hands.

Yugoslavia

Yugoslavia came into being at the end of World War I. Land from the dismembered Austro-Hungarian Empire was added to Serbia to create a south Slav kingdom in which the Serbs dominated some eight subject nationalities, of whom the Croats were the largest and most vociferous. Inhabiting the western half of the country, the Croats were Roman Catholic, while the Serbs, in the east, tended to be Orthodox Christians. After the German invasion and victory in April 1941, small parts of the country were hived off to Germany, Italy, Hungary, Romania and Bulgaria. German domination of the residue could not disguise the internecine war waged between different "Yugoslavs". A large, nominally independent Croatia was placed under the Ustaše, a Croatian separatist group led by Ante Pavelić, which unleashed a lengthy reign of terror, chiefly against Jews and Serbs. German rule in Serbia, through the puppet general Milan Nedić, was challenged by two resistance groups, the Četniks, led by the Serbian nationalist, Draža Mihailović, and the partisans under the leader of the outlawed communist party of Yugoslavia, Tito (Josip Broz). The partisans harassed Italian and German troops, engaging in sabotage, capturing towns, and tying down substantial numbers of occupying troops; when the odds were against them, they escaped to the rugged mountain areas. They attracted support from Croats and Bosnians revolted by the Ustaše atrocities, as well as from Serbs and Slovenes, giving them bases in all parts of the country. From the very start, they constructed a system of local government in areas they liberated. This confirmed the Četniks, representing the royal government-in-exile, in their hostility to the partisans whom they saw as no better than the invader; the result was civil war of a particularly bestial and brutal kind. Believing the widespread anti-German guerrilla activity to have been instigated by the Četniks, the Allies sent them aid in the hope that Mihailović would unite the opposition to German occupation. Only in 1943 did they accept that Mihailović was serving German interests while the partisans were fighting single-mindedly against German forces. The strength of Tito's partisans was evidenced when they were able to seize Italian arms and Yugoslav territory held by Italy, when Italy changed sides in September 1943. Their 250,000 soldiers controlled most of northwestern Yugoslavia, and in 1944 they were on the offensive, besieging German garrisons in several towns. On 20 October 1944 Belgrade fell to the partisans. Alone among occupied countries, Yugoslavia was liberated by indigenous effort, and Tito's prestige and popularity enabled him to rule for over 30 years after the war.

▼ Tito (Josip Broz, foreground right) was the Yugoslav communist partisan leader who fought successfully against both invaders and native rivals. He and his men lived and planned campaigns in hideouts in the rugged mountain regions of Yugoslavia, inflicting damaging guerrilla attacks on foes who often outnumbered them. From 1943, Allied aid strengthened his forces, which in 1945 liberated Yugoslavia.

The Soviet victory at Kursk in July 1943, and the western offensive based on the D Day landings in Normandy in June 1944, were only the beginning of the end of the "New Order". Liberation in both east and west brought rejoicing, but was achieved at a terrible cost to civilians as well as combatants. Towns in northern France were bombed by France's allies. The Warsaw uprising against German rule was denied Soviet aid by Stalin. Tenacious German military resistance virtually to the end condemned most occupied countries to suffer further destruction.

The new Europe

With the Treaty of Rome in 1957, a mere 12 years after the war, a "new Europe" was created in the west. Had the wounds of war already healed, enabling five countries which had suffered German occupation – "Benelux", Italy and France – to cooperate with a part of Hitler's former Reich? More realistically, the desperate economic problems faced by all of them, reinforced by American pressure, promoted economic cooperation. Further, the Franco-German rapprochement, the basis of the European Community, was welcomed as a device for tying part of a divided Germany into an international community, to prevent any future German aggression. In eastern Europe the formidable military victory of the Red Army ensured that countries which it had liberated from German thrall would be controlled by Moscow, to prevent a future invasion of the USSR from the west. Soviet authority ensured that in the east, too, several countries which had suffered German exploitation and barbarism should cooperate with a part of Hitler's former Reich. Perhaps Soviet pressure alone ensured that Czechs and Magyars and Poles would live peaceably as neighbors, not only with each other but also with a truncated German state, the German Democratic Republic. However much the experience of occupation was officially forgotten, there clearly has been no desire on the part of the former victims to pursue a reunification of Germany.

I led the peasants to Tito, without announcing their visit. Tito interrupted his work and spoke at length and warmly with them about the suffering of the people, about the prospects of ending the war, about postwar reconstruction. And we leaders admired the unpretentiousness and eloquence of our supreme commander. But we also waited impatiently for the conversation to end, our hungry eyes fixed stealthily on the roast lambs leaning against the beech tree.

MILOVAN DJILAS
CROATIA, 1943

THE FRENCH EXODE OF JUNE 1940

In June 1940 the population of northern and eastern France was gripped by a collective panic that came to be known as the "Exode" or Exodus. Some 5 million people, possibly more, fled from their homes to avoid the invading Germans; 2 million left the Paris region alone between 10 and 14 June. Some remembered the German occupation of northern France during World War I; others had heard of Nazi atrocities in Rotterdam and Guernica; most were simply swept along by the tide of humanity. Entire towns and villages were emptied of their inhabitants: it is estimated that the population of Chartres fell from 23,000 to 800, Troyes from 58,000 to 30. One Burgundy village was totally deserted apart from one family, which committed collective suicide.

This was no orderly evacuation. The panic of the population mirrored the disarray of the authorities. In Versailles the town hall provided no more than the following help: "The mayor invites the population to flee". The government itself left Paris on 10 June. Panic increased as the refugee columns were raked by German machine-gun fire. With France's roads clogged by fleeing refugees, military operations were hampered. But the Exode was essentially a response to defeat, not a cause of it. The refugees headed for what they presumed to be the safety that lay on the other side of the Loire, although by 19 June the last bridge over the river was down. Those who did get across swelled the populations of southern cities, creating administrative problems and social tensions: Bordeaux's population grew from 258,000 to 800,000, that of Cahors from 13,000 to 60,000.

The Exode was soon over. After the signature of an armistice on 22 June, the German authorities allowed most people to return home – except those from certain designated areas of the northeast, especially the so-called "forbidden zone" – and even provided petrol to help them do so. But the psychological effects of the Exode were more enduring. Members of families became separated and for months afterward newspapers carried advertisements from parents seeking the whereabouts of lost offspring. (A postwar French film, *Forbidden Games*, described the plight of a city girl who lost her parents in the Exode and ended up being cared for by a peasant family.) The sight of this civilian suffering had played a large part in Pétain's desire to sign an armistice; he portrayed himself as the protective father of a suffering nation. The Exode was an experience of trauma and dislocation for millions of people. They had felt abandoned by their leaders: the avuncular Pétain and his Vichy regime offered reassurance and authority. Families had been split up and children gone astray: Vichy stressed the importance of the family. Treasured private possessions had been lost or bundled ignominiously onto carts for public gaze: Vichy praised rootedness and the values of home and hearth. Pétain was a revered figure in 1940: the Exode offers one explanation why this was so.

▼▶ Desperate to escape from the invaders, families piled their belongings onto every conceivable form of transport: carts, bikes, barrows, etc. (below). Those fleeing by car ran out of petrol or became caught in queues. The novelist Antoine de Saint-Exupéry, viewing events from the air, described "roads black with an interminable treacle running on forever" (right). Children often became separated from their families (far right).

▼ **The map shows the movement of population from north and east to south and west, as well as the retreat of the government. One participant remembered that he had felt like a leaf swept along by a whirlwind. After such turmoil the cult of Pétain offered security. His benign countenance (below) was a ubiquitous presence. The regeneration of France – or National Revolution – that he offered took as its slogan: Work, Family, Fatherland.**

REVOLUTION
NATIONALE

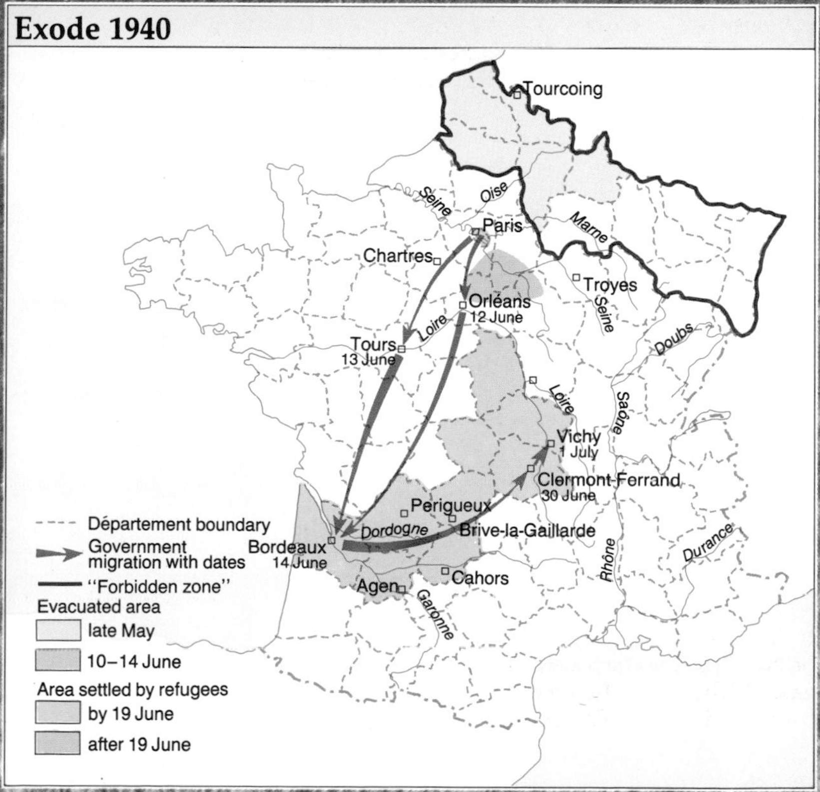

Exode 1940

Tourcoing
Seine
Oise
Paris
Marne
Chartres
Troyes
Orléans
12 June
Seine
Tours
13 June
Loire
Doubs
Loire
Saône
Vichy
1 July
Clermont-Ferrand
30 June
Perigueux
Brive-la-Gaillarde
Dordogne
Bordeaux
14 June
Cahors
Agen
Garonne
Rhône
Durance

Département boundary
Government migration with dates
"Forbidden zone"
Evacuated area
late May
10–14 June
Area settled by refugees
by 19 June
after 19 June

Datafile

In April 1941 the Nazis began to print millions of leaflets in the languages of the USSR in preparation for an attack in June. Meanwhile, British airmen had been dropping millions of leaflets on Germany. Paper was a weapon of war, and like other arms of war, propaganda caused the diversion of personnel and scarce resources from home civilian purposes. In the UK publishers were cut to only 37.5 percent of their prewar paper supply, and the number of new books fell from 14,094 in 1939 to 6,747 in 1945. Home listeners had less radio choice – propaganda to Europe burgeoned. Other forms of propaganda included posters and specially made cinema films.

▼ Statistics for radio sets in use are more suspect than most, particularly when national boundaries changed. But it is amply clear from those below how important the medium was in Western Europe, though listening in the East may have been harder. The BBC's own figures suggested that there were only some 50,000 sets in Greece at the end of the war, but this did not deter it from broadcasting in Greek for over 12 hours a week.

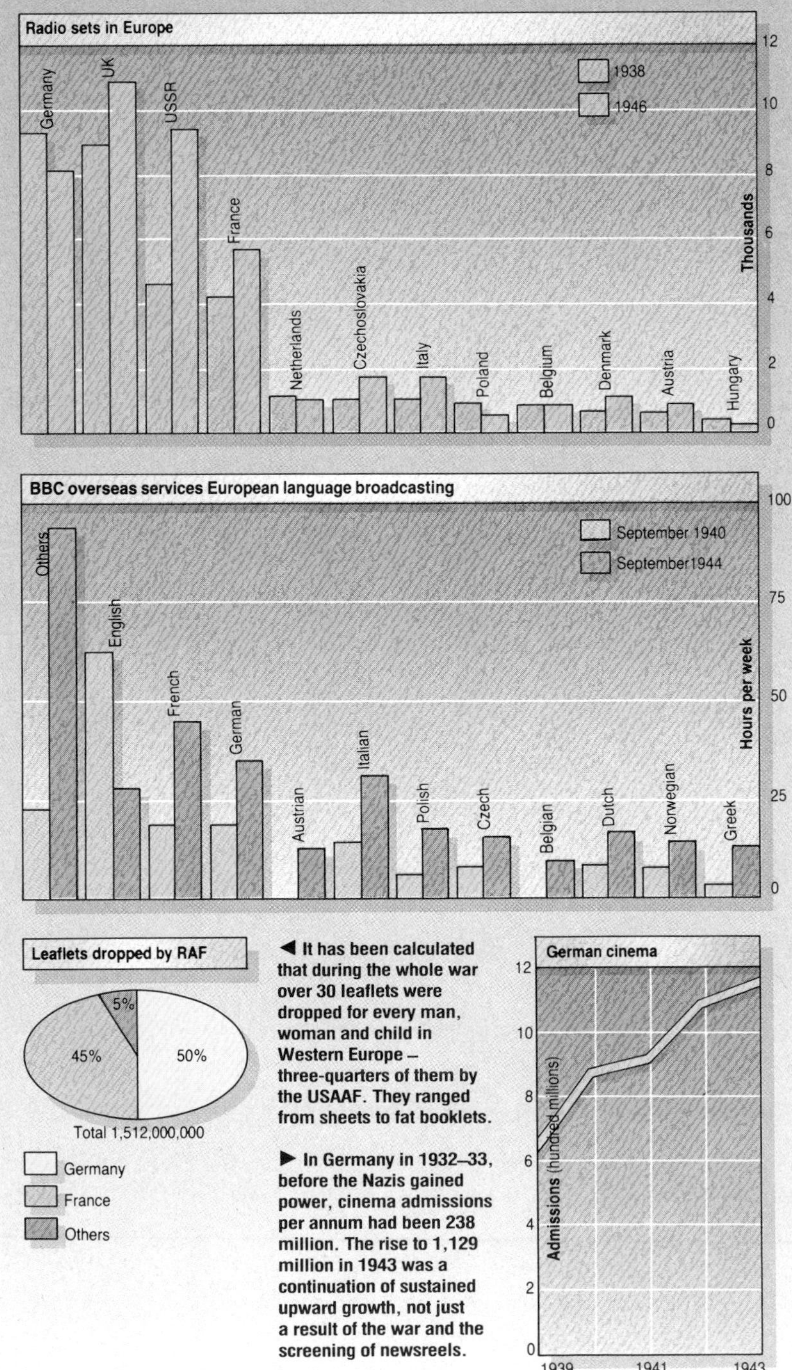

◀ It has been calculated that during the whole war over 30 leaflets were dropped for every man, woman and child in Western Europe – three-quarters of them by the USAAF. They ranged from sheets to fat booklets.

▶ In Germany in 1932–33, before the Nazis gained power, cinema admissions per annum had been 238 million. The rise to 1,129 million in 1943 was a continuation of sustained upward growth, not just a result of the war and the screening of newsreels.

If bombs failed to bring British, Germans and Japanese (till Hiroshima) to submission, neither did all the wiles of propaganda. From the earliest days of human society, secular and religious elites had visual, verbal and musical means to impress their power and moral superiority on subjects and on foreign peoples. Many celebrated classic books and artworks could fairly be described as propaganda. But 20th-century technology has vastly increased the scope for propagandists. British propaganda in World War I was believed, not least by Germans, to have set a new standard of effectiveness. The Nazis established a propaganda ministry, under Goebbels, in 1933. Mussolini followed suit in 1937.

When war broke out, all combatants used similar means. Propaganda fell into two broad categories. Some, dubbed "white" in the UK, told the truth, at least as the propagandist saw it. "Black" propaganda involved fakes and deliberate lies. Aims were likewise twofold. One was to sustain "morale" – the will to work and fight on – among home population and servicemen. In this, propagandists in all countries could claim success. Propaganda directed at enemy troops and populations and their "morale" never seems to have been decisive. Even the "propaganda of events" was not always potent. The UK struggled on in 1940, though this meant bankruptcy and loss of empire.

In propaganda's simplest and oldest form, a leader addresses his people. Roosevelt's "fireside chats" over the radio were famous. Churchill was essentially a parliamentary orator, while Hitler and Mussolini preferred huge rallies, often out of doors, but their speeches too were broadcast.

Posters are also simple. Like the newspaper press – tightly controlled in the Axis countries and the USSR, more flexibly handled in the UK and the USA, but everywhere a crucial propaganda outlet – posters could only be directed, in war, toward home populations or friendly or conquered nations. The British went in for homely, often humorous posters. German poster art was far more strident. Leaflets were another old device. But with aircraft they could now be scattered profusely over enemy territory.

Cinema was a vastly potent new medium, and newsreels and documentary films gave direct outlet to propagandist views. Fictional films were harder to use precisely. The function of entertainment might conflict with the clear expression of propagandist points. An extreme case which illustrates the ambiguity latent in the medium occurred when German and Japanese directors attempted to collaborate on a feature, *The New Earth*, about a Japanese hero's conversion back from democratic ideas. The Japanese director, Itami, objected that the German, Arnold Fanck, did not

PROPAGANDA AND THE ARTS

Propaganda media used in World War II

Goebbels and the organization of German propaganda

Propaganda and morale in the UK

Black GIs in the UK

The content of wartime films

Writers, musicians and painters

Francis Bacon

understand the Japanese mentality. Two versions went out, one for German consumption, the other for Japanese.

Radio was probably the single most important propaganda medium, at least in Germany and the UK. The Nazis, seeing its potential, had put on the market an excellent cheap *Volksempfänger* – "People's Radio" – and in six years before the war, the number of private radio sets in Germany had quadrupled. In the UK, too, they were commonplace. The medium's chief snag was that people could tune in to foreign propaganda.

Propaganda and art

At its most grandiose level, propaganda involved the proclamation of values for which nations fought. Here the arts and "high culture" were important. The pianist Myra Hess playing in a British documentary film represented liberal civilization; the Berlin Philharmonic orchestra stood for the preeminence of German music. While an Italian poster of 1942 projected the war as one for European culture against US barbarism, with a grinning black GI making off with the Venus de Milo, the Japanese sought to justify their New Order to other Asian peoples with the

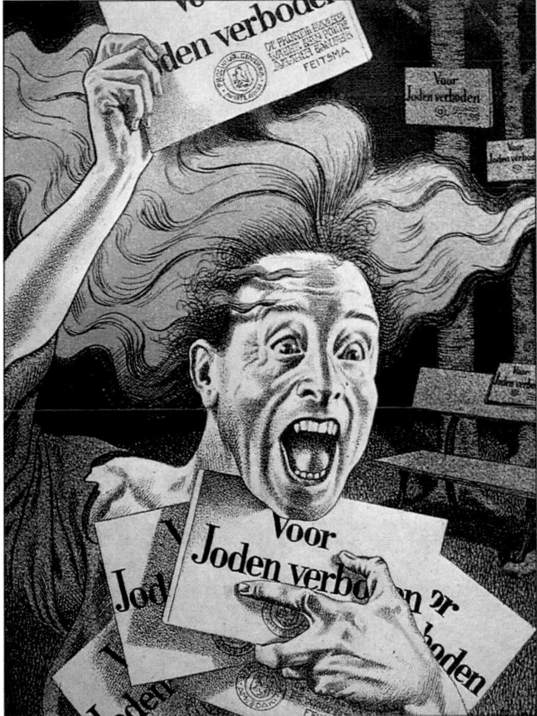

► A caricature of Goebbels drawn by L. Jordaan for the Dutch weekly *De Groene* (1941).

▼ In bombed London the *Evening News* is delivered to a vendor. British newspapers were restricted in size by paper shortage. They were also restricted in freedom of expression, mostly by self-restraint. Yet they were still eagerly purchased.

claim that "Nippon" culture, drawing on the ancient civilization of China and the ethical and esthetic traditions of Asia in general, but enhanced by the "scientific element", was "creating a new type of civilization, richer than any in existence."

At a lower level, no medium was too trivial to merit the attention of propagandists. Postage stamps were a well-established means of projecting "national" values. Allied and German "black" propagandists alike produced subversive imitations of enemy stamps – for instance, an Italian issue which showed Hitler and Mussolini face to face over the slogan "Due Popoli Una Guerra" was mimicked in a version which proclaimed "Due Popoli Un Führer". Amongst the innumerable matters which worried Goebbels were advertisements for the purchase of diamonds, and jokes in cabarets.

Propaganda blended, of course, with education. During the long siege of Leningrad, the city's children's magazine, *Koster*, missed only one issue – and that was broadcast: an index of the importance attached to indoctrination in the Soviet system. But young people everywhere found it hard to resist the popular culture of the USA. Swing music, Hollywood films, *Life* magazine and even chewing gum were mighty propaganda assets because, just as they were, without messages attached, they were infectiously attractive to foreigners.

▼ Approved art under Hitler dealt with a limited range of subjects, such as peasants with large families, purely "Aryan" allegorical nudes and, of course, marching men in uniform. Huge murals on such lines decorated public buildings. Lavish magazines displayed "Third Reich Art". For their 1937 exhibition of "Degenerate Art" (below), the Nazis selected the least successful and most experimental works of "Decadent Expressionism" and used them to show how brutally such sacred themes as womanhood and fatherland had been treated.

The master of German propaganda

As the tide of Nazi victories turned toward defeat, Dr Paul Joseph Goebbels mused: "In the hour of crisis at Dunkirk, Churchill showed admirable frankness in drawing the necessary consequences and telling the English people the absolute truth. At the time we did not understand this, but with these tactics, Churchill roused all the defensive resources of the nation. Our task today is to rouse the same defensive resources in the German nation." Goebbels, regarded by many, including himself, as the greatest propagandist of all time, was, up to a point, more respectful of the truth than most of his fellow Nazis.

During the war Germany's enemies believed Goebbels to be a "Propaganda Czar", omnipotent over the German media, a Mephistophelean genius, hideously assured in his brilliant lying. The truth was less remarkable. As in other spheres, control of news and opinion in the Reich was rather clumsily divided. Dr Otto Dietrich, head of the Nazi Party Press Department, ranked exactly equal to Goebbels, and did not get on well with him. Nor did Goebbels control the Division for Wehrmacht Propaganda, which presented the most important of all news, that from the battlefronts. Hitler himself frequently meddled with this. The army once reported that it had taken

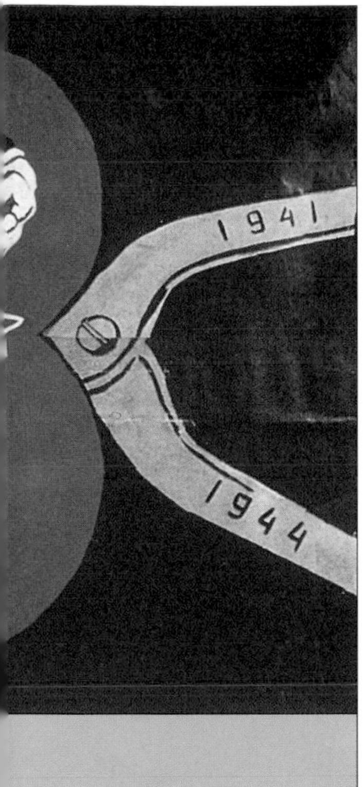

▲► Posters from three countries. Top: "Three years of war" from the USSR. Soviet poster art was particularly varied and effective in its sarcastic portrayals of Hitler and Goebbels. Top right: "The inhuman crimes of 'Gangster Pilots' expel the United States forever from civil contact", from Italy. The fascist authorities in Italy had been particularly sensitive about the depiction of Italo-Americans in US gangster films. Here they get their own back. Above, a typical German image of "Jewish Bolshevism" – "this threatens when we let up – fight to victory".

3,000 prisoners. Hitler insisted on an extra nought, then added, "Don't put 30,000 but 30,723, and everyone will believe that an exact count has been made." Though Goebbels lorded it over radio and films, and had much scope with periodicals – the "quality" weekly *Das Reich*, to which he himself contributed, had the remarkable circulation of 1.5 million – the historian Michael Balfour has concluded that "from at least 1942 until the closing months of the war" he had "lost control over Nazi policy towards the Press and over handling of news in general."

Yet, paradoxically, this was the time of his greatest glory. Whilst Hitler would not go near bombed cities, Goebbels endeared himself to their inhabitants by his visits. It was he who lost half a kilogram in weight when he called for total commitment to Total War in a long speech at the Berlin *Sportpalast* on 18 February 1943 which induced orgasmic fervor in an audience of 15,000. (Characteristically, he remarked afterwards that it had been "an hour of idiocy – if I'd told them to throw themselves out of the third-story windows, they'd have done it.") As Germany's fortunes declined, Goebbels' charisma increased.

Goebbels was born in the Rhenish town of Rheydt in 1897, the deformed son of devoutly Catholic petty-bourgeois parents. His puny frame and clubfoot debarred him from war service in World War I. Besides earning a doctorate in literature at Heidelberg University, he published an unsuccessful novel, before becoming the devoted master propagandist of the rising Nazi movement.

He never submerged his sardonic intelligence, nor a radical, antiplutocratic animus. Highly musical, he preferred Mozart to Wagner. A workaholic who concerned himself with the pettiest questions bearing on morale, he nevertheless had broader perspectives than many of his colleagues. Aided by terror-bombing of German cities and by the Allied policy of seeking unconditional surrender, Goebbels posed with success, in 1943, as the tribune of the German people, demanding sacrifice from the rich, and committed to Victory or Death.

German propaganda and the Soviet Union

Professor Hirt, head of the Anatomical Institute at Strasbourg University, wrote in late 1941 to Himmler's adjutant: "We have a large collection of skulls of almost all races and peoples at our disposal. Of the Jewish race, however, only very few specimens of skulls are available ... The war in the East now presents us with an opportunity to overcome this deficiency. By procuring the skulls of the Jewish-Bolshevik commissars, who represent the prototype of the repulsive, but characteristic, subhuman, we have the chance now to obtain scientific material." (Himmler obliged, though Hirt received not commissars but the skulls of specially murdered people, sent from Auschwitz.)

The British, presumed to be largely "Nordic", were never depicted as subhuman by responsible German propagandists. Slavs were more problematic. Some of them – Ukrainians, even Poles – hated Stalin so much that they would help Germany. Probably more Russian youths than Germans approximated to the ideal blond "Aryan" type. Communism, furthermore, was a hard creed to fight, when Nazis themselves indulged in antiplutocratic rhetoric. Many Germans had belonged to the communist party before the Nazis suppressed it. Goebbels personally supervised a successful "Soviet Paradise" exhibition in Berlin in April 1942, which claimed to show the appalling conditions in which Stalin's subjects lived. (He could not stop a characteristic Berlin joke going the rounds: "Why did they shut down the exhibition?" Answer: "Because the people of north Berlin wanted their belongings back.") Goebbels privately admired Soviet achievements and realized that a racialist line could be counterproductive. Germany needed to mobilize Eastern peoples on its own side. "In the long run we cannot solicit additional workers from the East if we treat them like animals within the Reich."

I DELITTI INUMANI DEI "GANGSTERS PILOTI" RADIANO
PER SEMPRE GLI STATI UNITI DAL CONSORZIO CIVILE

Such contradictions were transcended by conflating Bolshevism with Judaism. Propagandists attacked the "Jewish-Bolshevik Murder System". Jews had masterminded the alliance between communism and the US plutocracy. The forged "Testament" of Peter the Great revealing Slavic Imperialist aims was set beside the equally spurious antisemitic "Protocols of the Elders of Zion". Educational materials portrayed Stalin's regime as the puppet of Jews.

An SS pamphlet, *The Subhuman*, which was originally intended to serve as an introduction to Eastern peoples for German troops marching into Russia, portrayed these peoples as mongoloid animals. Goebbels tried to get it banned, at least in Germany. Far too late, in 1944, his propagandists began to make use of the Committee for the Liberation of the Peoples of Russia led by General Vlasov. The "subhuman" armies were by then on their way to Berlin.

Britain's radio war

The broadcasts from Bremen of William Joyce, "Lord Haw Haw", caused passing concern in Britain in 1939–40. An Irishman, formerly a member of Oswald Mosley's British Union of Fascists, Joyce attracted large audiences at first, but these tailed off as the war began in earnest. Joyce was a "white", truthful propagandist. The Germans also had fake "underground" stations purportedly operating on British soil and expressing pacifist, communist and Welsh or Scottish nationalist views. These had no discernible impact on British opinion.

But the British were not discouraged from mounting similar "black" operations. "Freedom stations" broadcast to Germany and occupied Europe. One of these, "LF", broadcast "news" to Germany which might have disruptive effect, describing, for instance, how SS men took a certain brand of sleeping pill to get through air raids, with the hope that listeners would demand these habit-forming drugs and become addicted. Sefton Delmer spoke on "Gustav Siegfried Eins" (GS I) as "Der Chef", a Prussian patriot disgusted by the depraved habits of top Nazis and SS men. His pornographic details had some appeal for German listeners; their serious purpose was to suggest and foster a rift between Nazi Party and the Wehrmacht. "Black" propagandists created dangerous rumors. Radio "G 9" mixed these with hard news. A sample from October, 1943: "It is announced that 12 U-boats have been sunk and 18 are missing." (True.) "In Hanau-am-Main the birth of an elephant has aroused excitement. 'Haruna' is known as the peace elephant because she had her first child in 1871 and her second in November 1918." (False.) "Sixty aircraft took part in a retaliatory raid over England during the night. London had its worst raid for two years." (True.)

A much higher proportion of truth to lies characterized the fake German forces newspaper which the USAAF dropped in hundreds of thousands daily behind the German lines late in the war – one of the relative successes of the leaflet campaign. British propagandists had been hampered by the RAF's refusal to set planes aside specifically for leafleting; the USAAF had a squadron of Flying Fortresses devoted to this task, and were the first to develop effective "leaflet bombs".

But the Allies' most useful propaganda weapon was the reputation which the British Broadcasting Corporation earned for its truthful reporting of news. It did not minimize British defeats, nor (deliberately) exaggerate British victories. The BBC was rewarded with large and appreciative audiences. Its "white" German Service broadcast mainly news (though it also presented "features" such as a soap opera chronicling week by week the life of a Berlin charwoman, and other humorous programs which sounded sympathetic rather than overbearing). The Gestapo estimated that the BBC's German audience was 1 million in 1943; by the autumn of 1944 it was reckoned to be 10 or 15 million, though many of these were outside the Reich.

The BBC's "V Campaign" was so successful that it got out of hand. In January 1941 the head of the BBC's Belgian service advised that they should spread the letter V, the first in the French word for Victory and the Flemish for Freedom. The idea caught on in France and Holland. The Morse signal for V was used, then the corresponding opening notes of Beethoven's Fifth Symphony. Chalked Vs appeared everywhere – on walls, roads, Wehrmacht cars – and schoolteachers were encouraged to call children to order by clapping the "V" rhythm. Soon "Colonel Britton", broadcasting Europe-wide on the English service, was coming close to incitement to sabotage. He was taken off the air, in May 1942, not long after the Dutch government-in-exile had complained that he was giving orders and making promises to their people without consulting them. Meanwhile the Germans had neutralized "V" by taking it up themselves. A huge V

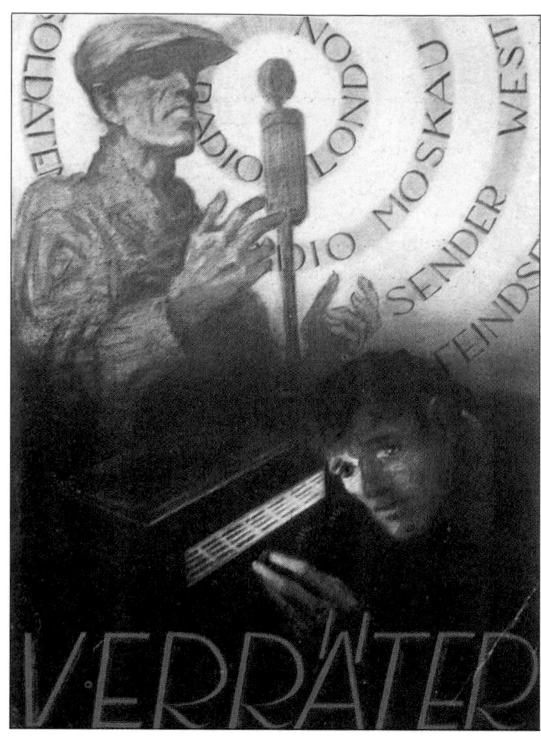

◀ "Traitor", a German poster of 1944 by Max Spielmanns, shows how concerned the Nazis were about listening to enemy propaganda broadcasts. But as the war turned against Germany on the Soviet front and Soviet forces captured huge numbers of prisoners, families in the Reich could only find out if menfolk were alive by tuning in to Soviet broacasts which from time to time listed names and even brought prisoners to the microphone.

▼ Most of Britain's inhabitants had long lived in towns. Nevertheless, the idealized rustic village was a potent symbol of national identity, of "England's Green and Pleasant Land". Southern counties which were still largely rural in aspect – Kent, Sussex and Hampshire – were on the "front line" in the "Battle of Britain", so the image of the English village could be related more directly, by juxtaposition and contrast, to German barbarism than could Yorkshire moors or Scottish coalfields.

Black GIs in the UK: a Public Relations Problem

Black US troops or GIs arrived in the UK in 1942. They multiplied the "colored" population tenfold and were camped in areas where people of African origin had never been seen. British civilians liked them very much. Unlike many white GIs they did not sneer at British standards of living. They loved children.

Churchill's war cabinet was uneasy about their presence. After attempts to prevent any black GIs coming at all had failed, the government accepted the US army policy of black-white segregation. The secretary of state for war (James Grigg) clearly thought that the British themselves should learn from Jim Crow. The colonial secretary (ironically) was the only minister to record his unease at the war cabinet's conclusion that "the people of this country should avoid becoming too friendly with coloured American troops." Brendan Bracken, the minister of information, in private accepted segregation wholeheartedly though he wrote a press article calling for the "social equality of coloured people".

In the violent affrays between white and black US servicemen which occurred at the rate of about four a week in the winter of 1943–44 – one, in Bristol, involved 400 GIs – British people usually sided with the blacks. When a black GI, Leroy Henry, was condemned to death by a military court on shaky evidence for the "rape" of a white woman, a popular tabloid, the *Daily Mirror*, launched an outcry which forced General Eisenhower to overturn the verdict. But the government said that British soldiers should "respect" the white American viewpoint.

▼ Lee Miller's 1944 photograph convincingly evokes the atmosphere of an English pub. The air of prosperous smartness about the black GIs is also credible. In September 1942 one Glasgow member of parliament complained bitterly that *black* US soldiers, like white ones, were paid more than five times as much as British privates. Oddly there was little or no connection between the arrival of black GIs and the craze for New Orleans-style jazz which hit Britain toward the end of the war.

Olivier's **Henry V** *had colour, movement, heralds and fanfares, music and above all language that stood my hair on end. Like all British films of the 1940s, this now seems to speak for a posh minority, but not then. We thrilled at the* **shared** *victory, the king's nocturnal conversations with his soldiers, the Robin Hoodness of it all...The four of us aesthetes went, on one of the few nights we were allowed out, to see the film for the third or fourth time in the nearby steel-town of Corby. Running through the wet streets for the last bus back to camp, we swore to be brothers forever, our boy-faces lit by flashes from blazing foundries.*

PETER NICHOLS

appeared in Paris on the Eiffel Tower, and a 9m (30ft) banner hung from the royal palace in Amsterdam proclaiming: "'V' for victory which Germany is winning for all Europe on all fronts."

Movies at war

In the interwar period and during the war cinema was increasingly popular everywhere in the world. In British India, for instance, the number of cinemas rose from 1,600 in 1938 to over 3,000 10 years later. Berlin alone had 400 cinemas in 1942, the whole of Greater Germany over 7,000. Goebbels of course gave this potent medium of film close attention. As resources grew scarce in 1942, such independent film companies as remained were nationalized to ensure that film was used in the most efficient way possible to meet demand at home and in occupied Europe. The bombing of cinemas was such a major catastrophe that Goebbels inspired massive efforts to restore them. Even in 1944 resources were made available for spectacular colored films, such as *Kolberg*, a story about German defiance of Napoleon, which was eventually screened in the last January of the war. Goebbels' "German Weekly Newsreels" continued in production until a few weeks before his suicide, though by the end nearly two-thirds of the Reich's picture houses had ceased to function.

In the UK some 25 or 30 million cinema seats were sold every week, implying that many people attended more than once. One young miner, indeed, boasted that he had seen 306 films in 1942, 382 the next year, and 430 in 1944. Most of these would have been American. The British cinema industry increased its prestige and popularity, with such features as *Next of Kin*, originally made to warn servicemen against "careless talk" but with such an attractive spy story that it was released to the public, and with a spectacular version of Shakespeare's *Henry V*, starring Laurence Olivier. But output was cut back under conditions of total war, while the great Californian Dream Factory continued to pour out features of every kind, from unashamed escapism to blatant propaganda.

Laurence Olivier's PRESENTATION IN TECHNICOLOR
of HENRY V
by WILLIAM SHAKESPEARE
A TWO CITIES FILM · EAGLE-LION DISTRIBUTION

▼ **The film which so thrilled the young RAF man quoted above was a faithful version of Shakespeare's play, which had originally served as nationalist wartime propaganda in the grim 1590s. It was lavishly produced, in Ireland, using technicolor.**

Cecil B. De Mille's Paramount studios, for instance, profited vastly from comedy films featuring the trio of Bing Crosby, Bob Hope and Dorothy Lamour on *The Road to* ... various exotic destinations. In another Hope movie, *My Favorite Blonde*, when Madeleine Carroll at last revealed that she was a British agent, the dimwitted hero retorted, "Too late, sister, I've already got an agent." But antifascism was taken very seriously indeed in the version of Ernest Hemingway's *For Whom The Bell Tolls*, which was Paramount's top hit of 1943, staring Gary Cooper and Ingrid Bergman. The latter also graced what is now the best-loved Hollywood war film, Warner Bros' *Casablanca*, where she played opposite Humphrey Bogart.

Two very famous Hollywood figures took controversial political stances. Charlie Chaplin was fervently anti-Nazi – his send-up of Hitler in *The Great Dictator* (1940) was adulated by the critics, less liked by the public. He was also pro-Soviet,

▲ A still from MGM's *Mrs Miniver:* an English parson officiates in a bombed church. Made in Hollywod, starring the very popular Greer Garson, it presented an idealized middle-class English family bravely at war on the "home front". Sophisticated British cinemagoers sneered. Less critical people wept copiously. A social survey studied reactions. One shop assistant said, "very sentimental and untrue to English life and yet it seemed to touch the spot". A male tax inspector found it "nostalgic and inspiring".

◀ A scene from Noel Coward's *In Which We Serve* (1942). Coward (front right) had been famous as the author of elegant stage comedies and debonair performer of his own light songs. He amazed his public by writing, producing and directing this somber film composed of flashbacks in the minds of crewmen from a sunken Royal Navy destroyer clinging to a raft off Crete. He himself starred as the captain.

The UK and America: the Story of Colonel Blimp

With the UK in desperate need of food and tools and munitions from the USA, Churchill's great speeches of summer 1940 were aimed at US opinion as much as British. Broadcasting to the USA was important war work. It was a great advantage that several distinguished US journalists in London were strongly anti-fascist. As London was bombed, the likes of Ed Murrow and Negley Farson helped to establish the image of an heroic city joking and working on under terrible assault.

As the USA turned from suspicious friend to difficult ally, British propagandists tried to exploit the positive and muffle the negative. Americans were highly susceptible to English landscape and "heritage" – the BBC favored for US consumption music evocative of rural England, like Julius Harrison's *Bredon Hill*, a rhapsody written, as the composer himself told American listeners, to celebrate a landscape representing "England at its oldest... the heart of Mercia... the country of Piers Plowman."

On the other hand, Americans had to be convinced that the obnoxious British class system was breaking down, and that GIs would not be fighting alongside an army run, as in George III's day, by class-bound aristocrats. Films were a key medium in this cause.

Before the war, the cartoonist David Low had created "Colonel Blimp" to typify the reactionary attitudes of the old-fashioned "officer class". At the height of the war, British critics of the government were prone to allege that the snobbish attitudes of "Blimps" in the army were seriously hampering the national effort. *The Life and Death of Colonel Blimp* (1943) was a major feature, filmed in rare and expensive technicolor by the gifted team of Michael Powell and Emeric Pressburger. It showed the old "Blimp" fading away and a "New Army", typified by a young officer risen from the ranks, at the helm of Britain's struggle. The ministry of information opposed it "strongly" and would not permit the release of Laurence Olivier from his service in the Fleet Air Arm to star in it. The war office asked for its suppression, partly on the grounds that German soldiers were represented as more realistic in war than British. Churchill called on Bracken, the minister of information, "to stop this foolish production before it gets any further. I am not prepared to allow propaganda detrimental to the morale of the Army." Bracken claimed, rather implausibly, that he had no power to halt *Blimp*. It has been suggested that he had deliberately created a clamor of controversy so as to convince Americans that the film was not official propaganda. This seems far-fetched. However, the fact that Churchill opposed the film being made, and that its export to the USA was prohibited for some time, did its reputation no harm when it crossed the Atlantic at last. Meanwhile, *Desert Victory*, an army film unit documentary which presented the British campaign in North Africa as a "people's war", was extremely successful in the USA, so much so that it excited jealousy among US filmmakers, who were still backward in documentary techniques. It was awarded an Oscar.

▲ Roger Livesey (right) as Colonel Blimp.

▲ **Chaplin's impersonation of Hitler ("Hynkel") in *The Great Dictator* (1940) was a financial success. He invested two million dollars of his own money and the film grossed five million. But many cinemagoers were disappointed when they saw it, perhaps because publicity "hype" for it had been excessive, perhaps because they were disconcerted when the beloved silent clown spoke on screen for the first time.**

declaring in 1943, "On the battlefields of Russia, democracy will live or die." His chief rival as the world's top laughter-maker was Mickey Mouse, created by Walt Disney. War came as a body-blow to Disney's cartoon studios: Europe had provided nearly half their market. *Pinocchio* and *Fantasia* (1940) and even *Bambi* did not initially recoup their costs, though Donald Duck won an Oscar for throwing a ripe tomato in *Der Fuehrer's Face* (1942). Disney faced financial disaster and his staff struck in protest against poor pay. The US government came to his rescue with a lucrative contract to make animated educational films for the armed forces, with such inspiring titles as *British Torpedo Plane Tactics*. Disney decided to put his own money behind his faith in strategic bombing. *Victory Through Air Power* only just broke even, but Churchill liked it and told Roosevelt so: FDR had it shown to the Joint Chiefs of Staff.

Germany ceased to import US films in 1940. Its film industry had to provide a range of products to match Hollywood's. As time went on Goebbels increased the ratio of "entertainment" to propaganda, though the former was not innocent of ideology; thus *Request Programme* (1940), which built on the success of a popular radio feature, presented a love affair between a German officer and a girl from whom he was repeatedly separated by duty. They were reunited through her radio request for the fanfare of the 1936 Olympics, where they had first met, and the happy ending was combined with a rousing chorus of the favorite warsong, "We Are Marching Against England." The propagandists in charge of German

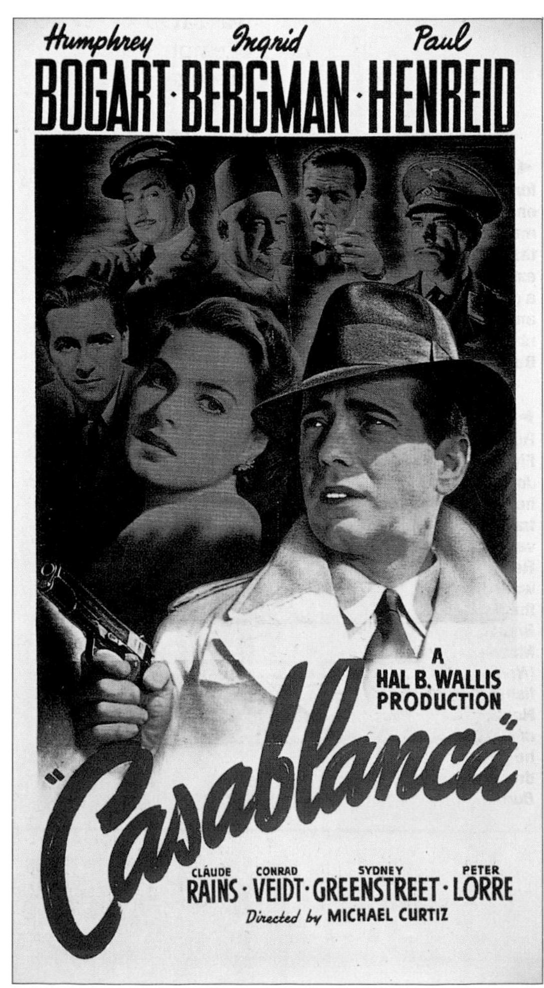

ROAD TO SINGAPORE
with
BING
CROSBY
DOROTHY
LAMOUR
BOB
HOPE
CHARLES COBURN · JUDITH BARRETT · ANTHONY QUINN · JERRY COLONNA
DIRECTED BY VICTOR SCHERTZINGER
A PARAMOUNT PICTURE

▲ Bing (left) and Bob (right) "hit the Road" in another hit. As Bob's ukelele suggests, they provided light relief for millions. For one British schoolteacher, the "craziness" and the songs of a *Road* movie "made it a real escape from reality".

cinema did not make the mistake of presenting the Nazi Party and its ideas directly. A common tactic was to produce historical films which associated political and cultural heroes from the past with Hitler and his regime, as in two films of late 1940: Horst Caspar starred as *Friedrich Schiller*, a young German genius struggling against reactionaries, Paul Hartmann as the patriotic *Bismarck* who, to safeguard his new German state, did a deal with Russia (like Hitler in 1939), and who was saved by "divine providence" from a murder attempt by an English Jew. The antisemitic hate propaganda of *The Jew Süss*, another lavish historical film with a galaxy of popular stars, prepared Germans for the "final solution" of the so-called Jewish Problem. Concentration camps were a British invention, as shown in Emil Janning's *Ohm Kruger* of 1941, in which heroic Boers struggled against the evil British Empire. "Enough of this humanity drivel!" cried the sadistic British commander, Lord Kitchener, using Boer women and children to provide a protective wall for his troops. But to an intelligent Martian, the outputs of Western film industries during the war would have looked very similar. Sentiment and suspense, transfiguration through death in battle against cunning and wicked national enemies, propaganda dressed as entertainment, gave pleasure to all sides.

◀ *Casablanca* won the Oscar for best picture of 1943. Its enduring charm still draws much from the revelation that the cynical American expatriate played by Bogart is a good anti-fascist at heart, and from the fresh European idealism projected by Ingrid Bergman.

▶ Here's a Nazi in trouble... Ronald Reagan and Errol Flynn in Warner's *Desperate Journey* (1942). Flynn, playing here an RAF bomber pilot trapped in Germany, was a vastly bigger star than Reagan. He saw very varied war service on celluloid – in the Naval Air Corps (*Dive Bomber*); in the Canadian Mounties chasing a Nazi (*Northern Pursuit*); as a fisherman helping the Norwegian Resistance (*Edge of Darkness*). At last in 1945 he joined paratroopers dropping on Asia in *Objective Burma*.

Writers in combat

For many writers in combatant countries, total war exacerbated a dilemma which was as old as the rise of the nation state and of "mass society". Should they devote their talents to public service – perhaps even stop writing and fight instead – or should they continue to follow their individual preoccupations and visions? Some writers were happy to make propaganda, like the leading Russian novelist, Ilya Ehrenburg, who turned full-time journalist and hatemonger. A number of writers of above-average talent had been attracted to Fascism before the war, and some were now ready to make their support noisy. The violently antisemitic French novelist L.-F. Céline was one who suffered after the Axis defeat – in his case, a spell in prison. Ezra Pound, one of the major figures in American Modernist poetry, a long-time resident in Italy, was arrested in May 1945 by American forces. He had made rabid attacks on Roosevelt, Churchill and the Jews over Rome Radio. He was adjudged insane and confined for 11 years in an asylum. Fellow-countrymen were still kinder to Norway's most famous living writer, Knut Hamsun.

This powerful novelist had won the Nobel Prize for Literature in 1920. His pro-Nazism is easy enough to explain. As a young man he had shared his generation's fascination with the writings of the philosopher Nietzsche. On his travels he had concluded that the USA and the UK were uncultured countries, perverted by industrial mechanization. He was attracted by the idea of a German Empire over Europe ruled by pure-blooded Nordic people, naively favored Nazism before the war, and equally naively took up an antisemitic position. His wife and two sons were active members of Vidkun Quisling's Norwegian Nazi Party. Hamsun's great age – he was 80 when war broke out – did not prevent him from vigorous public gestures after the German occupation of Norway in April 1940. "The Germans are fighting for us all", he assured his fellow-citizens at this time. When peace came, Hamsun was declared to be sane; he was also judged guilty of treason, fined heavily, but allowed to live out his days in freedom. He died in 1952.

The case of Jean Anouilh's *Antigone* shows how even an apparently "apolitical" writer could seem compromised in a context of occupation and resistance. Intelligent theater was a booming art form in occupied Paris. Updated versions of classic stories had long been a staple of the French stage. Jean-Paul Sartre showed in 1943 how this convention could be used to promote resistance. In *Les Mouches* (The Flies), Orestes' murder of his mother Clytemnestra and her lover the tyrant Aegisthus represented intellectual freedom striking a blow for society. *Antigone* appeared eight months later, in February 1944, and ran till

▲▼ J.B. Priestley (above), popular British novelist and playwright, made an enormous impression with his morale-boosting "Postscripts" to BBC news, from June to October 1940. With his homely Yorkshire accent and Socialist views, he generated the feeling that ordinary, decent British people could and must defeat Hitler and "create a noble future for all our species". The BBC halted a second series in 1941 on government instructions – Priestley was too leftwing. (Below) Ilya Ehrenburg, Soviet-Jewish novelist and propagandist.

Liberation and beyond. Anouilh in his plays liked to set youthful idealism against middle-aged compromise, and had in fact been inspired to write this work by the action of a young Resistance fighter who, in August 1942, had fired at a group of collaborationist leaders at a rally, and wounded Pierre Laval. But the German censor approved it, and collaborationist critics praised it highly. Was the reasonable, weary Cleon who absorbed Antigone's defiance perhaps to be identified, favorably, with Laval or Pétain? Some Vichyites in the audience certainly thought so. Word went round Resistance circles, *"N'y allez pas, c'est une pièce nazi."* The clandestine press denounced Anouilh, even comparing him to those Frenchmen who had enlisted in the SS. Fortunately, he emerged at last with reputation and honor intact.

At a juncture when timeless tales were liable to be received as topical allegories, it was very difficult to write directly, in plays or novels, about wartime societies, without making propagandist gestures. Little fiction from the war years is still worth reading. Writers' experiences would produce powerful work years or decades later, as famous novels by Joseph Heller and Kurt Vonnegut, Günter Grass and Heinrich Böll, Italo Calvino, Cesare Pavese, Evelyn Waugh and J. G. Ballard demonstrate. But the restraint upon honesty at the time was exemplified in 1944 when – in tolerant, unoccupied, and victorious Britain – several publishers turned down Orwell's anti-Stalinist beast-fable, *Animal Farm*. This was not an appropriate moment to criticize Britain's Soviet ally.

Russian writers, of course, had been used to worse restraints long before the war. Silence and suicide were proven ways of retaining honor. One major woman poet chose the latter recourse. Marina Tsvetayeva had returned to Russia from exile in 1939. She had been ostracized and could not publish. Her husband had been arrested and executed, her daughter sent to labor camp. Following the German invasion she was evacuated to the small provincial town of Yelabuga, and after ten days there she hanged herself. The great Anna Akhmatova had published nothing since 1932. Privately, she had completed her sequence *Requiem* for the victims of Stalin's purges of 1935–40, when her own son had been imprisoned. When war came, she experienced the bombardment of Leningrad before being sent to Tashkent, in Central Asia. But her gifts were now in demand as never before. She produced public poetry extolling the Russian people and their will to survive. *Muzhestvo* (Courage), in 1942, featured on the front page of *Pravda*, surrounded by war news and casualty lists. Meanwhile she attended to her longest work, *Poem Without a Hero* (1940–62), which she saw as essentially private.

In Britain, the cry was, "Where are the War Poets?" The young soldier poets who had died in World War I had become cult heroes. But their better World War II successors refused to oblige by providing gushes of patriotism or outraged responses to violence. Their often-cynical disillusionment was typified by the brilliant work of Keith Douglas, killed in Normandy in 1944. He wrote to a friend who had asked him to be more lyrical: "To be sentimental or emotional now is dangerous to oneself and to others." Reassurance and uplift were left to noncombatant poets, notably T. S. Eliot, whose wartime *Quartets* interwove "topical" and "public" resonances – bombed London features in "Little Gidding" (1942) – with "timeless" religious themes. Perhaps the most powerful British verse of the war years was written in Scottish Gaelic by Sorley Maclean, who fought in North Africa.

Major French poets who had been involved in the Surrealist movement between the wars were spurred, like Maclean, by explicit leftwing commitment. René Char served as "Capitaine Alexandre" in the Provençal resistance. Robert Desnos also worked for the Resistance and died in Theresienstadt concentration camp. Louis Aragon broke wholly away from Surrealist obscurity with six collections of wartime poetry addressed to the ordinary French reader – the Vichy censors eventually objected, and Aragon was published "underground". Paul Eluard moved in the same direction, toward a new directness.

Many major German and Austrian writers were already in exile when war broke out. Bertolt Brecht had fled in 1933. Via Switzerland, Denmark, Finland, Russia, he arrived in California in 1941. Along the trail he wrote masterpieces, such as *Mother Courage* (first performed in Zurich, 1941), which could not be seen by his German public. Fellow exiles on the West Coast included Thomas Mann and his brother Heinrich and the Austrian poet and novelist Franz Werfel. Stefan Zweig committed suicide in Brazil in 1942. The poets Jesse Thoor and Erich Fried wrote in London. Else Lasker-Schüler would die in Palestine.

Music divided

In 1933, on the eve of Nazi supremacy in Germany, the preeminence in Western musical culture of the German-Austrian tradition stretching from Bach through Beethoven to Wagner was taken for granted generally, though in fact the torch of experiment had now passed to France and the USA from German-speaking lands, where most of the chief innovators – Arnold Schoenberg, Alban Berg, Kurt Weill, for instance – had been Jewish. So were many supreme performers, and regular orchestral musicians. The irruption of Nazism from 1933 was followed by purges of Jewish players and official attacks on "decadent" experiment. Schoenberg left for the USA. So did leading conductors such as Otto Klemperer and Bruno Walter. Another distinguished conductor, Erich Kleiber, was not Jewish, but ended up in South America. The Hungarian conductor Antal Dorati, who quit Germany for the USA in disgust after antisemitic prejudice surfaced in his chamber orchestra, typified a host of gifted people. While important instrumental virtuosi such as the pianist Artur Schnabel also figured in this tragic exodus, convention so elevated the great conductors that they have figured from the 1930s till now as the prime symbols of music's relations with Nazism.

When war came, apart from the aged and self-infatuated composer Richard Strauss, neither Nazi nor anti-Nazi, whose conservative style

◀▲ Shostakovich (above) and Henry Moore (left, standing) were artists drawn into direct reaction to the war. Contrast Olivier Messiaen the French composer. A prisoner of war in 1940, he found a piano in his Silesian camp, and fellow inmates who played respectively violin, clarinet and cello. They gave his "Quartet for the End of Time" its première before prisoners in a camp washroom. But despite its title, this lyrical work projects Messiaen's enduring religious preoccupations, not a response to the war.

▼ There was little work for sculptors during the war. But as an official "War Artist" Henry Moore produced memorable drawings of shelters in London's underground, like the one below. Previously associated with Surrealism, he found that his sketching "humanised everything I had been doing", and realized that he had reached an "artistic turning point".

suited the regime well enough, Wilhelm Furtwängler was the only musical figure of great international standing left in the Reich. He used the power this gave him to defend Jewish orchestral musicians. Only once, in 1942, did he consent to conduct his Berlin Philharmonic orchestra at Hitler's annual birthday concert. Some top Nazis hated this favorite of Goebbels.

A young rival was at hand. Herbert von Karajan, barely 30 when war began, was, unlike Furtwängler, a Nazi Party member. (He said after the war that he had joined the Nazis in 1935 merely because membership was a prerequisite for his holding his first important post, as musical director at Aachen, but records survive which show that he had enlisted two years earlier.) Goering, political head of Prussia, controlled the Berlin State Opera where Karajan had a contract.

The British authorities had tried throughout to transcend musical nationalism. The leading young English composers, Benjamin Britten and

Michael Tippett, were both leftwing pacifists. (The latter spent two months in prison for insisting that music was the only "war work" that he was prepared to do. Yet during the war both won wider public recognition and applause for major works of nonconformist imagination, Tippett's oratorio *A Child of Our Time* (1944) and Britten's opera *Peter Grimes* (1945).

Prokofiev and Shostakovich

Paradoxically, the arts enjoyed high status and priority in the USSR under the rule of the brutal philistine, Stalin. The country's most famous composers, Sergei Prokofiev and Dmitri Shostakovich were evacuated from the front line and given the chance to work prolifically, though both had had censorship problems before the war, and both would suffer under Zhdanov's "antiformalist" cultural policy afterwards.

When the Germans invaded in 1941, Prokofiev was at once inspired to write an opera based on Tolstoy's epic of Russian resistance to Napoleon, *War and Peace*. This became his chief preoccupation for the rest of the war. Evacuated from place to place as the Germans advanced, he kept up a stream of other compositions – the ballet *Cinderella*, a Fifth Symphony, a string quartet, and much else, including the score for his friend Sergei Eisenstein's great film, *Ivan the Terrible*. Meanwhile, his efforts to stage *War and Peace* were frustrated at every turn. The first part proved very popular when it was first performed in Leningrad in 1946, but when part two had reached dress-rehearsal stage, it was suspended, for all its patriotic fervor.

Shostakovich had won popularity in Russia and later international acclaim for his opera, *Lady Macbeth of the Mtsensk District* in 1934 – but when Stalin saw it two years later it was condemned and taken out of circulation. Yet the composer was not essentially a radical modernist, and much of his work during World War II satisfied his own artistic personality as well as the demand for accessible, inspirational music.

Living in Leningrad, he volunteered for war service in 1941, but his eyesight was poor, and he was given firefighting duties at the Conservatoire. In December, he was evacuated to Kiev, where his Seventh Symphony was first performed in March 1942. It was a programmatic work depicting the tragedy and resilience under siege of the city of Leningrad, to which it was dedicated. An air-raid alert sounded during its first Moscow performance, but no one left for the shelter and the audience still refused to budge at the end, when it was widely applauded. Performances in the West soon followed.

Was art irrelevant?

In prewar Europe, "Modern" tendencies in art had been bitterly controversial. Such innovations as "Cubism" and "Surrealism" were associated with cosmopolitan, bohemian and "immoral" circles in the wicked city of Paris. "Expressionism", mostly a Germanic movement, was related to the "savage" art of Africa and the neuroses of the intelligentsia. "Constructivism" had had its short-lived epicenter in Russia after the Bolshevik Revolution of 1917. Hence moralists and traditionalists everywhere had inveighed against "nihilistic", "degenerate" and "Bolshevik" paintings and sculptures. Meanwhile the representation of alienation, anguish and upheaval had gone so far that few if any artists could find any fresh correlative to the unprecedented horrors of World War II. The Soviet Union had long imposed upon its artist "Socialist Realism", a profoundly regressive creed in which aristocratic and bourgeois heroic styles were conscripted to glorify the working man. The rise of Nazism had prompted a mass movement of German artists into exile. Such serious German artists as were not in exile survived only by "inner

Francis Bacon

Bacon (1909–) is a rare example of a major artist whose gift was nurtured rather than diverted by the war. Self-taught, he destroyed most of his earlier work. He was already 36 when he first attracted shocked attention. In April 1945 his *Three Studies for Figures at the Base of a Crucifixion* were first exhibited at a gallery in London's Bond Street alongside work by established British "modernists" – Henry Moore and Graham Sutherland – who seemed bland and complacent by comparison.

Anglo-Irish, born in Dublin, the son of a horse-trainer, Bacon had lived an apparently feckless life. His talent had been noticed as early as 1933 by a distinguished critic, Herbert Read, but for years thereafter he had subsisted on odd jobs and the proceeds of gambling. When war came, asthma ruled him out of military service. Then he returned to painting in earnest.

The "Three Figures", cramped in a harshly lit, strangely shaped, low-ceilinged space are half human, half animal, and suggest mutilation, anguish, hatred and gluttony. It would be too glib to say that they "respond" to the horrors of World War II. They bear some relationship to prewar European Surrealism, but stand outside that movement. A temperamental obsession with raw flesh, pain, grimace, scream, enabled Bacon to produce an artistic correlative for moral shock at bodies torn apart by bombs, men and women reduced to primal greed in concentration camps, Hitler in his bunker, Mussolini's body dangling from a meat hook in the center of Milan.

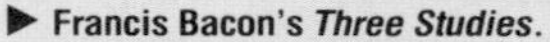

► Francis Bacon's *Three Studies*.

◄ The staider face of British wartime art. Meredith Frampton, a Royal Academician, portrays Sir Ernest Gowers, Senior Regional Commissioner for London, top man in the city's Civil Defence. It is perhaps a product of the ethos of the "People's War" that he is shown hard at work, unpretentiously dressed, with the ubiquitous British cup of tea, symbol of common sociability.

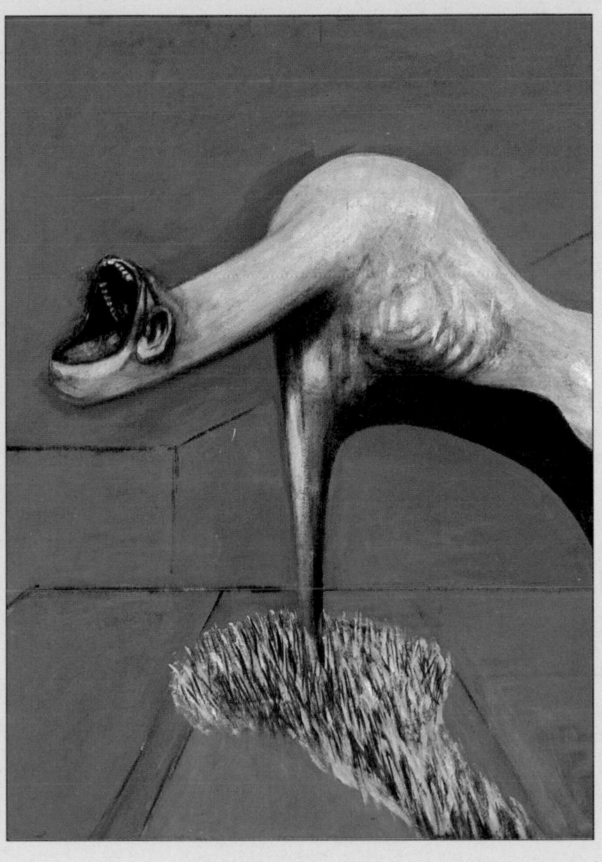

emigration", like Otto Dix, who confined himself to landscapes. Ironically, the reactionary impulse was strong also in liberal Britain, where it was almost spontaneous. Artists there had been slow to accept Continental innovations. "Unit One", formed by Paul Nash in 1933, had brought together leading British artists who were interested in European abstractionism and Surrealism – the major sculptors Henry Moore and Barbara Hepworth joined such painters as Ben Nicholson and Edward Burra. Yet Nash himself (1889–1946) had soon moved, in a quasi-Romantic spirit, to explore English landscapes, along with the younger Modernists, Graham Sutherland and John Piper.

All three worked during the war under the British government's "War Artists Advisory Committee" (WAAC). This aimed to reserve from other war service artists deemed capable of recording scenes of war and expressing the compelling emotions of wartime. These were either signed up for the duration, or employed on specific commissions. This scheme gave artists status and security – and flattered the public with images of itself – but it could not generate fresh artistic expression. Rather, the favored artists continued on preset courses. Sutherland and Piper lavished Neo-Romantic nostalgia on paintings of Britain's bombed architectural heritage. Nash hardly went beyond his own achievements in World War I, though his coldly brilliant paintings of air war and its effects exemplify his affinities with abstraction and Surrealism.

Picasso in 1937 had produced, in his painting *Guernica*, an unforgettable protest against fascist violence. After the war he responded to its horrors in his much less powerful *Charnel House* (1945). In between, he worked prolifically in occupied Paris, on still lives and pictures of women and landscapes, in styles which suggest his anguish over events. But these images are not widely remembered.

When the day of reckoning came for the horrors of this war, artists would not be important witnesses.

▼ ***Bending the Keel Plate*** **by Stanley Spencer, from his "Shipbuilding on the Clyde" series. This highly original English Christian visionary had mythologized, in quasi-naïve style, his home village of Cookham-on-Thames as a "holy suburb of heaven". In 1940 the WAAC commissioned him to paint shipyards. The camaraderie of Clydeside workers appealed to him. They too became creatures out of time.**

Datafile

Popular concern about war crimes is a 20th-century phenomenon, a natural product of mass politics and total warfare. Their trial and punishment was made exceptionally possible by the unconditional surrenders of 1945. The job could be done by the victors, with categories of criminality suitably enlarged to match the enormity of some of the things done. More numerous by far than unlawful methods or means of combat, the classic core of war crimes, were maltreatments of prisoners and civilians: for instance, starvation and slaughter of populations in occupied territory, forced labor there or by deportation and extraordinary neglect of the law on prisoners by Germany dealing with Russians, by Japan dealing with everyone. New as an international offence was "crimes against humanity", the best title that could be devised to describe Germany's treatment of the Jews, whether in its own territories or in conquered ones.

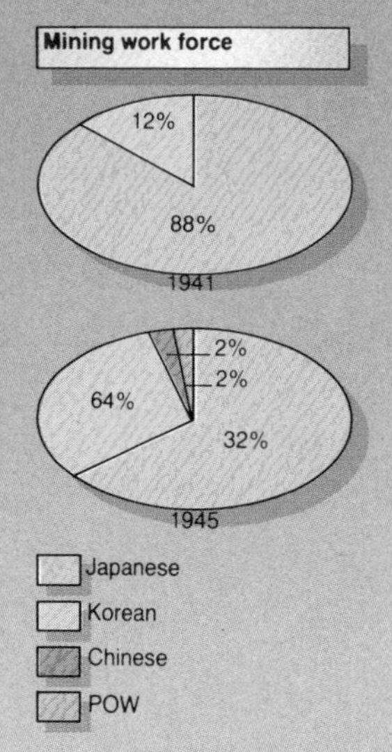

Civilians executed

Hundreds

15
10
5
0
1940 Years 1945

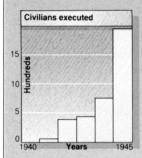

◀ Civilians executed in occupied Holland. The steep rise in the figures – all the steeper when it is remembered that "1945" here means less than half a year – vividly suggests how the atmosphere changed from the relatively easy-going first phase to the nervous and vindictive savagery of the last few months.

▲ Mines and quarries were a favorite scene of forced labor. Most of the work was unskilled, and the high rate of inevitable accidents and mortality was held to be of no account. Japan, chronically short of manpower for its armed forces, especially exploited prisoners and civilian deportees this way in the home islands.

▼▶ Statistics for the Nazis' attempted extermination of Europe's Jews can never be better than approximations, varying according to definitions (for example, many Jews in France were not French nationals) and border baselines (Poland, for instance, takes on quite different shapes according to date). But a cloud remains a cloud even if it has a fuzzy edge. Difficulty with precise figures cannot obscure the enormity of what was attempted and the scale of what was actually done.

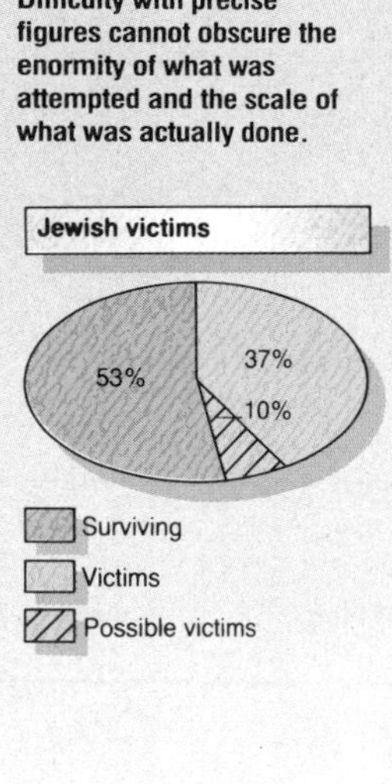

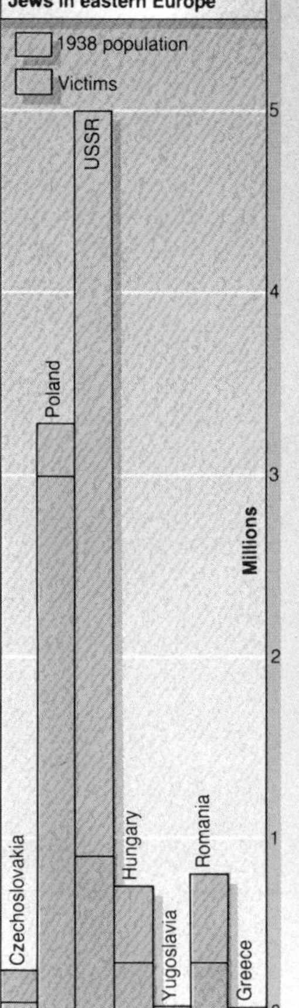

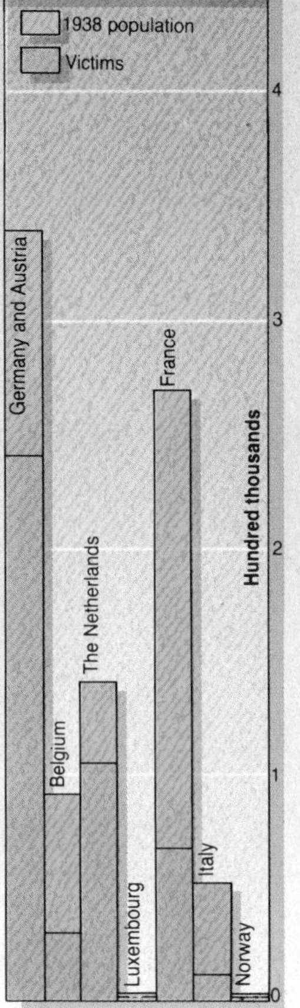

People who experience invasion, occupation and domination by enemy forces usually have a bad time. What happened during World War II was so bad, and often so atrocious, that many wartime acts of the Germans and Japanese and their allies were later categorized as criminal and punished. Some were clearly war crimes by the strict definition of the term: acts positively forbidden by the international laws of war. But the letter of those laws, as stated in the Hague Conventions of 1907 and the Geneva Conventions of 1929, did not cover all the vicious deeds that had to be dealt with after the war. New categories of crime were therefore formulated – crimes against humanity, notably the crime of genocide – so that the worst could receive the punishment and universal reprobation they deserved.

Attitudes to subject peoples

Germany and Japan alone need be considered. Their war aims included the exploitation of the human and material resources of the territories they intended to conquer. Both major aggressor powers sought to enlarge their economic empires. The Japanese fancied themselves as the hub of the Greater East Asia Coprosperity Sphere, the Germans as the middle mass of a Greater Germanic Estate. People in subject territories would be made to do whatever work and live in whatever degree of servitude might suit their masters' needs and convenience. This ordering of peoples and things came readily to many Germans and Japanese because they had clear ideas about their own racial superiority and consequent right, by their way of thinking, to dominate "inferiors". In the German case this idea extended to the extermination of some groups.

Occupation by an Axis power usually followed a standard pattern: first, the immediate seizure of useful materials and equipment; second, the gradual conversion of the economy to the occupier's exclusive benefit, with the occupied country bearing the costs of the occupation. Both Germans and Japanese were indifferent to the well-being of subject peoples, but there were important exceptions at each end of the conquerors' scale. At the milder end, a certain degree of consideration was shown to peoples felt to be "kith and kin": the Germans had such regard for the Danes, Norwegians, Flemings and Dutch, Estonians and Latvians. Peoples believed to be politically sympathetic also benefited: the Japanese, anxious to be welcomed as the liberators of Asians from white imperial rule, dealt leniently with Indonesians and Malays. At the other end of the scale were peoples viewed with such indifference or contempt that the reduction of their numbers by sickness and starvation was positively welcomed. The Japanese cared little for

WAR CRIMES

The viciousness of wartime occupiers' crimes

Japanese and German attitudes to other peoples

Treatment of slave laborers

Medical experiments

The Germans' attempt to exterminate the Jews

The postwar consequences of war crimes

what happened to Chinese whom they encountered throughout Southeast Asia as well as in China itself and were scarcely more concerned about the European whites in their prisoner-of-war and internment camps. The Germans had little concern for Slavs, whom their ideologues told them were too plentiful, and for whom the designated places in the Nazi Thousand Year Reich were those of serfs and slaves. The Jews were a special case altogether. (See p. 214.)

Slave labor and atrocities against civilians

Toiling in your own country for the benefit of the conqueror was one thing. Being taken to toil in his country was another. Germany was a populous and productive country, but the demands of an all-out, many-fronted war drove it to boost its work force with foreign labor. In the summer of 1939 foreign laborers constituted only about one percent of Germany's work force; five years later, about 20 percent – around 7 million foreigners in all. The great majority were there because they had been forced by one means or another, most of them nasty. Many hundreds of thousands of laborers were worked to death in mines, factories, and construction sites; most of the deaths were intentional. This was particularly the case with the millions of people who were processed through Auschwitz and Majdanek, which began as concentration camps providing and organizing slave labor and only subsequently acquired extermination facilities. All concentration camps in fact became work camps as time went on, and the SS's industrial concerns (part of their ramshackle empire-within-the-empire) relied integrally on them. Another form of enslavement – less intentionally cruel, but enslavement nonetheless – was forced labor in

▼ A Chinese patriot being prepared for execution by Manchurians who, with Koreans, often assisted the Japanese in the same way.

German industrial concerns, most notably Krupps and I.G. Farben.

In Japan 1.5 million despised Koreans were forced to do hard labor but did not suffer so badly. Only about 4 percent died on the job.

Many atrocities were committed against civilians involved – or believed to be involved – with the Resistance. International law concerning civilians under occupation was ambiguous. Nevertheless it did require that effort should be made to distinguish between passive civilians and those attempting to resist the occupier. Decent armies try to observe this distinction, but often the German armed forces and security services did not. Of these various organizations the SD, the SS and the parts of the army most influenced by them were the worst, but severity of treatment also varied according to region. On the Eastern Front terror was regularly resorted to in an attempt to produce a cowed quiescent civilian population. There and in the Balkans the killing of a single German was met with the execution of hostages in their hundreds, whereas in the west a handful often sufficed. The Wehrmacht normally killed only adults, but the SS was not so discriminating. In spring 1944 the northern Greek village of Klissura was the scene of a partisan action in which two German soldiers were killed; the SS's retaliation included the murder of 30 villagers over 60 and 38 under five. Other notorious reprisals, in central and western Europe, occurred at Lidice in Czechoslovakia and Oradour-sur-Glane in France. In retaliation for the assassination in a Prague suburb of the SS official Reinhard Heydrich, in May 1942, all the males of Lidice were murdered, the children were sent to a concentration camp and the women imprisoned – despite the lack of evidence that either village or villagers had anything to do with Heydrich's death. The killing of about 550 French civilians in Oradour (400 of them women and children burnt and blown up in the church) by the Waffen SS on 10 June 1944 was apparently intended to produce a general unresisting numbness throughout the regional population. Such atrocities could work,

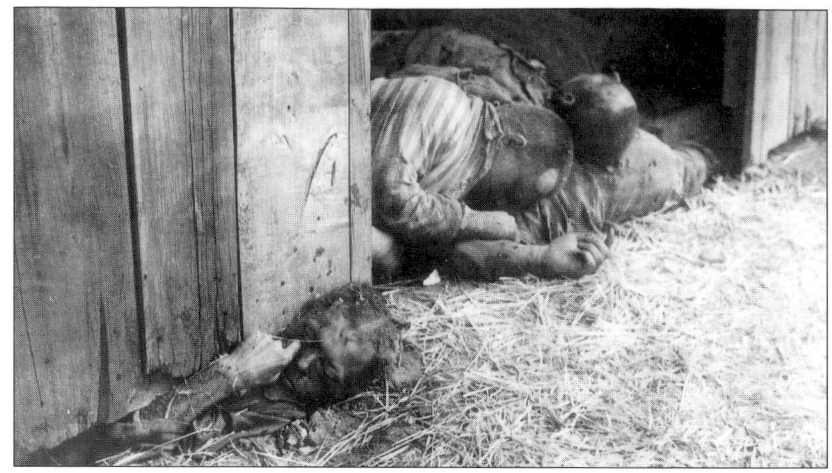

▲ This incinerated man was one of the 150 or so Russian, French, Polish and Jewish "political prisoners" burnt alive by the SS at Gardelegen (Germany). They were herded into a huge barn and made to sit down on petrol-soaked straw. The straw was set afire and the barn door locked.

◄ This picture is typical of those taken in the immediate wake of the Allies' arrival at sites of atrocity and horror in 1945. Highly indignant and largely incredulous at German civilians pleas of innocent ignorance, the victors often compelled local inhabitants to share in the exhumation of victims and photographed them with the evidence of what had been done in their name.

► Civilians murdered by Germans in the Crimea, discovered in early 1942. Slav peoples suffered bad times at the hands of the German invaders because of the low place accorded them in traditional German racialist theory (of which Nazi dogma was only an exaggerated version with a specially nasty antisemitic twist). Jews apart, Slavs were held to be the least valuable of noncolored races, a primitive and "animal" sort of people, fit only for serfdom within the new German empire.

Wartime Medical Experiments

Camps managed by the SS provided an opportunity for so-called medical experimentation of which SS doctors were quick to take advantage. Committed in any case by the code of their corps to subordinate all other feelings and principles to the supremacy of their race and nation, and counting among their number, as was to be expected, a high proportion of cranks and sadists, they did things which rank among the worst atrocities of the war and as shocking betrayals of medical ethics; eg maiming or, most often, killing victims by freezing and decompression.

The bulk of the evidence of these horrors emerged in 1947 during "The Doctors' Trial". Not necessarily the worst of the accused but destined to become by far the most celebrated was the one that got away, Dr Josef Mengele. He escaped to South America and died in the 1970s. Japan's counterpart atrocities had to wait many years before they were equally recognized. A handful of the army doctors who had done similar things to Chinese and other prisoners at "Unit 731" near Haerbin, Manchuria, were tried by a Soviet military tribunal at Khabarovsk in late 1949. But the bad faith engendered by the Cold War worked in two ways to dampen the effect; the Russians were concerned only to highlight the experiments in biological warfare (in which they suspected "the west" to have an unhealthy interest), and "the west" was unwilling to believe anything that came out of a Soviet court. Only in the 1980s did it become generally recognized that the findings of the Khabarovsk tribunal were essentially true, and that they were by no means the whole of the ghastly truth.

▲ This poster of the French "Institute for the Study of the Jewish Question" expresses the extravagance and fantasy of populist antisemitism. The age-old stereotype ghetto graybeard is used to symbolize ultra-modern international finance.

Kennzeichen für Schutzhäftlinge in den Konz.Lagern

Form und Farbe der Kennzeichen

	Politisch	Berufs-Verbrecher	Emigrant	Bibel-forscher	Homo-sexuell	Asozial
Grund-farben						
Abzeichen für Rückfällige						
Häftlinge der Straf-kompanie						
Abzeichen für Juden						
Besondere Abzeichen	Jüd. Rasse-schänder	Rasse-schänderin	Flucht-verdächtigt	2307 Häftlings-Nummer	Beispiel	
	P Pole	T Tscheche	Wehrmacht Angehöriger	Häftling Ia		

► A bureaucratic refinement from Dachau, close to Munich, the archetypal SS concentration camp from 1934 until its liberation by appalled American troops on 25 April 1945. Prisoners were elaborately classified and color-coded, with such results as are shown on the trouser-leg in the lower-right box.

but no civilized person could count a war won by such means as any sort of victory.

Destruction of the Jews

The wartime atrocity of the greatest magnitude was the Nazis' attempt to kill all European Jews. The Jews were not the only group the Nazis sought to exterminate. They also attempted to exterminate gypsies and killed about 250,000. They sought to exterminate all "degenerates" (by which they mainly meant homosexuals) and all communists they could identify. But Jews were their principal and most prized target.

Antisemitism in various forms was common throughout Europe, but German hostility to the Jewish people went far beyond ordinary antisemitism. The Nazis' paranoid world-picture allowed some sort of place for every other "race" (as they liked to put it) but none for Jews, whom they imagined to be not only "impure" but also actively and menacingly dangerous – to such an extent that even the sight of helpless Jewish children and feeble graybeards caused well-indoctrinated Nazis to shiver with a fear of what damage might be wrought by the hidden powers of "world jewry". They believed Jews to be potentially so dangerous that they could not even safely be enslaved. It sounds crazy, and indeed it was as absurd and vicious as could be, but that is what dedicated Nazis had in their heads. They had taught themselves to believe that the Jews constituted a colossal "problem", and from the winter of 1941–42 they devoted much effort to what they liked to call its "final solution", *die Endlösung*.

This came as the culmination of a variety of antisemitic measures of mounting monstrosity. Until the war in Europe began in September 1939 nothing was possible beyond causing German Jews to wish to emigrate as the Reich expanded. Grandiose plans were aired for the mass departure of the Jews to some distant foreign land (Madagascar was the favorite); while they remained on Reich soil their lives would be made miserable and a select few killed in concentration camps. The invasions of Poland, of the Low Countries and France in the west and most of the Balkans in the southeast in 1939–41 brought much mortality and anguish to the large, scattered Jewish populations that fell within the German grasp. German Jews were expelled eastward, Jews everywhere else were corralled into ghettos and labor camps, and murderous maltreatment was accorded to Jewish prisoners of war. But still nothing as grandiose as the "final solution" was heard of until the summer of 1941. It first appeared among the plans for the invasion of the USSR (which would begin with the invasion of the lands the USSR had annexed in 1939) and for the Germanization of the vast territories thus in prospect. These, it was thought, could not be properly "German" while Jews still inhabited them. They had to be "cleansed", made *Judenrein*. What was then to be done with all those Jews? In the minds of the Nazi ideologues and specialists in violence the idea began to materialize of systematically exterminating them; not just the Jews of eastern Europe but also, as opportunity permitted, the Jews of every other part of Europe.

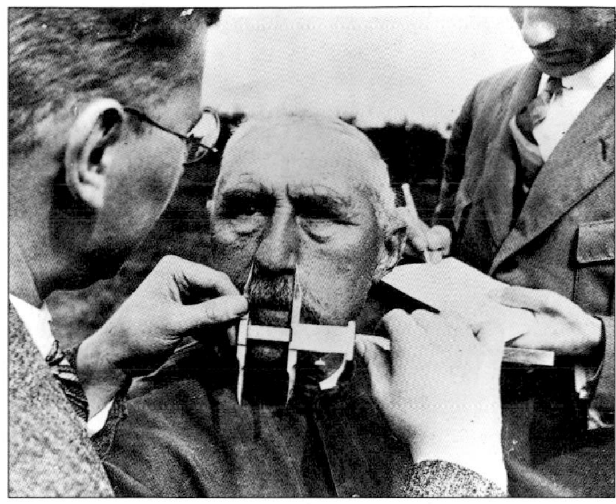

◀ **German officials measure a nose with a view to judging whether its owner was properly "Aryan" or not; an instance of the bogus-scientific absurdities into which there fanatics were commonly led.**

Work began on the Jews of eastern Europe in the wake of the German invasion. Spearheading it were the *Einsatzgruppen*, "special action groups" in the plain translation of their typically euphemistic title, but more accurately called murder squads: their main task was, simply, to massacre Jews and other listed undesirables. With the cooperation of the German and allied armies (some of whose more decent officers did not like what they saw or heard about and to their credit complained about it) and the often enthusiastic assistance of local police and militias, the *Einsatzgruppen* machine-gunned or otherwise did to death all the Jews they could lay their hands on, usually after first herding them into ad hoc camps and ghettos. There were innumerable scenes of horror. One of the worst was the filling of the Babi Yar ravine outside Kiev with nearly 40,000 dead or dying bodies.

Murder on an industrial scale

None of this was enough to meet the needs of the *Endlösung*. Much more planning had to be done. The work of the *Einsatzgruppen* and their local auxiliaries was patchy and piecemeal – and very messy. Killings so numerous and hasty were not easy to cover up, and the Nazis wanted to conceal their intentions from Jews still at liberty. Technology offered a less bloody and more secret means of mass murder: poison gas. Bureaucracy and efficient modern administration would surmount the other difficulties. A new master plan was developed, the details of which SD-leader Reinhard Heydrich explained to the heads of all involved agencies at a secret conference at Wannsee, a Berlin suburb, on 20 January 1942. The story would be spread that Jews were to be "resettled" eastward. The railroad systems of the whole continent would be geared to delivering trainloads of Jews (plus such others as were to be exterminated) to camps from which they would never emerge. Some would be sent initially to labor, the rest straight to newly constructed gas chambers. To handle the expected volume of business, additional camps with specialized extermination facilities would be constructed; the names of Chelmno, Treblinka, Sobibor, Belzec and Auschwitz now entered the gazetteer of Nazi infamy. The three camps at Auschwitz (Auschwitz I,

An open grave had been dug and they had to jump into this and lie face downwards. And sometimes when one or two rows had already been shot, they had to lie on top of the people who had already been shot and then they were shot from the edge of the grave. And Himmler had never seen dead people before and in his curiosity ... was looking in.

SS GENERAL KARL WOLFF
MINSK, 1941

▼ **A scene from the final hours of the Warsaw ghetto, conquered by the Germans at last, after four weeks of extraordinary resistance, in mid-May 1943. What little was left of it was then razed to the ground.**

Auschwitz-Birkenau and Auschwitz-Monowitz) were large industrial complexes. At its busiest period, handling Adolf Eichmann's big haul of Hungarian Jews in 1944, Auschwitz-Birkenau was managing to kill and cremate as many as 24,000 a day.

But death did not come only at that end of the railroad line. The journey itself and its staging posts were fraught with pain and terror, notwithstanding the Nazis' avowed intention to persuade victims that the very worst was not happening to them until that terrible moment of truth when gas began to come through the pipes in the "bath houses". The brutal round-ups and entrainments, the long-drawn-out journeys (food- and waterless in goods- and cattle-trucks), not to mention the steady starvation that had earlier marked life in the ghettos, brought death in many other ways, all by intention horrible. It speaks volumes about Nazi mentality that its *Endlösung*, although designed to kill its victims efficiently, had to terrify, humiliate and torture them as well.

The "final solution" did not wholly succeed. But the Nazis and their collaborators got fearfully close to their goal. About 300,000 of those in camps were still (just) alive when the Allied armies reached them in 1944 and 1945. Over a million more had not (yet) been "resettled". This was in part because the exterminators had not had enough time, in part because some Jews had managed to stay in hiding or had been hidden by courageous Gentiles, while a few of the thousands who had joined the partisans survived to tell their tale. It was, however, also because of the resistance to their deportation offered in varying measure by peoples and administrations all over occupied Europe. The rabid antisemitism sometimes shown by inhabitants of occupied countries, which was often of essential help to the occupier, was sometimes matched by ordinary decency and moral principle. One of the many weaknesses the Nazis identified in the Italians

▲▶ Disposing of the bodies was a problem the exterminators never solved entirely. The *Einsatzgruppen* relied mainly on burial, but that, whether in graves the victims were made to dig for themselves or in natural landscape features, tended to be doubly inadequate: there was not enough space, and they were difficult to cover up. To fill in this mass grave at Belsen demanded hard work. Crematoria ovens therefore became the preferred method. Tending the furnaces and removing the ashes (in which much tell-tale bone nevertheless remained) was done by prisoners themselves, as seen in the picture (right) from Dachau's tiny installation.

as allies was their lack of anti-Jewish viciousness. The Jews of Denmark (admittedly, only about 7,000 strong) were spirited away to Sweden on the eve of their roundup. The 48,000 Jews of German-dominated Bulgaria owed their lives to extraordinary demonstrations of public solidarity. It diminishes not at all the luster of that noble example to reflect that other peoples, no matter how much they might have wished to do likewise, would have found it formidably difficult. A preindustrial unpopulous nation with few railroads and many mountains, at the far corner of the continent, might get away with noncooperation but this was utterly beyond, say, the people of the Netherlands. Of their 140,000 Jewish fellow-citizens, three-quarters were killed. When the fighting at last stopped in 1945 over two-thirds – some 5.75 million – of Europe's prewar Jewish population was dead.

The perpetrators of these monstrous deeds never doubted that their actions were right for their nation and race (the only segment of humankind they cared about). But as defeat drew ever nearer in 1944 and 1945 they became worried about their effects on the feelings and temper of the likely victors. They therefore attempted to obliterate all traces of them. Demolitions were carried out whenever time permitted, the camp inmates were herded by road or rail away from the shrinking borders toward the interior of the expiring Reich. Not all were Jewish but most were. The last hundreds of thousands of Holocaust victims before the liberation were those who were shot, beaten, sickened or starved to death in course of this extemporized mass migration. But more than enough of camps and survivors was left for the world abruptly to learn, with a sense of moral shock greater for most Britons and North Americans than any experienced so far, the appalling facts about the concentration and death camps, and to take at last the full measure of the foe they had been combating. The impact of these revelations was twofold. They destroyed whatever small chance there was that public opinion would wish to show that "Magnanimity in Victory" recommended by the British prime minister Winston Churchill, and they rallied public opinion in victorious countries solidly behind the war-crimes trials which were soon to begin.

From war crimes to human rights

The British and North American publics had known little about the Holocaust and its associated horrors while they were being perpetrated. Allied governments, however, had known, and made plans for dealing with the perpetrators. Culpable individuals would be punished but they intended also to shape a new world order in which such antihuman horrors could never happen again. It is probably true to say that the present worldwide regime of concern for human rights would never have been launched without World War II's extraordinary and unprecedented denials of them. The charter of the International Military Tribunal for the Trial of German Major War Criminals, known to history from the place of its assembly as the Nuremberg Tribunal, proposed to hear indictments for three categories of crimes: crimes against peace (planning and waging aggressive war); war crimes (to include "deportation to slave labor" and "killing of hostages"); and – this was the one that prefigured the Universal Declaration of Human Rights – "crimes against humanity; namely, murder, extermination, enslavement, deportation and other inhumane acts commited against any civilian population. . .".

While the United Nations was formulating basic human rights legislation, war-crimes tribunals were established in the zones of occupied Germany and in every formerly occupied country. In the American, British and French zones between 1947 and 1953, 10,400 were tried; 5,025 were sentenced; 506 executed. It is largely from the superabundant evidence offered at the trials of, for example, Heydrich's successor Ernst Kaltenbrunner (before the IMT itself; the only SS/SD defendant there), the *Einsatzgruppe* commander Otto Ohlendorff (at a military tribunal in the American occupation zone) and Rudolf Hoess the commandant of Auschwitz (in a Polish court), that the war crimes of World War II – crimes against human rights in general, the Holocaust in particular – are so copiously documented.

With the limelight so much on concentration and death camps, little indignation was aroused in "the west" by what happened in German prisoner-of-war camps – which in any case, so far as British, Commonwealth and American servicemen were concerned, were relatively not so bad. Very different was the case with those taken prisoner by Japan. Not much was known of the conditions of life, labor and death in Japanese camps until Allied forces overran them in, mostly, the last weeks of the war, or even after it had been formally ended. The fury which followed the disclosure of those atrocities, kept hot through many years to come by the stories of the survivors when they got home, made the trials of the International Military Tribunal for the Far East (brief title, the Tokyo Tribunal) and the other war-crimes trials accompanying it even more engrossing to many in the victorious nations than their European counterparts.

AUSCHWITZ: THE NAZI DEATH FACTORY

The name Auschwitz will never be forgotten. It now refers principally to a small town in southern Poland (Oswiecim); in World War II it was the scene of crimes of unprecedented brutality: daily gassings of thousands of innocent victims; sadistic experiments carried out on humans; and prisoners systematically beaten, starved and worked to death.

It began with the construction of a concentration camp in May 1940, when Rudolf Hoess was appointed commandant. Expansion plans were laid early in 1941, to include a large synthetic rubber plant for I.G. Farben at nearby Monowitz (sometimes called Auschwitz III), to exploit prison labor, and a satellite camp at Birkenau (Auschwitz II). Then in the summer came secret orders from Himmler to make Auschwitz the main site for the extermination of the Jews.

Lessons learned from the earlier killing operations were to be perfected in one vast factory of death. The first gassings took place at the mortuary of the main camp in February 1942. However, operations were soon transferred to Birkenau for secrecy. Four large crematoria were constructed. Ovens were built inside the gas chambers to facilitate the disposal of the bodies. Zyklon B gas, used commercially as a pesticide, was adopted to increase efficiency.

Jews were rounded up from all over Europe and transported to Auschwitz like cattle. Their countries of origin included Poland, Germany, France, the Netherlands, Greece, Czechoslovakia, Hungary, Belgium, Italy, Yugoslavia and Norway, as well as the ghettos of the occupied East. On arrival they were segregated into those suitable for work and those condemned to immediate execution. There were heart-rending scenes as married couples were separated and mothers waved goodbye to their sons.

For those spared it was merely a stay of execution. Camp inmates were exploited as slave labor. Only overt brutality could keep such a system in operation. Long hours of physical work on inadequate rations sapped the prisoners' strength. Life expectancy was just three months. "Selections" regularly weeded out the unfit for the gas chambers. Others died of weakness and disease at work or in the overcrowded barracks. The calculated terror of the SS is revealed by Hoess himself: "They knew without exception that they were condemned to death, that they would live only as long as they could work."

The Auschwitz camps were discovered by the Red Army in January 1945. The camps' records had been destroyed by the Germans before their retreat, so the exact number of Auschwitz victims will never be known. But estimates range between one and two million.

It has been disputed whether Auschwitz was a unique historical phenomena. Comparisons have been made with other mass killings, such as massacres in Cambodia in the 1970s. However, the Nazis did not just murder their political opponents, but planned the elimination of an entire people on the grounds of race.

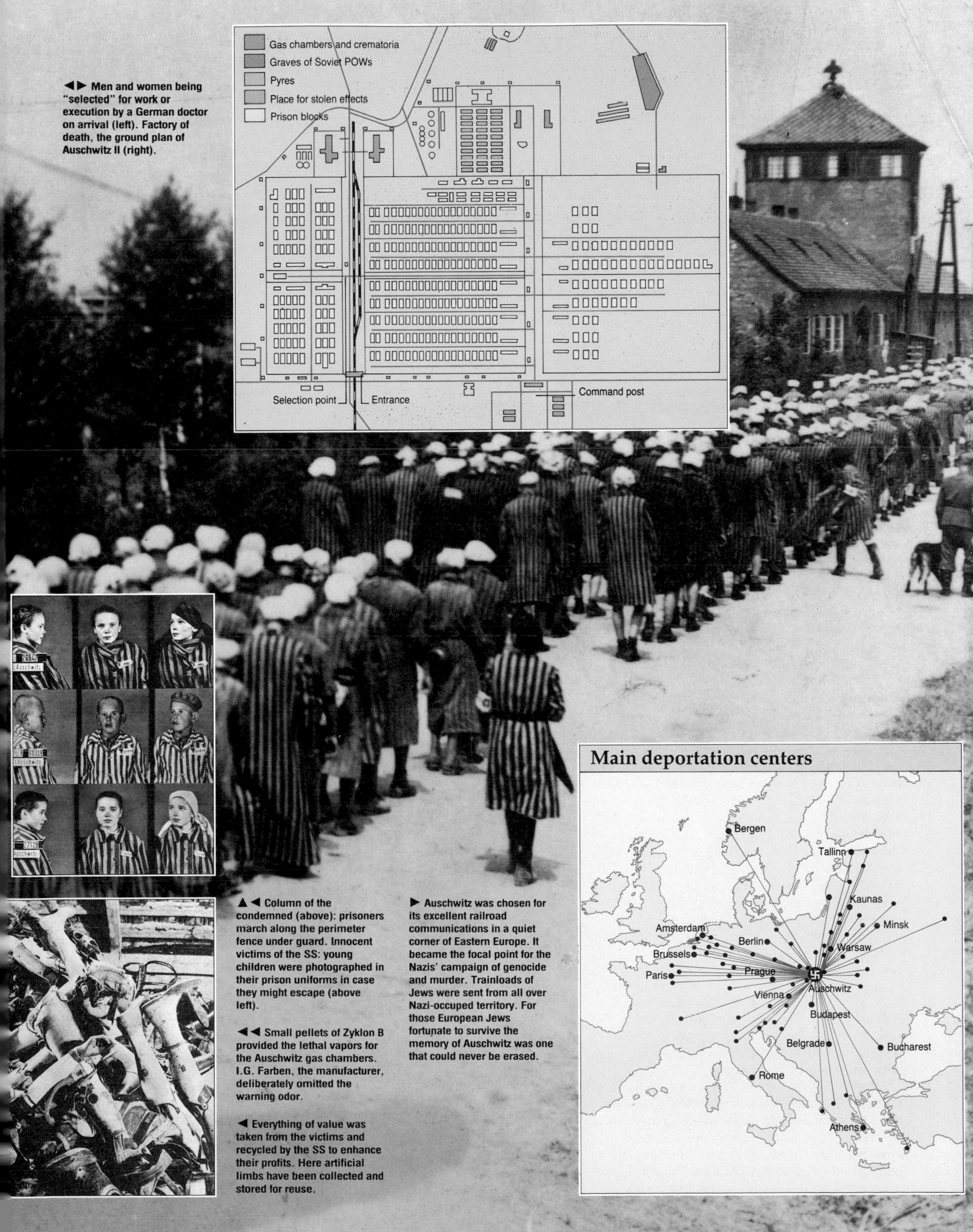

◀ ▶ Men and women being "selected" for work or execution by a German doctor on arrival (left). Factory of death, the ground plan of Auschwitz II (right).

▲ ◀ Column of the condemned (above): prisoners march along the perimeter fence under guard. Innocent victims of the SS: young children were photographed in their prison uniforms in case they might escape (above left).

◀ ◀ Small pellets of Zyklon B provided the lethal vapors for the Auschwitz gas chambers. I.G. Farben, the manufacturer, deliberately omitted the warning odor.

◀ Everything of value was taken from the victims and recycled by the SS to enhance their profits. Here artificial limbs have been collected and stored for reuse.

▶ Auschwitz was chosen for its excellent railroad communications in a quiet corner of Eastern Europe. It became the focal point for the Nazis' campaign of genocide and murder. Trainloads of Jews were sent from all over Nazi-occuped territory. For those European Jews fortunate to survive the memory of Auschwitz was one that could never be erased.

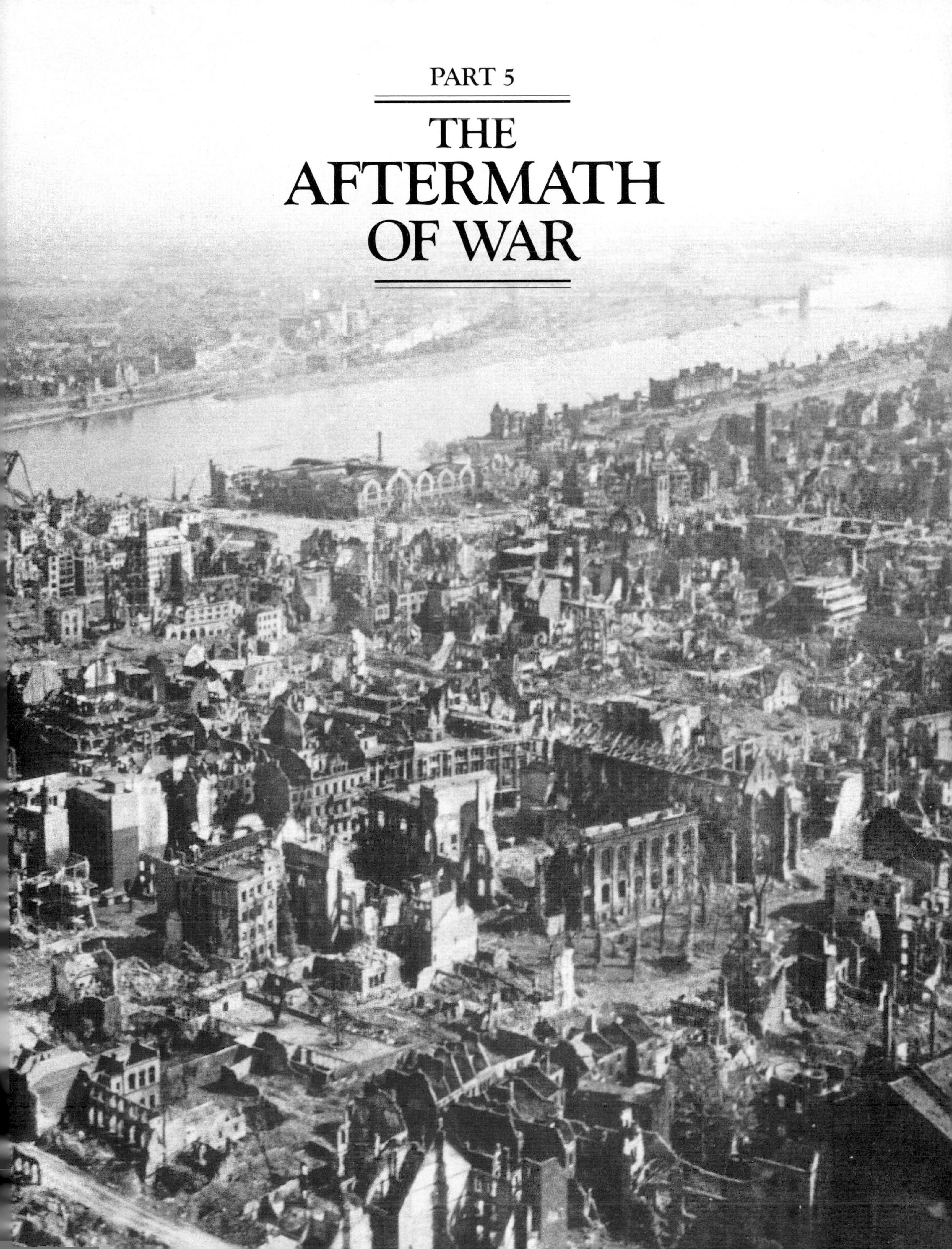

PART 5

THE AFTERMATH OF WAR

BIOGRAPHIES

Adenauer, Konrad 1876–1967

A christian democrat leader in the Weimar Republic, Adenauer was mayor of Cologne (1917–33) and a possible candidate for chancellor. He retired from politics when the Nazis came to power, but was restored as mayor of Cologne by the Allies in 1945. He led the christian democrat party (CDU) during the Allied occupation of Germany, and became the first chancellor of independent West Germany after the election of 1949, a post he held until 1963. As chancellor he presided over the revival of Germany as a first-class economic power in the 1950s, and was a signatory of the Treaty of Rome (25 March 1957), which established the European Economic Community. He negotiated German rearmament within NATO, on condition that German national sovereignty was restored. In 1958 he supported the deployment of tactical nuclear weapons in Germany to reduce the burden of its armed forces.

Attlee, Clement 1883–1967

Attlee joined the Labour Party in 1908, entering parliament as the member for Limehouse in East London in 1922. He held ministerial and cabinet posts in the Labour governments of the interwar years (1924, 1929–31), and from 1935 led the parliamentary Labour opposition to Ramsay MacDonald's National Government. In Churchill's wartime coalition he was deputy prime minister. After Labour's victory in the May 1945 election he became prime minister. Although described as "the little man" (by Ernest Bevin), Attlee was to preside effectively over those who enacted the wide-ranging social reforms of the Labour governments until October 1951. He continued to lead the Labour Party in opposition until its defeat in the 1955 election, after which he retired and was created Earl Attlee.

Ben-Gurion, David 1886–1973

A militant Zionist leader, Ben-Gurion believed force to be justifiable in establishing an independent Jewish homeland in Palestine. As leader of the Zionist labor movement he organized illegal immigration of Jews to Palestine after World War II, and from October 1945 to August 1946 coordinated a Hebrew revolt against British rule in Palestine. He delivered the declaration of Israel's independence on 14 May 1948. He was prime minister of Israel (1948–53 and 1955–63), and defense minister (1955–63). He created an army from the variegated Jewish liberation movements as a symbol of national independence, employing it against Israel's hostile Arab neighbors, for example against Egypt in 1956.

Bevin, Ernest 1881–1951

Born the son of a Somerset midwife, Bevin made his political reputation before the war as leader of the Transport Workers' Union. He was called into Churchill's wartime cabinet as minister for labor and national service. In the 1945–51 Labour government he served as foreign secretary and accompanied Attlee to the Potsdam conference (July–August 1945). A political realist, he aligned the UK with the USA after the war and supported Marshall Aid and an American military presence in Europe. His belief in western collective security led to the creation of NATO.

Churchill, Winston 1874–1965

Elected to parliament in 1900, Churchill held most of the important cabinet posts, including (as a Liberal) home secretary, first lord of the admiralty, minister of war and (as a Conservative) chancellor of the exchequer, before finding himself in the political wilderness from 1929 to 1940. He became the prime minister of a coalition ministry in May 1940. Having led the UK to victory in World War II his postwar career was something of an anticlimax. He and the Conservatives were defeated in the election of May 1945. He led the Conservative opposition until 1951, when he again became prime minister, holding the post until 1955. He died on 24 January 1965, and was given a full state funeral.

Cripps, Stafford 1889–1952

Cripps made his name as a lawyer before joining the Labour Party in 1930. He was solicitor-general in Ramsay MacDonald's Labour government in 1931, but militantly opposed MacDonald's National Government (1931–35). During the war he served Churchill as ambassador to the USSR (1940–42), as leader of the House of Commons, and finally as minister of aircraft production from 1942 to 1945. In Attlee's Labour government he served as president of the board of trade and then as chancellor of the exchequer (1947–50). He was director of the cabinet mission to India in 1946, which led to Indian independence. As chancellor he presided over a series of stringent measures of economic retrenchment. Illness forced him to retire in 1950.

Duclos, Jacques 1896–1975

Duclos was a communist party deputy before World War II, a wartime communist resistance leader and leader of the parliamentary communists in the postwar French Fourth Republic from 1946 to 1958. In 1947 the party was aligned with the Soviet bloc, and attempted to overthrow the constitution by a series of general strikes. Failure led to the isolation of the communist party in French politics, and when the government of Antoine Pinay tried to outlaw communism in 1952 Duclos was arrested on trumped-up charges. When a general strike to secure his release failed he reintegrated the party into the French constitutional structure, backing the government of the socialist Pierre Mendès-France in 1955 and standing as a presidential candidate in 1969.

Eisenhower, Dwight D. 1890–1969

During the war Eisenhower directed the Allied amphibious landings in North Africa (1942), Italy (1943) and Normandy (1944). After D Day he was supreme commander of the Allied forces in Europe until the end of the war. He became the first supreme commander of NATO in 1950. Elected by a landslide victory as Republican president of the United States in 1952, he held the office until 1961. His foreign policy was peaceful and defensively oriented. He ended the Korean War, and kept the United States out of entanglements abroad, by, for example, limiting the American commitment in Indochina and forcing the abandonment of the Anglo-French invasion of Egypt in 1956. He inaugurated the policy of detente with the Soviet Union, to try and end the Cold War. At home his administration laid the foundations for American prosperity in the 1960s.

Gandhi, Mahatma 1869–1948

Gandhi trained as a lawyer in England before taking up the cause of Indian independence during World War I. He was active in the Indian National Congress until 1934. He favored a policy of nonviolent obstruction of British rule in India, as seen, for example, in the campaign of civil disobedience in 1928, and the "Quit India" revolt of 1942. After the war he negotiated the granting of independence in 1947. However, he was unable to prevent the establishment of the separate state of Pakistan and the conflict between Hindus and Moslems after independence. He was assassinated by a Hindu extremist on 30 January 1948.

Ho Chi Minh 1890–1969

Ho Chi Minh received a western education in France where he began agitating for Vietnamese independence from France. He led the communist party in French Indochina from 1930, founding the Viet Minh (League for the Independence of Vietnam) in 1941. At the end of the war he set up a communist regime in Vietnam in opposition to the puppet regime of the pro-French prince Bao Dai. He received support from China in his struggle against French imperialism, enabling him to establish an independent communist state north of the 17th parallel after the victory at Dien Bien Phu in May 1954. He remained secretary-general of the communist party until 1959, and then became head of state, continuing to influence policy from behind the scenes.

Hoxha, Enver 1908–86

Educated in France, Hoxha was active in the Albanian communist party before the war. He was appointed party secretary when Tito reorganized the Albanian partisans in 1941, and became head of the Anti-Fascist Council of National Liberation in May 1944. After German withdrawal in November 1944 he established communist political ascendancy. Hoxha organized the state on Stalinist lines, concentrating power in his own hands; at one time he was premier, foreign minister, defense minister and commander in chief of the army. Independent and isolationist, under his rule Albania remained backward, agriculturally and economically underdeveloped. His relations with Tito were close until Yugoslavia broke from the communist bloc. He then aligned Albania with Stalinist Russia, until Khrushchev's policies led him to move toward China in the 1960s.

Hull, Cordell 1871–1955

A Democratic senator and secretary of state before the war, Hull supported American aid to the Allies. During the war he planned the establishment of a postwar international peace-keeping force, which evolved into the United Nations in 1945. He was awarded the Nobel Peace Prize in 1945 as "father of the United Nations".

Jiang Jieshi 1887–1975

Jiang (formerly known as Chiang Kai-Shek) trained as a soldier in Japan, and participated in the Chinese republican revolution of 1911. He rose in the nationalist army, seizing power in the Guomindang in the Guangzhou coup of 20 March 1926. He was military dictator at the head of a national government from 1928 until the Japanese invasion of 1937. He led resistance to the Japanese from 1937 to 1945, while husbanding his forces for the inevitable civil war with the communists. After losing this war he retreated in 1949 to the offshore island of Taiwan. With American aid, he turned Taiwan into a formidable economic power, with a view to the reconquest of mainland China.

Jinnah, Mohammed Ali 1876–1948
Jinnah studied law in England, and was active in the Indian National Congress in the 1920s. In the 1930s he became converted to the need for a separate Moslem state to be created when India achieved independence. He split from the National Congress in 1940, and as head of the Moslem League negotiated after the war for a separate state of Pakistan when independence was granted. He was the first governor-general of the state of Pakistan created in 1947, and is venerated as the father of his country.

Khrushchev, Nikita 1894–1971
A Ukrainian coalminer, Khrushchev joined the communist party in 1918. He rose to power in the Ukrainian party as a tool of Stalinist repression in the 1930s and was its head from 1938 to 1949. After Stalin's death in 1953 he won the power struggle with his rival Malenkov. He was party secretary from 1953 and also premier from 1958. He began the process of destalinization with his "secret speech" at the 20th party conference in July 1956, calling for a return to "Leninism". Abroad, Khrushchev asserted the Soviet Union's control over its eastern European satellites by the creation of the Warsaw Pact in April 1955, and by the repression of the popular uprising in Hungary in October 1956. He favored rapprochement with the West, attending, for example, the Geneva summit in 1955, until the U2 spyplane crisis in 1960. His attempt to deploy Soviet ballistic missiles in Cuba led to the crisis of 1962. He fell from power in 1964.

Kim Il Sung 1912–
Kim Il Sung joined the Korean communist party in 1931, and led Korean resistance to the Japanese in the 1930s. In 1945 the Soviet Union backed his establishment of a communist regime in North Korea. He was chairman of the Korean workers party from 1948, premier from 1948 to 1972, and was president and head of state from 1972. He invaded democratic South Korea in 1950 in an attempt to extend communist power. Although defeated, the reunification of Korea under communist government remained an objective of North Korea.

MacArthur, Douglas 1880–1964
MacArthur was the US army chief of staff from 1930 to 1935, then military adviser to the Philippines. He lost the Philippines to the Japanese in 1942, but as commander in chief of American land forces in the Pacific he defeated the Japanese with his strategy of "leapfrogging". He received the Japanese surrender in 1945, and was supreme commander, Allied Powers in Japan from 1945–50, overseeing the democratization and economic development of Japan. He commanded the United Nations forces in the Korean war from 1950–51, saving South Korea by an ambitious amphibious landing at Inchon and advancing to occupy North Korea. His over-ambition brought China into the war, and poor relations with President Truman led to his dismissal in April 1951.

Mao Zedong 1893–1976
A founder member of the Chinese communist party in 1921, Mao went on to lead it in its long political and military struggle with the Chinese nationalists. He led the "Long March" retreat from the nationalists in 1934–35, setting up an independent soviet state based at Yan'an. He fought against the Japanese during World War II, and in a civil war from 1946–49 conquered China from the nationalists. He was party chairman in the People's Republic of China from 1949–76. His plan for economic expansion, the "Great Leap Forward", failed in the 1950s, and his "Cultural Revolution" of 1966–76 had a retrograde effect on China. He was the sponsor of communist popular movements throughout Southeast Asia.

Marshall, George C. 1880–1959
General Marshall was chairman of the Allied joint chiefs of staff committee. He made an unsuccessful mission to China to try and reconcile the communists and nationalists in 1945. He was secretary of state in the Truman administration from January 1947. He believed that America had a duty to revitalize Europe after the war, to save it from communist influence and give an outlet for American trade. Under the "Marshall Aid" plan, 13 billion dollars were loaned to Europe to relieve starvation and fund economic regeneration, for which he won the Nobel Peace Prize in 1953. Ironically, it increased the rivalry between the USA and the USSR, deepening the Cold War.

Mountbatten, Lord Louis 1900–79
A cousin of King George VI, he served his early career in the navy. In April 1942 Churchill appointed him head of combined operations, where he was responsible for the Dieppe and St Nazaire raids, and for the early planning for the Normandy landings. In August 1943 he assumed supreme command of British forces in Southeast Asia. Raised to the peerage in 1945, he served as last viceroy and first governor general of India until June 1948, overseeing the granting of Indian independence. He returned to the navy and completed his career as chief of the defence staff, 1959–65.

Nehru, Jawaharlal "Pandit" 1889–1964
Educated as a lawyer in England, Nehru came under the influence of Gandhi on his return to India and took up the cause of Indian independence. He was elected president of the Indian National Congress in 1929, declaring India's goal to be complete independence. In pursuit of this goal he suffered imprisonment by the British nine times, but independence was eventually achieved in 1947. He was the first prime minister of independent India, which he governed on democratic socialist principles. Independent and nonaligned in his foreign policy, he led India in war against Pakistan over Kashmir in 1948, and against China in 1962.

Rhee, Syngman 1875–1965
An activist for Korean independence from the 1890s, Rhee was president of the "Korean government-in-exile" in the United States from 1919 to 1939. In 1945 he was placed at the head of a "provisional government" in Korea by the USA. After a campaign of terror and the elimination of his main rival he was elected first president of the Republic of Korea in 1948. He was backed by the United Nations when South Korea was invaded by the communist north in 1950. His dictatorial totalitarian regime was overthrown by a popular uprising in 1960 and he returned to exile in America.

Schuman, Robert 1886–1963
A member of the French national assembly from 1919, Schuman was active in the resistance during the war. In the short-lived governments of the Fourth Republic he served as minister of finance (1946), premier (1947 and 1948), foreign minister (1948–52) and minister of justice (1955–56). In 1950 he developed the "Schuman Plan" for European economic and military unity and Franco-German rapprochement. From this evolved the European Coal and Steel Community, established in 1952, and the European Economic Community set up in 1958. He was first president of the EEC assembly from 1958–60.

Stalin 1879–1953
As an active revolutionary communist from 1901, Stalin became a member of the first politburo after the revolution of November 1917. He defeated his rival Trotsky in the power struggle after Lenin's death, and in the interwar years imposed his own totalitarian regime, based on terror and repression, on Soviet Russia. He survived the crisis of the German invasion of June 1941, and led the USSR to victory in the war. His search for security after the war brought him into conflict with his wartime allies, the USA and the UK. He supported the postwar partition of Germany and the creation of a series of satellite communist states in eastern Europe. This produced a "Cold War", and an "iron curtain" (Churchill's term) across Europe. At home repression continued up to his death.

Truman, Harry S. 1884–1972
As vice-president, Truman succeeded as Democratic president on Roosevelt's death in April 1945. He remained president until 1953. His lack of experience of foreign policy meant that he missed the chance to establish a permanent understanding with the Soviet Union after the war. Instead he promulgated the "Truman Doctrine" in defense of national freedoms in Europe, and supported the Marshall Plan as an extension of this doctrine. Faced with the spread of communism he began the development of the hydrogen bomb in response to Russia's development of the atomic bomb, and brought America into the Korean War to oppose the spread of communism in Asia. Truman's achievements include the creation of NATO and support for the establishment of an independent Jewish homeland in Palestine.

Tito 1892–1980
Tito learnt communism and Stalinism as a prisoner of war in Russia during the revolution and later as an exile there in the 1930s. He became general secretary of the Yugoslav communist party in January 1939. He led a nonparty partisan alliance against the Germans during the war, being marshal and commander of the armed forces in the government for the liberation of Yugoslavia from 1943. With British and Soviet support he triumphed over his domestic enemies at the end of the war, ruling Yugoslavia on autocratic Stalinist principles until his death in 1980. He resisted the influence of the USSR, resulting in Yugoslavia's expulsion from the Cominform (Information Bureau of the Communist Parties) in 1948. Efforts by Stalin's successors to get him to join the Warsaw Pact failed. Instead, as a communist neutral, he received financial support from the USA.

Ulbricht, Walter 1893–1973
Ulbricht was a leader of the communist party and member of the Reichstag in the Weimar Republic. He fled from Germany when the Nazis came to power, but returned as an administrator of Soviet-occupied territories in 1945. He refounded the German communist party, and became deputy prime minister of the German Democratic Republic (East Germany) on its establishment in 1949. He was general secretary of the communist party from 1950 and head of state from 1960. He tied East Germany closely to the Soviet bloc, remaining implacably hostile to West Germany, and resisting destalinization in the late 1950s.

Datafile

A dominant theme of the postwar period was the drive for security: it was common to both superpowers, but resulted in anxiety rather than reassurance. The approach of the USSR was to build a protective cordon around its borders, largely by fostering coups in neighboring countries. The USA, of course, is surrounded by a moat; in the immediate postwar period, before the coming of the intercontinental ballistic missile, its main concern was the security of Western Europe, but it was also concerned about any area which appeared to be menaced by communism. Rather than taking over governments, it attempted to build up allies which could defend themselves.

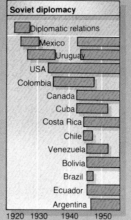

Soviet diplomacy

Diplomatic relations
Mexico
Uruguay
USA
Colombia
Canada
Cuba
Costa Rica
Chile
Venezuela
Bolivia
Brazil
Ecuador
Argentina
1920 1930 1940 1950

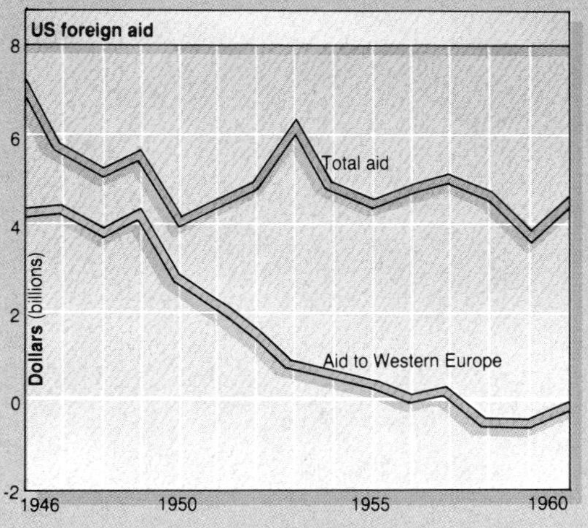

◀▲ **After the immediate postwar years, the proportion of American aid which went to Western Europe declined rapidly, that continent being replaced by the Western Hemisphere, the Far East and the Pacific and, especially, the Near East and South Asia. Soviet diplomacy had made few inroads in the Western Hemisphere before the war. The respectability conferred on the USSR by its wartime cooperation with the USA enabled it to become better established in the USA's "backyard".**

▶ **The rapid demobilization of the American armed services after the war, which was required by US public opinion, is manifestly clear in the graph, which only begins to move upward again with the advent of the Korean War. Between 1953 and 1960 the number of personnel dropped again by one-third.**

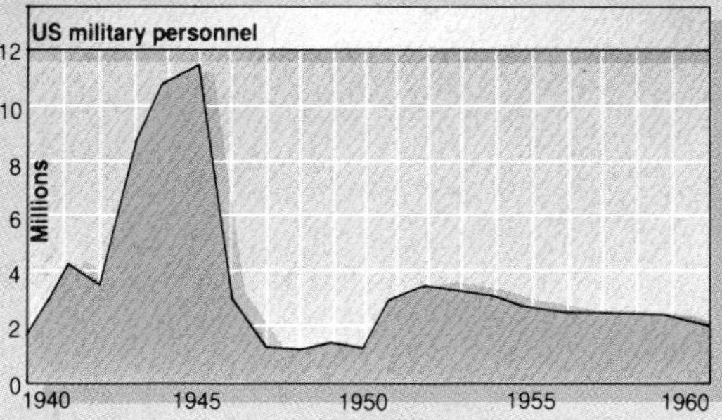

Chronology: the superpowers and Germany

1945

May
Berlin surrenders to USSR

June
Allied Control Council and Soviet Military Administration in Germany (SMAD) formed

July
Western Allies take over their sectors in Berlin; Interallied Military Kommandatura set up to administer Berlin

1946

April
Socialist Unity Party (SED) formed in Soviet zone. Paris Conference about Germany reaches no decisions

1947

January
Economic administrations of British and American zones merged as Bizonia

March–April
Moscow conference reaches no decisions about Germany

1948

February–March
Conference in London between the Western Powers and the Benelux countries recommends union of the Western zones

March
Allied Control Council ceases to function, due to differences between USSR and Western Powers over London Conference

June
Soviet troops blockade Berlin; airlift by Western Powers begins; USSR quits the Berlin Kommandatura; USSR and Eastern Bloc condemn the London conference

July
Western Powers start drafting constitution for Western zones

November
Appointment of provisional government for East Berlin ends unified Berlin administration

1949

May
USSR calls off Berlin blockade

May–June
Allied foreign ministers meet in Paris, but reach no decision as to Germany

September
Government of Federal Republic of Germany formed

October
Foundation of German Democratic Republic; SMAD dissolved

As long ago as 1835, the French political scientist Alexis de Tocqueville, in his *Democracy in America,* had predicted the rise of the superpowers: "There are, at the present time, two great nations in the world which seem to tend toward the same end, although they started from different points: I allude to the Russians and the Americans .... The Anglo-American relies upon personal interest to accomplish his ends, and gives free scope to the unguided exertions and common sense of the citizens; the Russian centers all the authority of society in a single arm: the principal instrument of the former is freedom; of the latter servitude. Their starting-point is different, and their courses are not the same; yet each of them seems to be marked out by the will of Heaven to sway the destinies of half the globe."

In 1939 this predominant position still lay in the future – there existed several great powers as well as others with pretensions – but by 1945 the position was transformed. The sheer scale of their mobilization since 1941 had lifted the USA and USSR into a league of their own. Although the UK still hoped to play the role of a great power, it would be very much as a junior player, compared to the USA and the USSR; as for the other major countries, defeat had, for the time being, taken them almost entirely out of the game. The superpowers stood apart, linked by a common drive for security, and a determination to apply the lessons of history in their reconstruction of the postwar world.

Discussions between the two had begun early in the war, and throughout the war the USA, and particularly President Roosevelt, hoped that the two countries could perpetuate their alliance to ensure peace; on the other hand, the USSR, and particularly Stalin, assumed that they could not. The expectations of both were rooted in history, and both prepared to restructure the world in ways best suited to their aspirations. During the war they had of necessity to work together to defeat the immediate foes, but by 1947 it was clear even to the USA that their goals were in conflict. By 1949 much of the industrialized world was grouped into two more or less formal and hostile blocs.

Soviet foreign policy: expansionism or ideology?

For years political scientists have argued over the question of what drove the USSR: the historical Russian drive for expansion, or a newer drive to make the world safe for communism. The truth is probably a combination of the two. The driving principle of the Russian state had for centuries been to seek security through expansion. Stalin took over this axiom of Russian behavior and added an ideological twist: security meant control of spheres of influence, made up of states on the

THE WORLD OF THE SUPERPOWERS

The effects of wartime mobilization on the USA and USSR

The USSR's need for security in the postwar world

Stalin's advances in Asia and Europe

The USA and plans for new international organizations

The origins of the Cold War

Soviet border firmly under the control of local communist regimes, which were in their turn firmly under the control of the USSR. If this control could be extended beyond the immediate border states, so much the better.

Recent history meant that the Soviets had a special fear of a resurgent Germany, which implied a particular need for control there as well as in all interlying states. Stalin, in short, thought specifically in strategic and military terms, rather than primarily in ideological ones: for example, he was willing to sacrifice the strong possibility of a communist regime in Greece, which was not on his border (nor in Central Europe), in exchange for Western acquiescence in Soviet control of Bulgaria and Romania, which were.

Stalin's drive for security through territorial expansion began in 1939, with the takeover of the three Baltic republics (Estonia, Latvia and Lithuania, which had been jumping-off points for

▼▶ On 9 May Moscow celebrated Germany's defeat (below). Stalin and Truman (right) met at Potsdam in July together with Churchill to settle Germany's future.

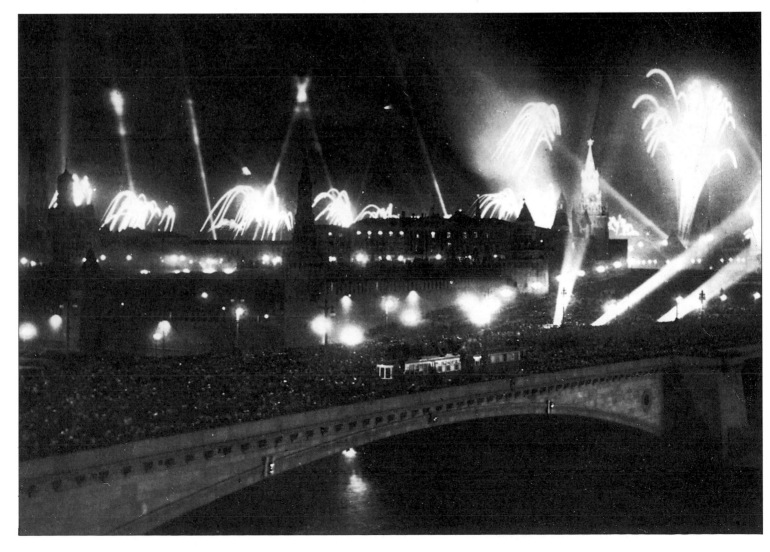

the anticommunist invasion of the Bolshevik Republic in 1918) and the move against Poland, which had formed part of the czarist empire. During the war the emphasis was, naturally, on Europe. But Stalin perceived Soviet interests as extending beyond Eastern and Central Europe, as was demonstrated by his activities toward Japan, Turkey and Iran. From Japan the USSR claimed the return of lands and privileges granted away in 1875 or lost by the Russian defeat in the Russo-Japanese War of 1904–05: the Kuril Islands, the southern part of Sakhalin Island, the lease of Port Arthur as a naval base (important as a warm-water port) and the restoration of economic privileges in Manchuria. From Turkey Stalin demanded the northeastern districts of Kars and Ardahan, and military bases in the Dardanelles straits and on the Aegean Sea. He presumably expected that Turkey would yield to pressure and would not be supported by the preoccupied Western powers. But in this case Stalin failed. He failed in Iran as well, although here it was the cunning of the Iranian Premier Qavam as-Saltaneh, rather than the tough talk of President Truman, which seems to have been crucial. Iran had been occupied by the USSR and the UK in

The Superpowers and the Bomb

At the end of World War II, the USA stood alone as a nuclear power. The government wanted to keep it that way, but the attempt to monopolize the secret proved fruitless. In August 1949 the USSR stunned the West by exploding its own atomic bomb. It had been assumed by the Americans that the Soviets were incapable of building their own bomb, and when the atomic spy Klaus Fuchs confessed shortly thereafter that he had been passing on all he knew to the USSR, the Soviet bomb was ascribed to his treachery. But the Soviets had begun their own research in 1939, and accelerated it in 1943. Further, the Soviets had the benefit of the official report of the Manhattan project.

The UK, too, decided to build a bomb: this was partly because it felt cheated by the USA, which it had freely provided with the information necessary to build one: but beyond that, the UK government believed that this was the only way to deter nuclear attack and keep its position as a major power. However, it was not until October 1952 that the UK exploded its own bomb. By that time the USA and USSR had moved on a generation, and were working on the more powerful Superbomb or hydrogen bomb. The USA exploded its first H-bomb in November 1952, the USSR following in August 1953. The UK finally caught up in May 1957. By this time, other powers were at work: behind the UK were France and China and then Israel, South Africa and Iraq. In short, wartime scientific developments and postwar rivalries left the world under threat of annihilation.

▼ "Ivy Mike", the first American hydrogen bomb, was exploded at Eniwetok atoll on 1 November 1952. With power equivalent to 10.4 million tons of TNT explosive, the cloud stem pushed upward 40km (25mi), while the cloud itself went up 16km (10mi) and spread out about 160km (100 mi). At 65 tons, the bomb was too heavy to deliver: portability came with "Bravo", in the spring of 1954.

◄ Rebuilding work in Leningrad, 1946. The Germans had besieged Leningrad for almost three years, devastating much of the once-beautiful city. It, along with Stalingrad, and much of the USSR subject to German invasion, had to be rebuilt after the war. To replace Soviet industrial plant destroyed by German guns, Soviet troops stripped their zone of occupation in Germany.

order to safeguard the flow of supplies to the Soviets, but by 1946 the USSR had still not evacuated its troops. They only did so in May of that year because they mistakenly thought that they had an ally in the Iranian premier.

The USSR and European communism

These were specific examples of Soviet probes for possible weak spots in various states' defenses, a wholly traditional Russian habit. However, of more general geopolitical concern was repeated evidence of the Russians' dualist or Manichaean view of the world: communist and capitalist, with war between the two camps inevitable. Two landmarks were the Duclos letter of April 1945 and the setting-up of the Cominform in September 1947.

In April 1945, once the defeat of Germany was clearly only a matter of time, a letter signed by the French Communist Party leader Jacques Duclos was published in the journal *Cahiers du communisme*. It expressed Stalin's view that the postwar world would see total enmity between capitalist and communist states, and foreign communist parties were ordered to purge those members who advocated "peaceful coexistence and collaboration in the framework of one and the same world".

Nevertheless, during the period from 1944 to 1947, communist parties in Western Europe were encouraged to enter into coalitions with "bourgeois" governments; the tactical considerations behind this maneuver, however, were subsumed within a greater strategy, articulated by Andrei A. Zhdanov in a speech at the founding conference of the Communist Information Bureau (the Cominform) in September 1947. The point of the Cominform was formally to group together the communist parties of the East European states plus those of Italy and France, and the point of Zhdanov's speech was to justify its foundation ideologically. This he did, as one historian described it, by emphasizing "the enduring and irreconcilable antagonisms between the capitalist and Communist systems".

This, then, was the driving theme behind Stalin's postwar foreign policy. Whether it arose from already developed Soviet plans, or was partly a reaction to Western moves, are problems which will not finally be answered unless Soviet archives are one day open to scrutiny – and perhaps not even then.

The US vision of the postwar world

The USA was as concerned for postwar security as the USSR, and as driven by history. As one historian has written, "Lessons of the past significantly conditioned Washington's plans for peace", and here the reference is to the post-World War I period. Germany in 1918 had not believed itself defeated: this time, the enemy would be thoroughly defeated and then disarmed. President Roosevelt and Secretary of State Cordell Hull, among others, believed that economic barriers had led to the Great Depression, which had contributed to the march to war: this time the USA would insist on the lowering of barriers to trade and investment, which implied the convertibility of currencies and the elimination of tariffs. It was believed that American failure to join the League of Nations had contributed to the collapse of international order: a new international organization to maintain peace would be established. Finally, the limitations on self-determination after World War I had contributed to the political instability of the interwar period: this time the principle should be applied more completely, which would mean, among other goals, the decolonization of the British, French and Dutch Empires.

In short, the USA thought in global terms, and indeed, since mid-way through the war the official American position had been that American security in the postwar world could be secured only by a fundamental restructuring of the international order. However, in the immediate postwar years, the USA thought less in military terms than did the USSR, or else it would not have demobilized so quickly. Rather (except for the atomic bomb), it looked more toward economic instruments to enforce policy, and only in 1949, with the establishment of the North Atlantic Treaty Organization (NATO), did it accept completely the need for a strong military capability as well.

Planning for the postwar years began during the war, and by the end of 1945 the major organizations were in place. During the early years of the war there was public discussion in the USA over whether or not it should encourage the establishment of another collective security organization on the lines of the League of Nations:

▼ Ruler of the USSR for almost 30 years before his death on 5 March 1953, Josef Stalin was a worthy and bloody successor to the czars. A former theology student, he worked for the Party from his teens. He spearheaded the prewar forced collectivization of agriculture, purged all actual or potential opponents, led the USSR to victory in the war and created an Eastern European empire.

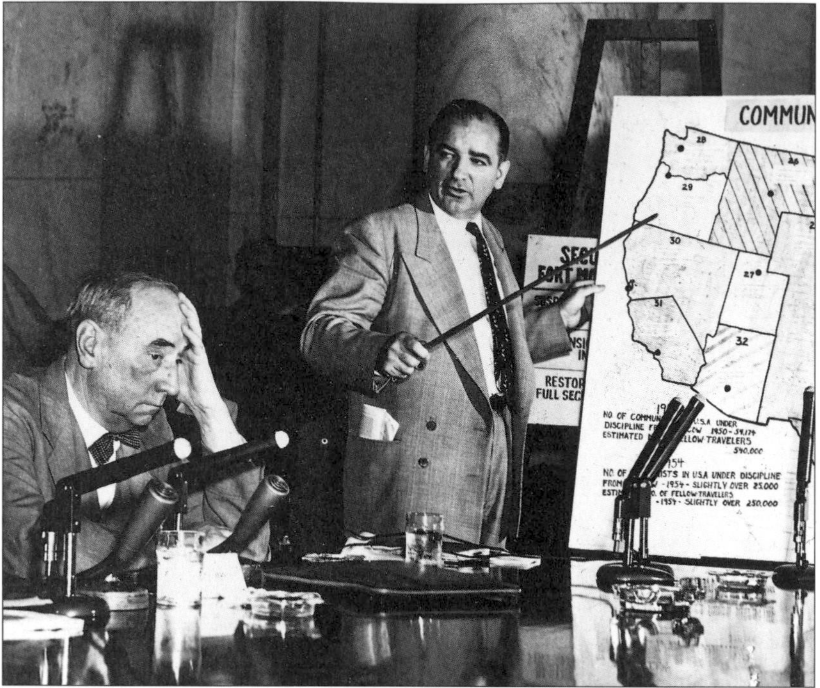

▲ **Against the background of communist victories in Europe and Asia, Joseph McCarthy, a member of the US senate and a vicious character assassin, rode to fame and power early in the 1950s on a wave of American fear about domestic communist subversion. His technique was to accuse individuals of spying for the USSR, or subverting American democracy, or knowing people who did, or belonging to groups which might not be sufficiently strongly opposed to those who did. He denounced these individuals and groups to US senate subcommittees, where he was protected by congressional privilege against legal action by the guiltless individuals whose lives he had ruined.**

by the summer of 1942 public-opinion polls showed that three out of four Americans favored this. The state department proceeded with planning for such an organization, with the significant modification that four powers, the USA, the USSR, the UK and China, would have a veto over its activities. This was necessary in order to secure the agreement of the USSR and of the US senate. At the Moscow Foreign Ministers' Conference on 1 November 1943 the above-named four states agreed the necessity of establishing an international organization, open to all peace-loving states, to maintain international peace and security. The general structure of such an organization was worked out at a conference of the four held from August to October 1944 at a mansion called Dumbarton Oaks in Washington, D.C.; however, there was disagreement on the voting procedure for the planned Security Council, which was only settled at the Yalta Conference in February 1945, when the principle of the veto was decided upon. A conference of 50 nations ensued, from April through June 1945 at San Francisco, in which the United Nations Charter and the Statute of the International Court of Justice were drafted. The UN finally came into existence on 24 October 1945.

In July 1944 forty-four nations met at the United Nations Monetary and Financial Conference at Bretton Woods, New Hampshire, USA. The outcome of their discussions was the establishment of the International Monetary Fund, whose main aim was to expand international trade by stabilizing exchange rates, and the International Bank for Reconstruction and Development, often called the World Bank, which aimed to provide loans for development purposes when private capital failed to do so. The Articles of Agreement of both organizations came into force on 27 December 1945. In the case of the IMF and the World Bank, however, the USSR was not a member: the Soviets did not want to relax trade barriers in areas under their control.

The origins of the Cold War

Through the media of the UN, the IMF and the World Bank, the world now entered into a new era of collective security and international trade, both of which were expected by the Americans to help ensure peace. Yet in the postwar period the outlook appeared to be dominated by the growing and implacable hostility of the USSR. Thus American policymakers had to decide upon both strategy and tactics: the strategy became the "containment" of the USSR, and the tactics depended upon the use of America's economic power. Early in the postwar years the US government believed that it had only limited resources with which to fight any external threat, and indeed, that the USSR was luring the USA into bankruptcy. Therefore the USA would adopt a policy of defending "strong points": those areas which had a military-industrial capacity, sufficient raw materials and secure lines of communication with the USA. Dealings with the USSR were to be conducted with "patience and firmness".

By the end of 1947 the USA had decided that there was minimum possibility of cooperation between the two superpowers, a decision which

► **The United Nations embodied the hopes of those who believed – against all the evidence of the short and failed career of the League of Nations – that world public and diplomatic opinion could be mobilized against an aggressor. However, the UN, founded in the San Francisco opera house on 24 October 1945 (right) as a result of decisions made during a series of wartime conferences, went further than the League in its realistic assessment of the need for the great powers to have their power recognized. Each was given a veto. Further, the UN was also given several humanitarian adjuncts, such as the International Children's Emergency Fund and the Relief and Works Agency.**

► The postwar wealth of the USA was legendary, a combination of a superabundance of food, an industrial sector which had developed extensively during the war, an abundant labor force – and the fact that no rebuilding of the country was required, since it had escaped entirely unscathed from the fighting. Public spending on armaments merely fueled the economy. In short, Americans had never had it so good.

seemed more than justified by the communist coup in Czechoslovakia in February 1948, followed by that in Hungary in June. To prevent the communist system from spreading even further, the USA more urgently set about "containing" the USSR, but from 1947 to 1949 the weapons remained economic. For one thing, the Americans believed that the Soviets had no immediate intention of starting a war: they lacked an atomic bomb, and they were achieving their aims in Eastern Europe without fighting the West. American demobilization could continue and, indeed, there was a limit to what military forces could achieve in peacetime: thus economic aid to certain countries would achieve more security per dollar than would military aid. But beyond that, and in contradistinction to the Soviet approach, the Americans believed that their own interests, and those of global stability generally, would be best served by the emergence of independent and self-sufficient centers of power overseas.

The best-known expression of American aid to rehabilitate strong points is the so-called Marshall Plan, but economic containment was to move decisively in the direction of military containment in the working-out of the implications of the Truman Doctrine. Early in 1947 the UK, which among the Western powers had taken the predominant role in Greece and Turkey, informed the USA that its own economic weaknesses forced it to end its military and financial support for those countries. The USA had to decide quickly whether or not to assume the UK's role: the result was a request by President Truman to Congress in March 1947 for 400 million dollars in aid. The immediate threat was perceived to be that of an armed communist takeover of the two countries; but the need for a quick response by Congress forced the administration to justify its request in more global terms. On 12 March 1947 Truman announced that "it must be the policy of the United States to support free peoples who are resisting attempted subjugation by armed minorities or outside pressures", thereby appearing to commit the USA to resist Soviet expansion anywhere in the world. Historians have differed as to whether this was a "fundamental point of departure for American foreign policy in the Cold War", or whether it merely marked the culmination of "patience and firmness". That aid to Greece in particular was voted to stave off what appeared to be a communist military threat must link it to later moves in military as well as economic containment.

A year and a half later discussions began between the USA and various European nations which culminated in the agreement to establish NATO in April 1949. In other words, the USA began to assume the development of a bipolar world, and to work on that assumption and in short, had to recognize that, contrary to President Roosevelt's hopes, the "Grand Alliance" was not going to cooperate to maintain peace. Rather, it was split asunder, so that by the early 1950s, former allies had become enemies and former enemies had been rehabilitated as allies, and the world was firmly in the grip of an increasingly menacing Cold War.

REVENGE AND JUSTICE

It was predictable that many among the victorious nations – especially in those which had experienced enemy occupation – would wish to punish the authors and beneficiaries of their woes. Less predictable was the resolve that such punishment should be judicially determined. But vengeance and justice are bad bedfellows. The forms which vengeance took in the wake of the Allied armies' arrivals were inevitably nasty. That justice took rough forms in the confused interim between liberation and the restoration of regular government was not surprising. What was surprising was that the settlement of accounts nevertheless had some fair-minded and discriminating aspects.

In a class of its own was the expulsion of Germans from the east European lands where their name now stank. Some of them had perhaps long lived there, others had only recently been lodged there in place of evicted Slavs. Now the tide turned, the USSR and Poland moved their frontiers westward and Czechoslovakia became again what it was before "Munich". Huge numbers of German-speakers had (or felt they had) no choice but to flee westward into the Allied-occupied zones: 1.5 million from the Sudetenland, about 7 million from beyond Poland's new western frontier, and so on.

Rough and questionable justice was meted out by lynch-mobs and people's courts directly after the departure of occupying forces. These were hazardous weeks for persons reputed to be traitors and collaborators, many of whom were summarily and often savagely dealt with before regular courts could consider their cases more dispassionately. The damning word "collaborator" was too readily attached to people who had reckoned that they could best serve their country by cautious cooperation with the occupier. Many relatively innocent men and women were unjustly done to death, and opportunity was freely taken for settling all sorts of old scores.

Justice was better served in the multitude of courts, civilian and military alike, which were soon trying alleged war crimes right across Europe from the Pyrenees to the Urals and all over East and Southeast Asia. Their quality of justice was of course conditioned by ideology and circumstances, but over-severity and bias were certainly not universal. Leniency and acquittals became more frequent as time went by. The International Military Tribunals which crowned the whole unprecedented business had mixed aims: to punish the guilty, but also to put on record the criminality of the men, organizations and regimes responsible, in the victors' view, for causing the war and for making it so horrible. In some of these aims they succeeded. "The Nuremberg Principles" asserted the superiority of international over national law and the personal responsibility of even the highest-placed breakers of it. The traditional law of war was reaffirmed and new human-rights law was helped by the creation of a new category of "crimes against humanity".

►▲ The Tokyo courtroom (right): 28 defendants face the bench of judges (one from each of 11 victors) with ample space for translators, press and camera men. At Nuremberg the USA, USSR, UK and France tried 22 defendants. Goering (above) was the most important Nazi tried, and one of the sharpest. The proceedings were free enough for him to be able to give the prosecution lawyers some sticky moments.

►▼ Vengeance and cruelty are more conspicuous than justice as the corpses of Mussolini and his mistress hang upside down at a Milan filling-station and a batch of Belgian women who have been going with German men are publicly humiliated. Below, May 1945, refugees in Berlin.

THE FOUNDATION OF ISRAEL

Hitler's policy of genocide had the ironic effect of giving a new political and humanitarian meaning to the cause of Zionism. Before the outbreak of World War II, the creation of a Jewish state in Palestine, to which all Jews would have the right of return, was a political impossibility, due to Western indifference and Arab hostility. The Holocaust transformed the impossible into the necessary, by demonstrating, especially to the American public, the force of the Zionist argument that without a state of their own, Jews would be permanently vulnerable to persecution and mass murder.

World War I had left the former Turkish territory of Palestine as a League of Nations mandate under British rule. In the interwar years Zionist organizations sponsored Jewish immigration. On the eve of World War II, the British mandatory authorities were caught between Arab and Jewish nationalist movements. In 1935 the Peel Commission recommended the division of Palestine into three states, one Arab, one Jewish, and one under the British mandate. The Jewish population would occupy roughly one-fifth of the territory of mandatory Palestine, and would be in no position to dominate the Arab majority. This was unacceptable to militants on both sides. The Arab revolt of 1936–39 – a series of riots and attacks on British and Jewish targets – bloodily reaffirmed their determination to preserve Palestine as a predominantly Arab homeland, in the face of increased Jewish immigration and settlement. In contrast, the Jewish Agency (Zionist executive), led by the future first prime minister of Israel, David Ben-Gurion, took a more cautious line and accepted the principle of partition on pragmatic grounds. Some rightwing militants rejected this decision and formed independent armed cells.

World War II transformed the entire political context of the Arab–Jewish conflict in Palestine. The liberation of Nazi concentration camps and a series of war crimes trials disclosed the enormity of the "Final Solution", and created a consensus that a home had to be found for the survivors of the Holocaust. In addition, the economic burdens of warfare made it inevitable that the UK would reassess its overseas commitments, including its responsibility as the mandatory authority in Palestine.

Faced by armed revolt from both Jews and Arabs, the British authorities handed the problem over to the United Nations. On 29 November 1947 the United Nations voted to partition Palestine into an Arab and a Jewish state, keeping Jerusalem as an international entity. This solution was accepted by the Zionists and rejected by the Arabs. A bitter armed struggle followed. Over the next 18 months, Jewish force of arms prevailed against numerically superior Arab forces. The state of Israel was proclaimed on 14 May 1948, and recognized by the United States and most of the Western world. The Zionist dream had been realized, but the underlying conflict with Palestinian Arabs remained unresolved.

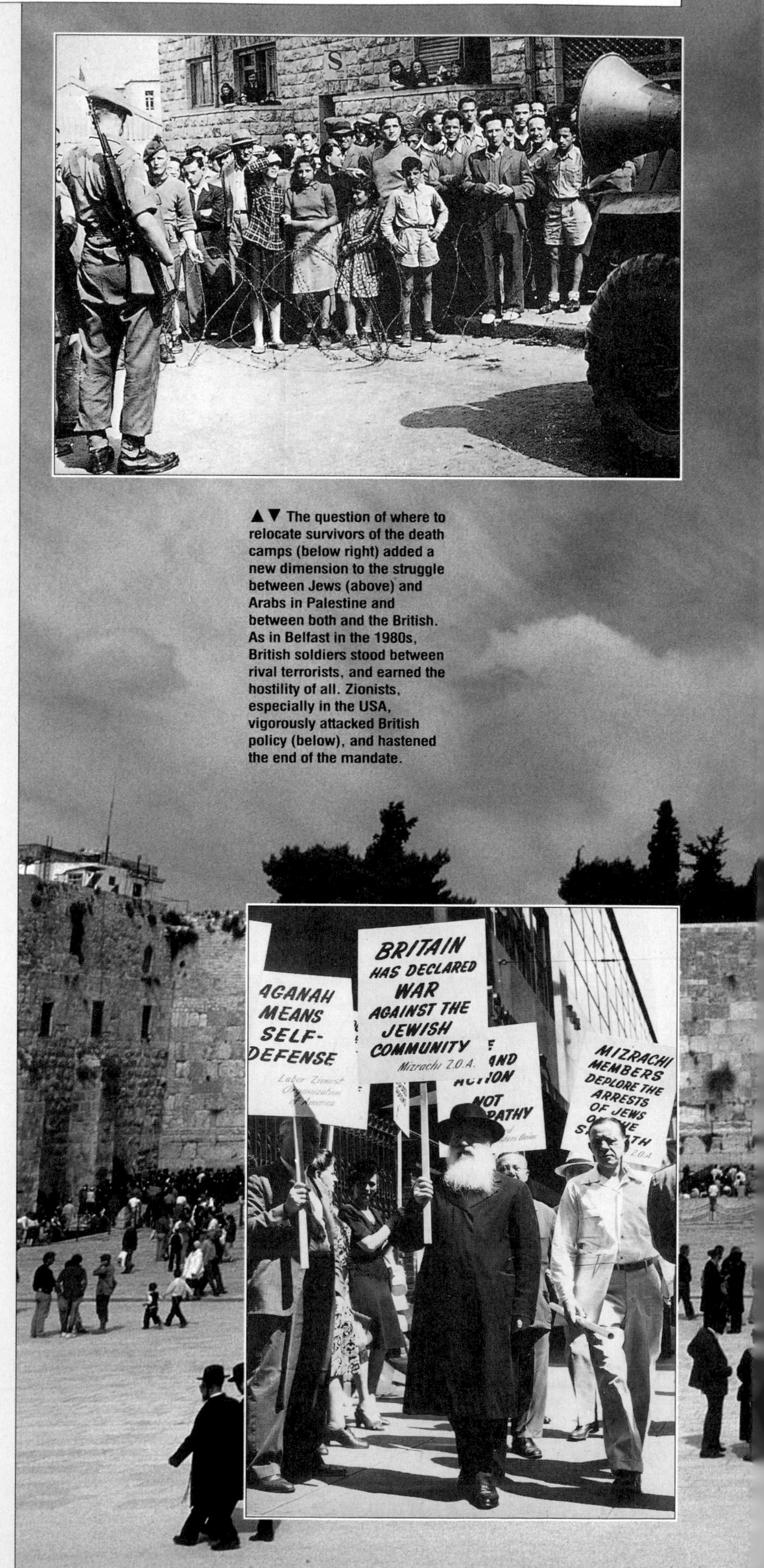

▲▼ The question of where to relocate survivors of the death camps (below right) added a new dimension to the struggle between Jews (above) and Arabs in Palestine and between both and the British. As in Belfast in the 1980s, British soldiers stood between rival terrorists, and earned the hostility of all. Zionists, especially in the USA, vigorously attacked British policy (below), and hastened the end of the mandate.

▼ Jerusalem stands at the heart of Jewish history and nationalism. Jews have worshiped for centuries at the Western Wall of Solomon's Temple, and throughout the world observant Jews still turn to Jerusalem during their devotions. After 1947 Jordan held the Old City of Jerusalem, but in 1967 it was captured by Israel and annexed to the Israeli state.

► David Ben-Gurion (1886–1973) was chairman of the Jewish Agency for Palestine from 1935 to 1948, and first prime minister of Israel. Palestinian Arabs lacked a leader of his vision, and thousands fled or were expelled during the 1947–48 war. The camps in which they and their descendants (below right) have lived since then symbolize their tragedy.

Datafile

Asia responded in different ways to the consequences of war. Japan rose from the ashes of defeat and achieved astonishing economic progress from the late 1950s, to be followed in the 1970s and 1980s by the economic achievements of Taiwan, Singapore and South Korea. China's backward economy enjoyed rapid economic growth in the post-Mao era. Indonesia was afflicted by internal struggles.

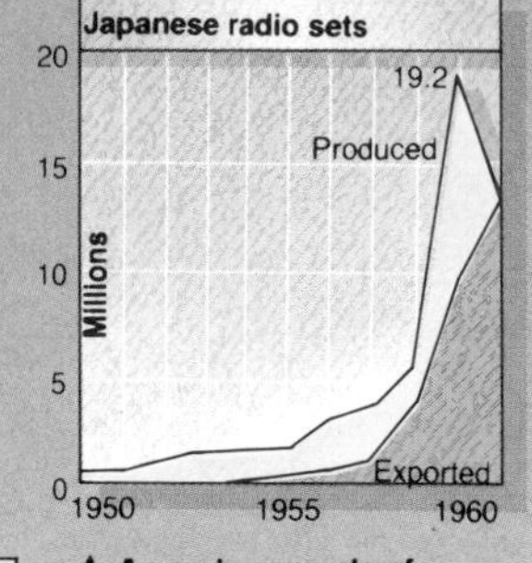

▲ An early example of Japanese commitment to electronics is seen in these statistics portraying the swift growth in the manufacture and export of radio sets. Note that the real takeoff occurs in the late 1950s with significant exports. This is particularly interesting in that exports in total only quadrupled in the same period.

Casualties in Malaya

Thousands

25 20 15 10 5 0

Terrorists

Security forces

Civilians

1948 1950 1955

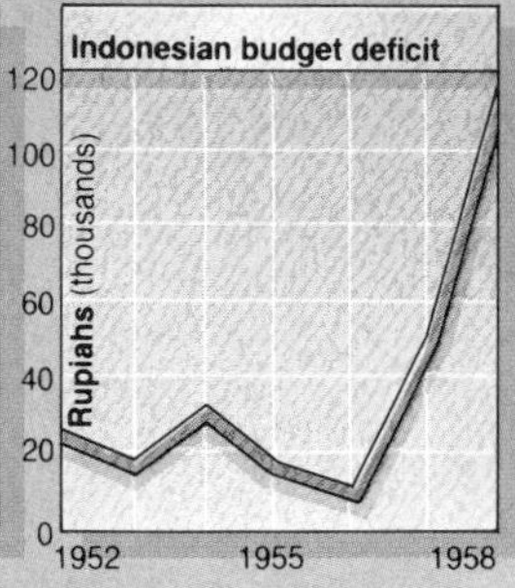

▲ Indonesia's economic experience after gaining independence proved disappointing. Budgetary deficits were high and then escalated in the late 1950s. In Malaya casualty figures during the communist insurgency demonstrate that even at its time of greatest impact, 1951–52, the insurrection had not obtained sufficient support.

▶ Mao's determination to develop the Chinese economy is illustrated by oil usage. At first China was heavily dependent on imports from the USSR but these diminished from the late 1950s. During the 1960s Chinese production expanded and Soviet imports ended during the acrimony of the Sino-Soviet split of the mid 1960s.

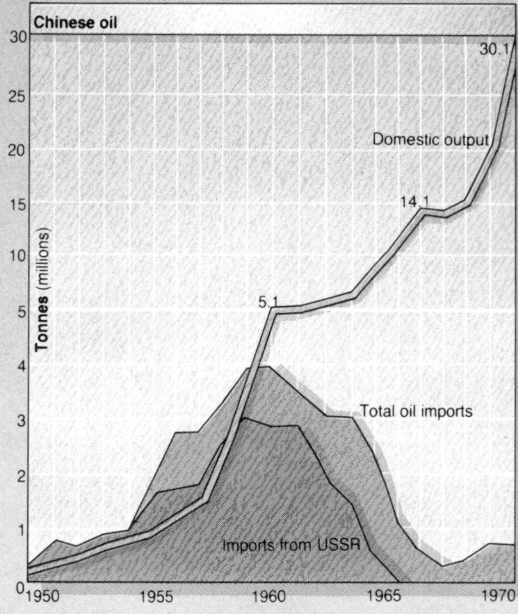

Chronology: occupation, decolonization and civil wars

1945

August
USA and USSR agree to divide Korea at the 38th parallel; Ho Chi Minh declares the independence of Vietnam

September
Japan makes its formal surrender; the Allied occupation begins under MacArthur; British forces intervene in Indochina and Indonesia to assist the French and Dutch

1946

July
The USA relinquishes authority in the Phillipines; civil war starts in China

1947

August
India and Pakistan are given independence as dominions

1948

January
Burma becomes an independent republic

June
Emergency declared in Malaya as communist party fights British

July
Republic established in S. Korea and constitution approved for the Democratic People's Republic (North Korea)

1949

July
France recognizes Bao Dai as head of state in Vietnam

August
The Netherlands recognizes the Republic of Indonesia

October
Mao Zedong proclaims the People's Republic of China

1950

June
North Korea invades the south; UN forces respond

October–November
China intervenes in Korea

1951

July
Armistice talks in Korea

1952

April
Occupation of Japan ends

1953

October–November
End of Korean war

1954

July
Vietnam divided

1957

August
Independence of Malaya

The long-term results of the Pacific war were paradoxical. Japan had gone to war, first with China, then with the USA and the UK, with two ambitions: first and principally to expand its own power, but secondarily – the cloak under which the Greater East Asia Co-Prosperity Sphere was promulgated – to put an end to European hegemony in Asia. In 1945 Japan was utterly and catastrophically defeated. Yet within a very few years of that defeat both goals had been to a large degree accomplished.

Freed of the distracting burdens of military imperialism, Japan made a spectacular economic recovery to become, within a generation, the dominant economic power in Asia, and gained a far more powerful presence on the world scene than it could ever have attained by military means. In addition the British, French and Dutch empires which had ruled, respectively, India and Malaya, Indochina and the East Indies were all more or less rapidly wound up.

A third outcome of the war, however, which neither the American victors nor the Japanese vanquished had wished for, was the spread of communist rule over large parts of Asia. This ensured a continuing American interest in the region. Just as it did in Europe, the onset of the Cold War in Asia caused a reversal of wartime alliances. In Europe, within four years of the death of Hitler, Western Germany was transformed by NATO into an ally against the former Ally, the USSR; similarly in Asia, America's former enemy Japan was quickly turned around to form a bulwark against the threat from China. Twice, in Korea and in Vietnam, the Americans went to war to try to stem the communist advance.

The legacy of World War II only began to recede in the late 1980s, as a less polarized and less ideological world began to emerge. By now former colonies had long been independent; and the anti-Western appeal of communism had begun to wear off; the capitalist prosperity of Japan, South Korea, Taiwan, Hong Kong and Singapore offered an alternative model. China was emerging from revolutionary isolation into greater openness and a more relaxed self-confidence, even considering a more friendly relationship with the Soviet Union, while the focus of American alarm had shifted to the Middle East. The postwar power vacuum had been filled. The three major powers which ultimately emerged from the upheaval of 1937–45 – communist China, democratic capitalist Japan and independent, democratic but nonaligned India – had matured into stable and unthreatening members of the world community. The Pacific basin, with its immense populations and increasingly advanced technology, now began to assume a centrality in world affairs that looked set to

ASIA REBORN

The long-term results of the war in Asia

American occupation of Japan and its consequences

Imperial possessions in Asia gain their independence

Communist victory in China

America's attempts to resist the spread of communism in Asia

◀ Japanese women were eager to take advantage of gaining political equality in 1947. Here women are seen casting their votes.

grow. The death of Emperor Hirohito in 1989 finally drew a line under the war in the Pacific.

The recovery of Japan

The Allied occupation of Japan, under the command of General Douglas MacArthur, was a remarkable success story. It was essentially American in character and aimed to eliminate militarism, introduce American-style democracy and encourage the emergence of a prosperous "western" nation. Yet MacArthur understood the importance of applying reforms in such a manner as to carry the Japanese with him. In particular, impressed by Emperor Hirohito's frank acceptance of responsibility for past failures, he rejected the idea of trying him with other accused leaders as a war criminal. Various generals, admirals and bureaucrats were tried and some were executed, but Hirohito was merely stripped of his divinity and left in place as a ceremonial head of state to provide a necessary element of continuity.

The new democratic constitution, introduced in 1947, created a bicameral legislature with ultimate authority vested in the lower house. Women were given the vote. A drastic land reform was implemented which ended the remaining power of the landlords and created a stable peasantry. Attempts to destroy the powerful *zaibatsu* (financial combines) were less successful. Trade unions were encouraged and grew so rapidly that serious labor disputes occurred. At first this new assertiveness of labor was welcomed, but as the Cold War developed industrial militancy was suppressed.

From late 1947 the nature of the American occupation changed. The original radical zeal gave way to a more pragmatic conservatism in which the emphasis was placed on economic revival. With the outbreak of the Korean War Japan was increasingly seen by the USA as an important ally against the spread of communism. At San Francisco in 1951 the United States and Britain concluded a formal peace treaty with Japan: the settlement was generous on internal arrangements but involved the loss of all the territorial gains made by Japan since 1895. The USSR and China did not sign it. The American occupation formally ended in April 1952.

The end of empires

The ending of the European empires was most dramatically symbolized by the British withdrawal from India. Unlike most of the rest of Southeast Asia, India was never conquered by Japan. But Indian nationalism was encouraged by the spectacle of European vulnerability revealed by the war. The Indian nationalist leaders, Gandhi and Nehru, were wary of the Japanese, but saw that the war would greatly increase the pressure on Britain to concede independence. Even a Conservative government in London would have had difficulty in staying long after 1945. The Labour government had no wish to linger. Finding the mutual suspicions of Hindus and Moslems too deep to resolve, prime minister Clement Attlee and Mountbatten, the last viceroy of India, accepted the case for partition without further

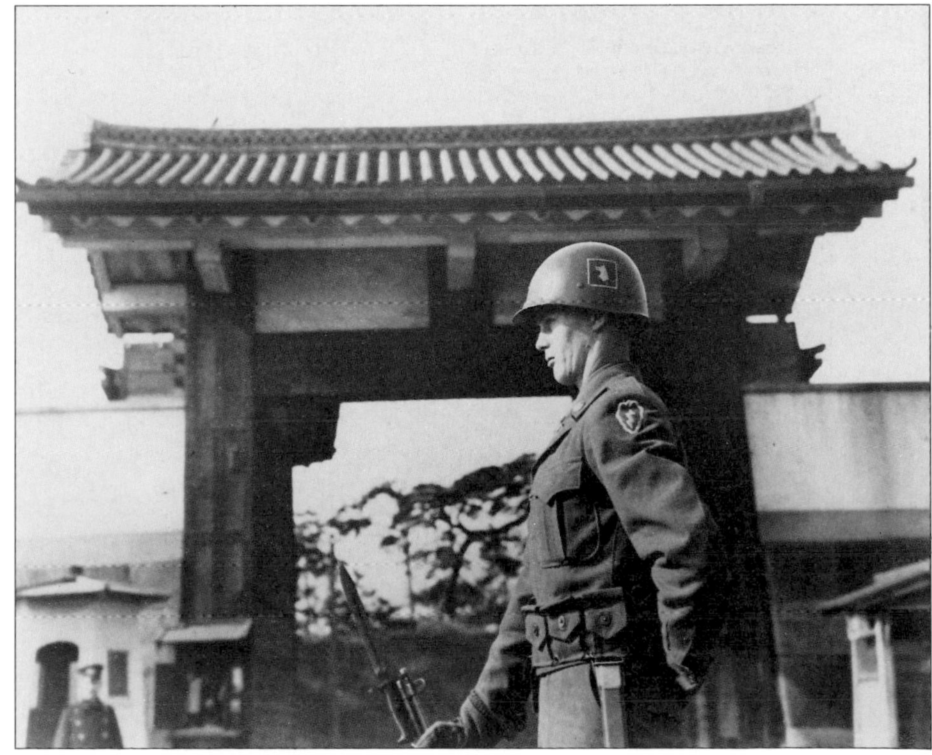

▼ An American soldier on guard duty in Tokyo provides a symbol of the reality of defeat; and the Japanese entrance makes an ironic contrast with the GI. Despite some unpleasant incidents, relations between occupiers and occupied were generally amicable.

The Independence of India and Pakistan

The attainment of independence within the Indian subcontinent afforded the most spectacular example of postwar decolonization. The British contained nationalism during the war but the postwar British prime minister, Clement Attlee, and his Labour government realized that it would be impossible to maintain the Raj without committing large-scale resources. To supervise the transition to independence they appointed Lord Louis Mountbatten viceroy.

He had vague socialist sympathies and had displayed shrewd political judgment when heading Southeast Asia Allied Command from 1943 to 1946. The chief issues in Indian independence were the demand for the creation of a separate Muslim state of Pakistan, the fate of the princely states and the timescale. The leading Muslim, Mohammed Ali Jinnah, had originally supported the Indian nationalist Congress Party but broke from it in 1940 and developed the Moslem League so as to secure a separate Pakistan. Churchill had told him in December 1946, "You only have to stand firm and demand your rights not to be expelled from the British Commonwealth and you are bound to be accepted".

In February 1947 Attlee promised that power would be transferred by June 1948. Mountbatten handled skillfully the daunting problems he had inherited. The cabinet told him to obtain a unitary solution so as to permit the pursuance of a unified defense policy. This was impossible, because the demand for Pakistan was too strong. The destabilizing trends were so powerful that Mountbatten could see that Britain must leave rapidly. Nehru, President of the Congress Party, wanted a united India but reluctantly accepted that partition must occur.

The dominions of India and Pakistan were established on 15 August 1947 and the princely states were compelled to adhere to either state. Independence was accompanied by communal slaughter on a large scale – between one quarter and half a million perished. Independence came too quickly; but it was a great achievement to implement it on an agreed basis for such a vast area.

◀▲ The greatest single event in decolonization is shown (above) as vast crowds assembled in Delhi in August 1947 for Independence Day and to watch the Indian flag being raised at the Red Fort. This picture indicates fervent yet controlled enthusiasm. Elsewhere many perished in savage intercommunal slaughter. The principal personalities in achieving independence (left) were Gandhi (right) and Nehru. The saintly and devious Gandhi was the chief strategist of independence and Nehru was the first prime minister, charged with the task of converting aspiration into reality.

delay. India and Pakistan became independent states within the British Commonwealth in August 1947, to be followed in 1948 by Ceylon. Burma, following its liberation from Japanese occupation, also received its independence in 1948, but remained outside the Commonwealth.

Elsewhere in Asia decolonization was not so swift but just as certain. In Malaya the Japanese had fostered nationalist feeling against the British but had severely oppressed the Chinese population, with the result that Chinese communists were prominent in the wartime resistance. After the war they turned their British-supplied arms against the British themselves and launched a long guerrilla insurgency which lasted from 1948 to 1960. It failed for lack of support from the majority Malay population. Malaya became independent in 1957. (Singapore joined Malaysia in 1963 but chose independence in 1965.)

In the Dutch East Indies, too, nationalism had been too far stimulated by the war to be suppressed after it. The Dutch government tried briefly but accepted the inevitable in 1949 and Indonesia became independent in 1950. Of all the European powers, the French fought hardest to restore their rule in Indochina. Here again nationalism was identified with communism. British forces helped the French in occupying Saigon and outmaneuvering Ho Chi Minh's proclamation of a progressive republic in August 1945. Ho retreated to Hanoi to rebuild his forces while France endeavored to impose a solution giving limited autonomy to Vietnam, Cambodia and Laos while retaining the reality of control itself. But French military strategy was inept and ill-placed to counter a guerrilla movement enjoying peasant support. The bankruptcy of French policy was demonstrated by the humiliating surrender of Dien Bien Phu in 1954. The Geneva conference the same year, chaired jointly by the UK and the Soviet Union, agreed on the division of Vietnam into two zones, supposedly to be reunified through free elections.

Communist China

In China the principal effect of the Japanese war was to assist the recovery and steady advance of the communists. Mao Zedong had assumed the leadership in 1934–35 during the Long March. Far abler and more ruthless than his predecessors, Mao remolded the party in his own image during the next decade. He warned that the Chinese road to socialism would be long and that "formula Marxism" should be rejected. In order to attract noncommunists alienated from the Guomindang, he was anxious to allay fears that a communist victory would be accompanied by extreme radical policies. But the identification of nationalism with communism was demonstrated in the party's waging of guerrilla warfare against the Japanese. The Guomindang government by contrast was moribund, unable to oppose the Japanese effectively and facing a deteriorating economic situation marked by rampant inflation. On paper the Guomindang was superior, with a much larger army than the communists. When the civil war resumed in 1946, foreign observers expected the Guomindang to prevail. But the foundations of Guomindang rule were rotten and Jiang Jieshi a discredited leader. With the moral disintegration of the Guomindang the communists achieved a rapid victory in 1948–49. The People's Republic of China was proclaimed on 1 October 1949.

At first pressure of circumstances compelled Mao to rely on the USSR for support but he regarded Moscow with suspicion. The negotiations that led to the Sino-Soviet treaty signed in February 1950 were difficult and Mao resented Stalin's domineering attitude. The Guomindang's survival on the offshore island of Taiwan (Formosa) and vehement American anticommunism

▲ The bitter struggle in Indonesia is illustrated in the photograph and poster. The poster demands liberty for the East Indies and urges people to strive for this. The reality of conflict is depicted with four Indonesians (left) and one Chinese (right) detained by Dutch forces in 1947 after capturing Tegal, a port in northern Java. The five men were accused of responsibility for a number of murders.

◀ A symbolic removal of the old order in Indonesia. Mobile cranes are removing a large Dutch angel of peace from a plinth in the former Wilhelmina Park in Jakarta in June 1950. The statue had been erected in 1882 and was to be replaced with a mosque.

prevented any rapprochement between the USA and the new China, while the Chinese were themselves too negative toward the west to exploit the opportunity created by British recognition in January 1950. Facing the daunting problems of governing a country devastated by decades of conflict, Mao might possibly have applied more pragmatic policies, both domestically and internationally, had it not been for the Korean War and the Chinese decision to intervene. But Mao then imposed a harshly repressive internal regime and any likelihood of China's pursuing an outward-looking foreign policy was doomed for 20 years.

The Korean War

Communism gained a foothold in Korea in August 1945 as a result of the Soviet occupation of the northern half of the peninsula, above the 38th parallel, while the United States occupied the south. With Russian help, Kim Il Sung soon emerged as an able and astute communist leader. A fervent nationalist, Kim was determined to see Korea united, independent and communist. In the south the Americans blocked radical reforms and promoted the veteran conservative nationalist, Syngman Rhee – a wily operator who used the Americans for his own purposes. He resembled Kim in the depth of his patriotism, but was fanatically anticommunist. Kim and Rhee competed in shrill rhetoric, each threatening to unify Korea by force after their respective states gained independence in 1948. The situation in Korea was extremely unstable.

With the mounting tension of the Cold War between 1948 and 1950 the United States was bound to act against any perceived communist aggression. When North Korea advanced across the 38th parallel on 25 June 1950, President Truman of the USA called for "police action" by the United Nations under American leadership. A Soviet boycott of the security council enabled the USA to secure condemnation of North Korean aggression. The UN was thus committed to war, although most of the forces provided were American.

The first year of the conflict was marked by astonishing reversals as first one side and then the other seized the advantage. Each side showed remarkable initiative in different phases – in the original North Korean attack, in the massive Chinese intervention, in General MacArthur's brilliant operation at Inchon, and in General Ridgway's success in restoring the UN position in 1951. The latter part of the war was characterized by interminable armistice talks. They ended in 1953 without a political settlement, leaving Korea still partitioned.

Asia an ideological battleground

American determination to resist the spread of Chinese communism, however, was only strengthened. In the 1950s Southeast Asia was seen by the USA as vulnerable to the "domino theory" of progressive subversion. After 1954 the USA took over the French role in attempting to block communist expansion into South Vietnam, leading in the 1960s to a large-scale, brutal and ultimately unwinnable war against Ho Chi Minh's jungle-skilled guerrillas. Vast resources were expended, Vietnam itself was devastated and American society was bitterly divided before the Americans, in 1975, finally accepted defeat. Communist regimes were established in reunified Vietnam, Cambodia and Laos. With the Americans gone, however, they quickly fell into renewed fighting among themselves. Meanwhile President Nixon, taking advantage of continuing antagonism between China and the Soviet Union, had achieved in 1973 a startling American rapprochement with China, which after the death of Mao became steadily more open to the west and gradually emerged as a major force on the world stage.

▲ A meeting between General Douglas MacArthur, head of the Allied occupation in Japan, and President Syngman Rhee of South Korea reflects American power in East Asia. Rhee visited Tokyo in October 1949. In the center is Jean MacArthur, the general's second wife. South Korea achieved independence in 1948 but was heavily dependent on the USA for military and economic support. Rhee knew that MacArthur favored a tough policy aimed at defeating the Chinese Communists.

► Henri Cartier-Bresson's photograph of the communist army occupying the former Guomindang capital of Nanjing in June 1949 suggests the dedication and tenacity of the communist troops. Awe and inquisitiveness can be seen among the faces in the crowd.

▲► Two faces of the Vietnam war. Above, a Vietnamese child, burned by napalm, is held by his father, following an American attack on Vietnamese guerrillas. Right, Australian troops about to embark in helicopters.

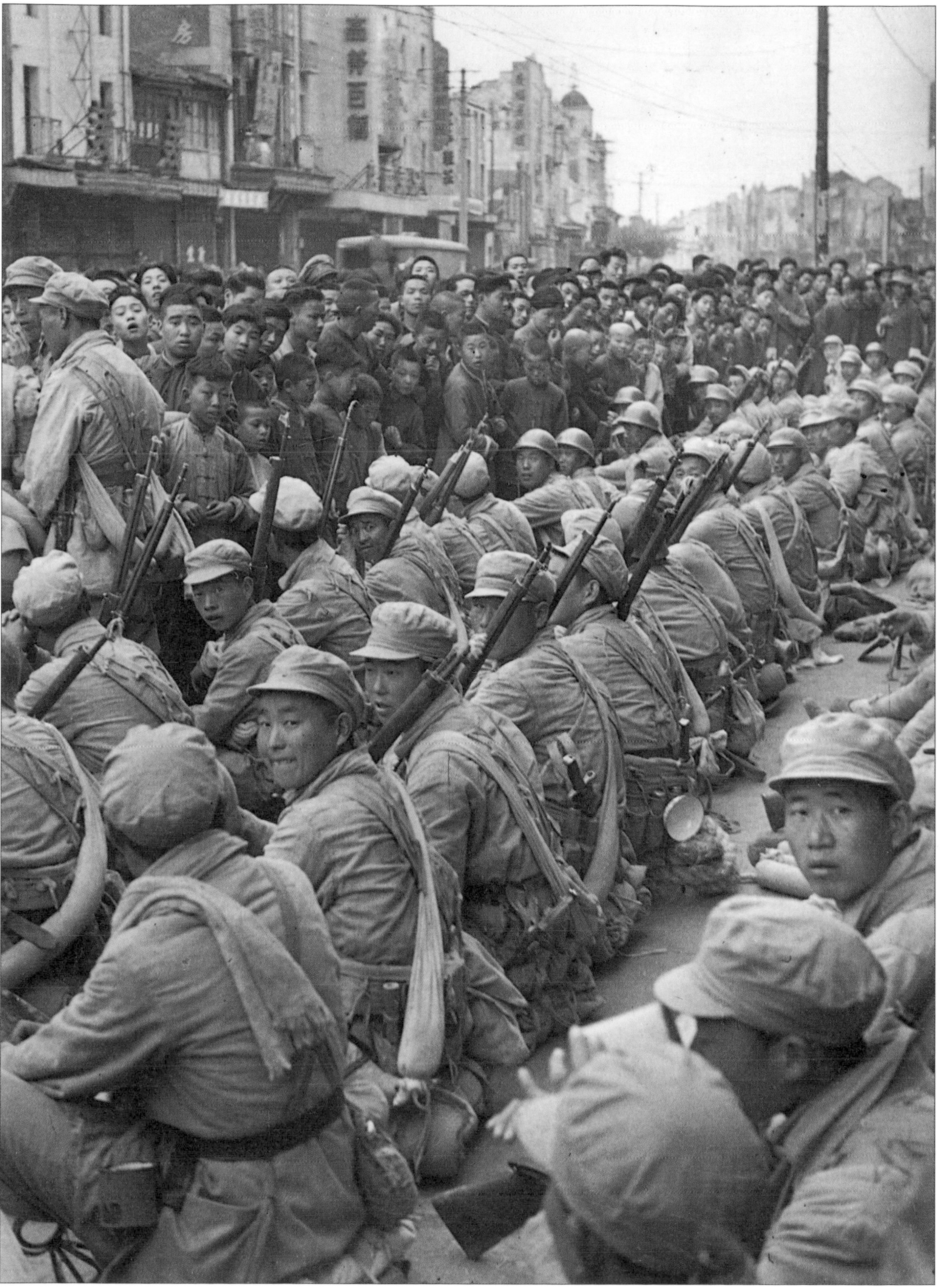

Datafile

For a decade after the war, many thinking Europeans must have felt themselves to be pawns in the game of the superpowers. Indeed, it was akin to the Great Game of imperialism in Africa during the 19th century: the adherence of countries to one side or another was sometimes desirable not for their own sake, but to deny support to the other side. The postwar period saw Europe congeal into two separate alliances.

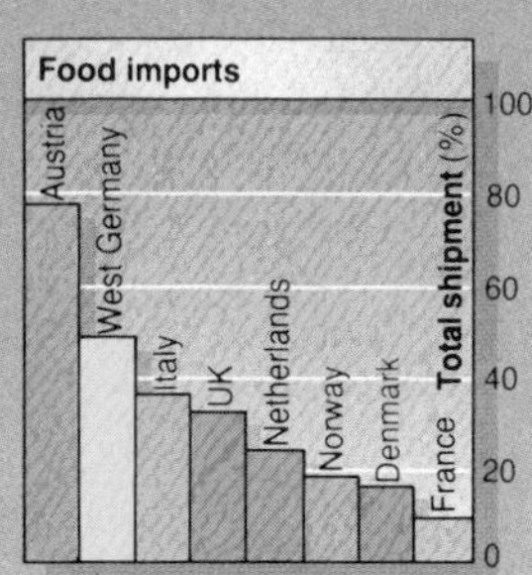

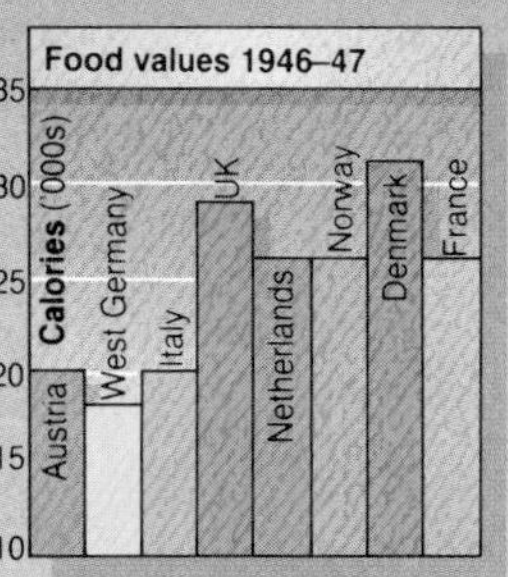

▼ **In both steel and electricity production, the proportional increases in Hungary and Poland, unlike those in Yugoslavia, are notable: the two former countries suffered depopulation, especially Poland, which lost land and people to West Germany, while Yugoslavia was not nearly so ravaged by the war.**

▼ **The proportion of West European exports supplied by the UK was bound to drop as the German industrial sector came back on stream after the war: between 1945 and 1950 UK exports as a percentage of the total of Western European exports fell by 27 percent, while West German exports increased by 103 percent.**

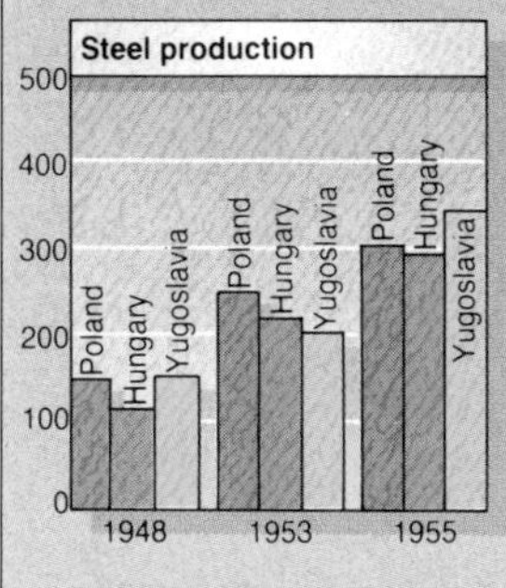

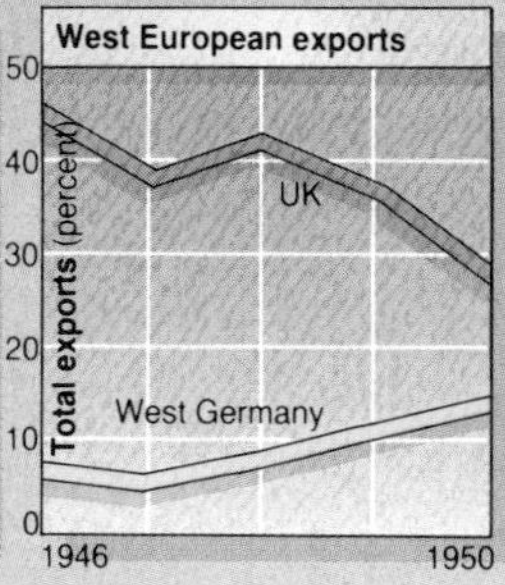

Electricity production

500
400
300
200
100
0
1948 1953 1955

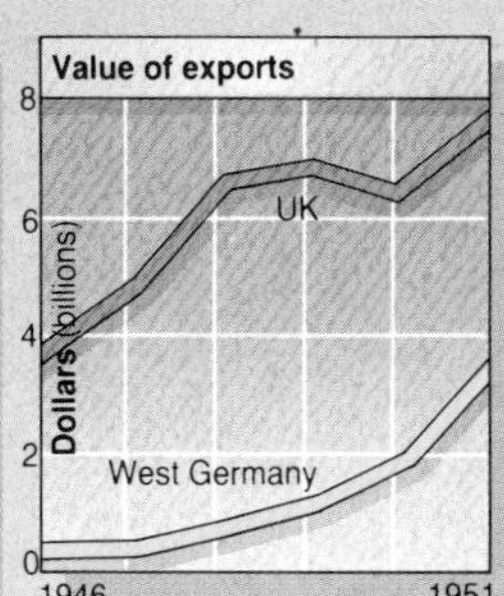

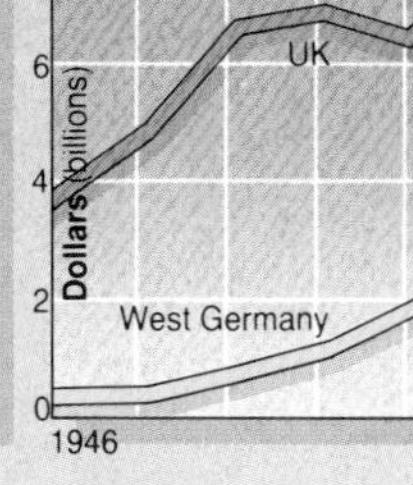

▲ **In 1946 the European nations appeared to influential members of the American government to be teetering on the verge of famine. A priority, therefore, of Marshall Aid was to provide food. It is clear from the two charts that those countries which most needed food in 1946–47 (measured by calorific intake per person) were those receiving a significant proportion of their aid in food by 1949–50 (top). The provision of food meant that scarce dollars could be spent on machine tools and raw materials to enable Europe to restart its industrial sector; by the end of Marshall Aid, the usual intra-European trade of food for manufactured goods has resumed.**

Chronology: the fall of the Iron Curtain

1944

October
Communists start civil war in Greece; ends February 1945

1945

May
Communist groups begin work in Soviet-occupied Germany; return of President Beneš to Prague

June
Formation of National Unity government in Poland pro- and anti- communist

November
Small farmers' party leader Tildy forms coalition in Hungary but communists hold key posts; Soviets force communist-dominated government on King Michael of Romania; Tito's national front party forms government in Yugoslavia

1946

January
Communists gain "electoral" victory in Albania

September
Bulgaria proclaimed a republic; communist- dominated fatherland front forms government

1947

January
Government intimidation leads to communist victory in Polish polls

February
Communists destroy Hungarian small farmers' party, and overthrow the Nagy government (May)

August
Communists dissolve political opposition in Bulgaria, and hang its leader, Petkov (September); General Markos proclaims Soviet Republic of Greece, and the civil war recommences

October
Polish opposition leader Mikolajczyk flees

December
Romania proclaimed republic

1948

February
Communist coup forces Beneš to put communists in charge of Czech government; defenestration of opposition leader, Jan Masaryk (March)

August
Defeat of communist revolutionaries in Greece

1949

October
German Democratic Republic proclaimed

One of the most enduring legacies of World War II has been the division of Europe. At the heart of that division is the partition of Germany, potently symbolized by the Berlin Wall (see pp. 246–47); but more than geography has separated Eastern Europe from Western Europe. Briefly, the years from 1945 through 1948 saw those states bordering the Soviet Union lose their independence one by one, primarily through coups by communist parties supported by Soviet armies of occupation. By September 1948 negotiations had begun which were to lead to the formation of the North Atlantic Treaty Organization (NATO) in April 1949. Its counterpart, the Eastern European Mutual Assistance Treaty (the Warsaw Pact) was agreed in May 1955, just five days after the Federal Republic of Germany gained its sovereignty. With the signing of a treaty between the German Democratic Republic and the Soviet Union in October 1955, which likewise granted sovereignty to the former Soviet zone, the crystallization of the two Europes was complete.

The wartime Allies' plans for postwar Europe

From the earliest days of American and Soviet involvement in the war, the members of the Grand Alliance had different ends in view for Europe. Roosevelt's USA government believed that the way to prevent future wars was for the principle of self-determination to be more fully applied in Europe. The implication was that free elections should be held, but this ran directly counter to Soviet plans for Europe. Stalin wanted first of all to consolidate the Soviet recovery of lands which had been part of the pre-1917 Russian Empire, and secondly, to ensure that there would be no hostile countries on his borders – free elections would almost certainly produce several governments hostile to the Soviet Union. The UK hovered uneasily between. Churchill realized that a refusal to give the USSR a free hand in Eastern Europe in 1939 had encouraged Stalin to turn to Germany. Conscious of this, as of the realities of geography and power, he was inclined to concede more to the Soviets than Roosevelt thought American public opinion would stand, in exchange for Soviet agreement to refrain from meddling in areas of British interest. It was this tension between American hopes of self-determination and the Soviet intention to ensure secure borders which ensured the breakdown of the Grand Alliance.

By means of the Soviet-Nazi Pact of 1939, Stalin had gained Germany's agreement to the Soviet takeover of Finland, which had been an autonomous grand duchy within the Russian Empire from 1809–1917; of the Baltic republics, Estonia, Latvia and Lithuania, whose freedom from Russia had been gained only in 1918; of

EUROPE DIVIDED

The wartime Allies' plans for Europe after the war

The Yalta and Potsdam conferences

Communist coups in Eastern Europe

NATO and the creation of the Federal Republic of Germany

The role of Marshall Aid

Postwar Europe: two armed camps

eastern Poland, which had belonged to Russia from the partitions in the 18th century until Poland received its independence in 1919; and of Bessarabia, which had belonged to Russia from 1912 until 1918, when it had seceded to Romania. Stalin made it clear to the Western powers that he wanted their confirmation of Soviet possession, but Roosevelt managed to postpone this until the end of 1943. In November, however, at a conference in Tehran, Roosevelt felt driven at least to imply to Stalin that the USA would not challenge the Soviet position in Poland and the Baltic states, although Finland was a special case (and, indeed, ultimately retained its independence).

It was not only that Roosevelt was reluctant to agree to these specific postwar arrangements: in contradistinction to Stalin, Roosevelt hoped to postpone political discussion until after the war. For one thing, the self-determination of peoples would be hard to arrange while the fighting continued; furthermore, military victory was the overwhelming priority, and if the Allies began maneuvering to gain postwar advantage, the acrimony this would cause could only benefit the enemy. There were also domestic political reasons for Roosevelt's preference for postponing such discussions: during most of the war he governed in tandem with a very conservative, Republican-dominated Congress, and bipartisan agreement – at least in the early years of the war – covered only the need to defeat Japan and Germany.

Churchill was more inclined to negotiate with Stalin, since he was acutely aware of the UK's probable postwar weakness, and the duration of American interest in Europe was problematical. In October 1944 the UK and USSR concluded the secret so-called "percentages agreement": Soviet influence in Bulgaria and Romania would be 90 percent, as opposed to 10 percent for the UK; in Hungary, Soviet influence would be 80 percent and the UK's 20 percent; in Greece, British influence would be 90 percent and Soviet 10 percent; and in Yugoslavia each would have 50 percent influence. This was Churchill's attempt to save he what could of Eastern Europe for the West: the capitals of Bulgaria, Romania and Yugoslavia were already occupied by the Soviet Red Army. The USA, however, refused to back the agreement, since it would be politically

▼ Soviet soldiers in Vienna in 1945. Austria remained divided and occupied for 10 years. The Viennese feared Russian soldiers more than their Western counterparts, partly because of their propensity to loot.

impossible to sell such an apparently cynical policy to the American people. It was notable, however, that Stalin held to the agreement over Greece, refusing to help the Greek communists when their uprising in Athens was suppressed by the British in December 1944.

The Allies and Poland and Germany

By early 1945 it was becoming impossible for the USA to ignore the Soviet insistence that the Allies set out the lines on which the postwar occupations would be established: the defeat of Germany was clearly a matter only of months, and arguments over Germany – and Poland – took up much of the time at the Allies' conferences at Yalta in February 1945 and Potsdam in July 1945. In early 1945 public opinion in the West still favored postwar cooperation amongst the three main allies, and indeed, Roosevelt and Churchill still hoped that the wartime unity could be maintained in some form. Therefore the two Western leaders gave Stalin much of what he wanted in Poland: while insisting that free elections be held, they withdrew their support for the Polish government-in-exile in London. In March the Red army liberated Poland, and in June 1945 a Provisional Government of National Unity was formed, comprising both the anticommunist government-in-exile and the procommunist Polish Committee of National Liberation, which had been formed in Moscow and had subsequently moved to Lublin. By July 1945 the agreed Polish border had moved westward: whereas at Yalta the Western allies had urged that Poland's western border follow the Curzon Line (proposed after World War I), at the Potsdam conference in July 1945 the three leaders agreed that it should instead follow the line of the Oder-Neisse. The point was that the USSR did not intend to return Polish land seized in 1939, and Poland was to be compensated at the expense of Germany. Thus, one-fifth of prewar Germany became part of Poland, and the Germans expelled by the Poles later constituted nearly 25 percent of voters in the FRG.

The discussions over Germany at Yalta and Potsdam were easier, since the Allies all agreed on the need to render it harmless for the foreseeable future. At Yalta they agreed that Germany would be divided into zones of occupation, with France invited to take a fourth zone. At Potsdam this policy was confirmed; they also agreed that the economy was to be decentralized in order to break up excessive economic concentration as exemplified by cartels and other monopolistic arrangements, although the country was to be treated as a single economic unit. This, however, would require wholehearted cooperation by the Allies; and this cooperation soon began to break down: the outcome for Germany was be its division into two sovereign states.

The period from the end of the war to the end of 1947 saw the increasing division of Europe into two camps, as the USSR consolidated its direct control over the states of Eastern Europe. Curiously, however, the first two states to become communist-controlled, Yugoslavia and Albania, evaded Soviet control, primarily because they did not share common borders with the USSR. The Yugoslav Tito had been the leader of the largest and most effective resistance movement in Europe; on the liberation of Belgrade by the Red Army in 1944 a People's Republic was proclaimed, the king deposed and Tito became prime minister in 1945. While he concluded economic and military agreements with the various Eastern European countries, he refused to accept direction from Moscow, and Yugoslavia was expelled from the Cominform in January 1948. Tito remained a thorn in Stalin's side, and relations with the USSR were only normalized in May 1955, although Yugoslavia refused to join the Warsaw Pact.

Albania as well refused to join the Soviet-directed military alliance, although it was probably more driven by intense xenophobia than was Yugoslavia. After the Italian invasion of Albania, the king of Italy had asumed the crown of Albania. With the defeat of Italy, the Albanian King Zog was deposed, and in 1946 a republic was declared. Enver Hoxha, the communist leader, assumed control.

At Yalta, the three Allies had pledged to help any liberated states to form interim governments which were to be broadly representative of all democratic elements, and they further pledged that free elections would be held in these states as soon after liberation as possible. In the Eastern European states, however, these activities were

◄ Enver Hoxha (above), the leader of the Albanian communist party from 1945 until his death in1985, was also a fierce nationalist. In 1961 Albania was thrown out of the Warsaw Pact and COMECON (the Soviet trade organization) for backing the People's Republic of China against the USSR. Jacques Duclos (below), founder-member of the French communist party in 1920 and effective leader of the French communist resistance effort during the war, was for some years communist parliamentary leader.

▼ These youths throwing stones at a Soviet tank were part of the wave of riots and strikes which swept East Berlin and parts of Soviet-occupied East Germany in June 1953, after Stalin's death. Workers were protesting against increased production quotas and police repression. The Soviets suppressed the uprising and enjoyed the cooperation of the East German leadership in reestablishing control. The German Democratic Republic remained one of the USSR's staunchest allies.

◄ Marshal Tito, prime minister and then president of Yugoslavia, addresses a crowd of Slovenian partisans. The son of a peasant, Tito spent his life as a communist political organizer. Thanks to his wartime leadership of the National Liberation Front of partisans, he was acclaimed the national leader soon after the liberation of Yugoslavia. A nationalist as well as a communist, he resisted postwar Soviet attempts to determine Yugoslav policies.

carried out under the eyes of the occupying Soviet forces, and during 1947 and early 1948 all these states fell under the control of the various communist parties. Poland was the first: the Provisional Government of National Unity had been formed from pro- and anticommunists in June, and in the elections of January 1947 the government bloc gained 382 out of the 444 seats in the Sejm; the atmosphere, however, increasingly became one of police terror, and in October 1947 the anticommunist leader fled to London.

Hungary enjoyed free elections in 1945, with the result that the small farmers' party received 57 percent of the vote and the communists and social democrats each received 17 percent. In the 1947 elections the communists emerged as the largest single party with nearly 23 percent of the vote. By skilful political maneuvering they secured the most important government posts; they soon exercised absolute power.

In Romania a government was formed in November 1946, from the communists, the liberals, the social democrats and the Plowmen's Front (a peasant communist party). A year later this government compelled King Michael to abdicate and on 30 December 1947 proclaimed a People's Republic. A popular democratic front won overwhelming majorities at subsequent elections, and the country was soon under communist control. Soviet troops remained in Romania until July 1958, at first to maintain lines of communication between the USSR and Soviet troops in Austria, and after July 1955 as part of the Warsaw Pact. The USSR had recurring difficulties with its Romanian ally, which for much of the postwar period insisted on following an independent foreign policy line.

Much more amenable to Soviet policies was Bulgaria, which owed its very statehood to Russian support in 1878 and 1909. When the German army withdrew in 1944 there was a coup which brought to power the fatherland front, a coalition of communists (support for whom had been growing between the wars), the agrarian party and the social democrats. In 1946 the monarchy was abolished by referendum, and elections held to the Grand National Assembly, which the communists won by 277 to 187. The following year the agrarian party was suppressed and its leader hanged, by order of the communist leader, and after July 1948 no members of the political opposition were allowed to sit in the Assembly.

The USA had been watching these changes with growing alarm. It feared that the same might happen in France and Italy, both of which had large communist parties. This was an important stimulus to proposals for economic aid for Europe. Aid was offered to all European countries, East and West, although the USA assumed that the Soviet Union would not participate. Czechoslovakia, however, first accepted, and then was forced to withdraw by Stalin: it was the fate of Czechoslovakia in February 1948 which rang alarm bells in the US congress and encouraged the passage of the Marshall Aid bill. After elections in 1946 the communists emerged as the largest party, but a growing food crisis made them fear the results of the forthcoming June 1948 elections. The communist leader forced a crisis, "hostile elements" were purged from the parties which formed the government coalition – and the moderate foreign minister, Jan Masaryk, somehow fell to his death from his window. Thereafter there was a single slate of candidates for election.

The Czechoslovakian coup eliminated any

The Marshall Plan

In spring 1947 the USA became increasingly concerned about the ability of the European nations to recover from the war. It feared that, in their weakened state, they would be prey to an expansionist USSR, through invasion or through the domination of communist-led domestic coalitions. The USA also feared that it might plunge into a recession if a thriving Europe did not exist as a focus for American exports and private investment. The Europeans were suffering from a "dollar famine": most of what they wished to import had to be paid for with dollars, but they had very few exports with which to earn those dollars. The main point of the Marshall Plan, called after the secretary of state, George C. Marshall, was to supply dollars to Europe.

The USA insisted that the 16 participating nations (Austria, Belgium, Denmark, Eire, France, Greece, Iceland, Italy, Luxembourg, the Netherlands, Norway, Portugal, Sweden, Switzerland, Turkey and the UK – Czechoslovakia had accepted the American invitation, but had been forced to withdraw by the USSR) present a unified plan for recovery, rather than drawing up separate lists of requests. This was partly to keep the total amount requested within reasonable bounds; a second, strategic, purpose was to force the 16 to integrate their economic recovery programs. The USA believed that if Europe could create a free-trade area on American lines, recovery would be facilitated and worldwide multilateral free trade, a prime goal of the American government, would be achieved.

In 1947 those countries deemed most at risk from internal communist threats, Austria, France and Italy, received 522 million dollars in interim aid; once the Economic Co-operation Act was passed in April 1948, the European Recovery Administration was set up and missions sent to all 16 countries: the flow of aid could then begin. From 1948 to 1951 the USA granted more than 13 billion dollars. The results were impressive indeed: Europe's gross national product jumped by more than 32 percent, while agricultural production grew by 11 percent and industrial production by 40 percent above their prewar levels.

▲▶ The poster (above) celebrates the American goal of removing barriers to trade amongst European states. Food was important (right); but machine tools catered for the future.

▲ Nato chiefs of staff, of the UK, the United States and Belgium, in conference on defence, December 1950.

◄ London housewives receiving ration books in May 1951. In some respects living conditions were more difficult after the war than during it. In Britain, food rationing was tightened up, while clothes rationing continued until 1949. Memories are green of the day in 1951 when sweets ceased to be rationed, but food rationing only ended completely on 3 July 1954, when Churchill told the housewives of Britain to throw their ration books into the fire.

illusions which Western governments might have had in respect of Soviet intentions, and heightened their fears of new Soviet moves; one consequence was new attention paid to rearmament. The Soviets did indeed make a move, and this a dramatic one: on 24 June 1948 they imposed a complete blockade on the Western zones of Berlin, which lay 175 km (110 mi) inside the Soviet zone. The attempt was clearly to force the Western powers out of Berlin and bring it under the complete control of the USSR. The West reacted by instituting the Berlin airlift, whereby supplies were flown into the city night and day. Eventually Stalin lifted the blockade in May 1949. Berlin remained, however, a focus of tension between Eastern and Western blocs; this culminated in 1963, with the building of the wall.

FOR EUROPEAN RECOVERY
UNITED STATES OF AMERICA
FIRST CARGO OF
CARIBBEAN SUGAR
SHIPPED UNDER MARSHALL
AID

Postwar Europe: two armed camps

Meanwhile negotiations had begun in September 1948 between the Western European democracies, the USA, Portugal, Canada and Denmark, toward the establishment of a military alliance. This was partly in response to Soviet moves in Europe, partly from disappointment with the UN, where the Soviet veto was hampering the work of the security council. These countries established the North Atlantic Treaty Organization or NATO: the treaty states that an armed attack on one is an attack on all, and that each will help the country attacked by "such action as it deems necessary". It was signed on 4 April 1949, linking Western Europe in an anti-Soviet bloc.

The seemingly permanent nature of the division of Europe was underlined the following month when the three Western zones of Germany were joined together in the Federal Republic of Germany (FRG); in response the Soviets in October 1949 signed a treaty with the German Democratic Republic (GDR), previously their zone of occupation. This situation held until 1955, when the Western Allies agreed that the FRG should be allowed to rearm and join NATO, at the same time becoming a sovereign state. In October the USSR signed a treaty with the GDR, recognizing it as sovereign, although the Western powers refused to extend such recognition.

Austria, which had also been divided into four occupation zones at the end of the war, was reunited and made neutral. The USSR had always insisted on linking an Austrian settlement with that of Germany, but by 1955 Nikita Khrushchev, the new Soviet leader (Stalin had died in March 1953), wanted to improve relations with Europe and agreed to withdraw Soviet troops from Austria, in October 1955 – the first voluntary Soviet withdrawal from Central Europe. But this presented a problem in logistics: stationing Soviet troops in Hungary and Romania had been justified by the need to maintain communications between the USSR and its troops in Austria. This justification no longer remained, but the Soviet wish to retain troops in Eastern Europe did.

This, plus renewed Soviet apprehension about Germany after their failure to prevent its rearmament and entry into NATO, plus the general need to safeguard their interests, led to the formation by the Soviets of the Eastern European Mutual Assistance Treaty, known as the Warsaw Pact. Signed five days after the GDR attained full sovereignty, the Treaty linked all the East European states except Yugoslavia under a unified military command. Thus the USSR moved, like the USA, to a multilateral defense pact, albeit one which was rather more tightly controlled by the dominant power than was NATO.

The outcome of World War II in Europe was, first, a diplomatic revolution, as former enemies became allies and former allies the enemies, and secondly, the freezing of the new configuration of the powers into two armed camps. Though each camp contained members who were not keen on marching in step (in particular Yugoslavia in the East and France in the West), for a generation the picture remained largely unchanged: two sides locked into a balance of terror.

BERLIN: THE FRACTURED CITY

During the war the Allies agreed that, after they had defeated Hitler, they would divide Germany into four zones of military occupation (US, Soviet, British and French). Berlin had of course been a prime target for the Allies, because of its enormous political and psychological importance as the German capital. As a mark of continued Allied cooperation, Berlin would become the seat of the Allied Control Council, which was to be in charge of the military occupation. The capital of the former Reich was itself also to be divided into four occupation zones, becoming a "special area" under joint Allied control, although, ominously, it lay deep in the Soviet zone. The Americans, British and French kept to this extraordinary plan – despite some high-level misgivings – with only the haziest idea of what it might imply, and without proper agreement on access. The Russians reached the city first, in April 1945. Not until summer 1945 did the western powers enter their agreed sectors.

After the defeat of Nazi Germany the wartime alliance quickly ran into trouble. Relations between the USSR and the western powers worsened. Four-power cooperation in Germany and Berlin became unworkable and by 1948 the western powers despaired of reaching agreement on the German problem. They now felt they were conducting a holding operation against communism in Europe and began to plan a democratic constitution and a new currency for their western zones.

This panicked the already suspicious Russians, who closed land access to Berlin in June. The West's response was bold and spectacularly successful. For nearly a year they supplied by air the essential supplies for two million West Berliners. When the blockade was lifted the partition of Germany was consolidated: the three western zones were combined as the Federal Republic of Germany (FRG), with West Berlin as a *Land* with special status, while the Russian zone became the German Democratic Republic (GDR). Divided Berlin signaled the superpowers' intent to cede no further ground. For Britain and France, their continuing roles as occupying powers in Berlin enhanced their status as European Great Powers. In 1955 the FRG joined NATO, and the GDR the Warsaw Pact.

During the 1950s the Russians strove to defuse this "unexploded mine" in their security zone. They suppressed East German demands for unity and the Soviet leader Nikita Khrushchev railed against a divided Berlin, threatening again western access to the city. But US President John Kennedy proclaimed that any diminution of access or threat to western protection would undermine the security of the whole free world. The East Germans responded to this and to the hemorrhage of German refugees to the west by building a wall round the western sectors in 1961, a symbol of the crude partition of the city. Its jarring presence is a reminder that both east and west had wanted unification only on their own terms: division was for both a lesser evil than unity in the other's camp.

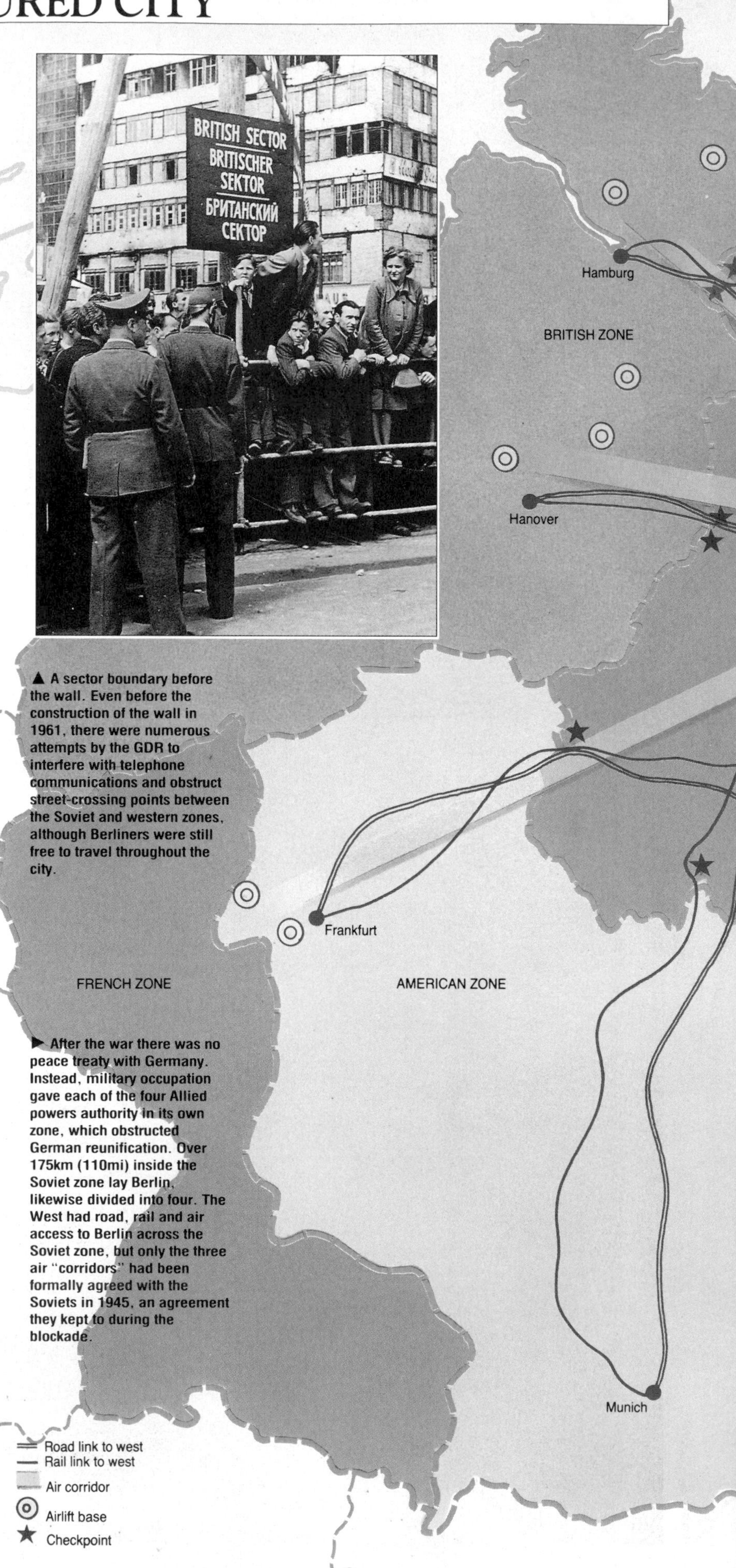

▲ A sector boundary before the wall. Even before the construction of the wall in 1961, there were numerous attempts by the GDR to interfere with telephone communications and obstruct street-crossing points between the Soviet and western zones, although Berliners were still free to travel throughout the city.

▶ After the war there was no peace treaty with Germany. Instead, military occupation gave each of the four Allied powers authority in its own zone, which obstructed German reunification. Over 175km (110mi) inside the Soviet zone lay Berlin, likewise divided into four. The West had road, rail and air access to Berlin across the Soviet zone, but only the three air "corridors" had been formally agreed with the Soviets in 1945, an agreement they kept to during the blockade.

► RAF Avro transport planes at Gatow, Berlin. Coal, food and machinery were ferried daily by air to West Berliners during the blockade – the biggest airlift in history. At its height, one plane was landing in Berlin every two minutes. The airlift had to continue whatever the weather. The Americans even ran "Operation Little Vittles" – dropping sweets and goodies for the children.

▼ By 1945, Berlin – the former center of National Socialist power – was a wasteland, with buildings destroyed, coal, electricity and water shortages and only meager food supplies. Allied military troops kept within their allocated sectors lacking a common purpose.

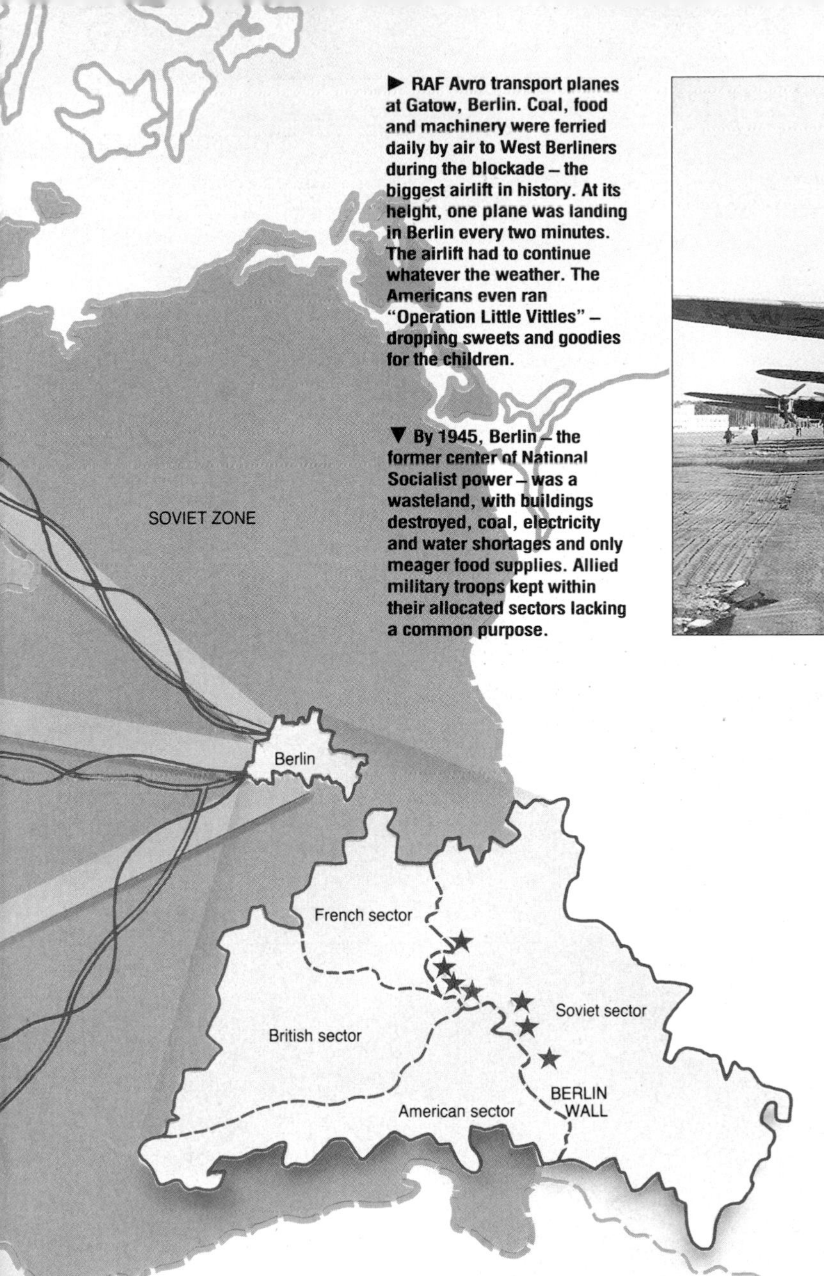

▼ "Hurrah, we are still alive" shouts the sign as news that the Soviets had lifted the blockade reaches ordinary Berliners. The people of West Berlin had held firm against the Soviets' war of nerves.

▲◄ The first fatality at the Berlin Wall (above). Would-be escapee Peter Fechner is carried away by a border guard. The wall has claimed the lives of nearly 100 men and women seeking their families and their freedom in the west. The wall is now over 160km (100mi) long and has been "improved" by the addition of trenches, iron girders, alarm fences, guard dogs and watchtowers (left).

FURTHER READING

THE THEATERS OF WAR

Japan's War in Asia

Crawley, J.B., *Japan's Quest for Autonomy, National Security and Foreign Policy, 1930–38* (Princeton, NJ, 1966).

Dallek, R., *Franklin Roosevelt and American Foreign Policy 1932–45* (New York, 1975).

Li, L., *The Japanese Army in North China 1937–41* (Tokyo, 1975).

Ogata, S.N., *Defiance in Manchuria* (Berkeley, Calif., 1964).

Peattie, M.R., *Ishiwara Kanji and Japan's Confrontation with the West* (Princeton, NJ, 1975).

Pelz, S.E., *Race to Pearl Harbor* (Cambridge, Mass., 1974).

Hitler's War in Western Europe

Bond, B., *France and Belgium 1940* (London, 1975).

Chapman, G., *Why France Collapsed* (London, 1968).

Guderian, H., *Panzer Leader* (London and New York, 1952).

Horne, A., *To Lose a Battle: France 1940* (London, 1968).

Young, R., *In Command of France* (Cambridge, Mass., 1978).

Japan's War in the Pacific

Blair, C., *Silent Victory: The U.S. Submarine War Against Japan* (Philadelphia, Pa., 1975).

Fuchida, M., and Okumiya, M., *Midway: The Battle That Doomed Japan* (Tokyo, 1968).

Lewin, R., *The American Magic: Codes, Ciphers and the Defeat of Japan* (New York, 1982).

Lundstrom, J.B., *The First South Pacific Campaign: Pacific Fleet Strategy, December 1941–June 1942* (Annapolis, Md, 1976).

Mayo, L., *Bloody Buna* (New York, 1974).

Stewart, A., *Guadalcanal, World War II's Fiercest Naval Campaign* (London, 1985).

Thorne, C., *Allies of a Kind: The United States, Britain and the War Against Japan, 1941–45* (London, 1978).

Empires at Bay

Kiernan, V.G., *European Empires from Conquest to Collapse* (London, 1982).

Mandel, E., *The Meaning of the Second World War* (London, 1986).

Thorne, C., *The Issue of War: States, Societies and the Far Eastern Conflict of 1941–45* (London, 1985).

Tsou, T., *America's Failure in China 1941–50* (Chicago, 1962).

The Eastern Front

Dunnigan, J.F., (ed.), *The Russian Front: Germany's War in the East, 1941–45* (London and Melbourne, 1978).

Erickson, J., *Stalin's War with Germany*, Vol. 1, *The Road to Stalingrad*; Vol. 2, *The Road to Berlin* (London, 1973, 1985).

Larionov, V., Yeronin, N., *et al., World War II: Decisive Battles of the Soviet Army* (Moscow, 1984).

Seaton, A., *The Russo-German War 1941–45* (London, 1971).

Shtemenko, S.M., *The Soviet General Staff at War 1941–45* (Moscow, 1985).

Zhukov, G., *Reminiscences and Reflections* (Moscow, 1985).

The War in the Mediterranean

Carver, M., *Tobruk* (London, 1964).

Graham, D. and Bidwell, S., *Tug of War: The Battle for Italy 1943–45* (London, 1986).

Jackson, W.G.F., *The North African Campaign, 1940–43* (London, 1975).

Nicolson, N., *Alex* (London, 1973).

Strawson, J., *The Italian Campaign* (London, 1987).

The Second Front in Europe

Ambrose, S.E., *Eisenhower: The Soldier* (London, 1984).

D'Este, C., *Decision in Normandy* (London, 1983).

Hamilton, N., *Monty: The Field Marshal* (London, 1986).

Keegan, J., *Six Armies in Normandy* (London, 1982).

Lewin, R., *Hitler's Mistakes* (London, 1984).

The Defeat of Japan

Belote, J.H. and W.M., *Typhoon of Steel: The Battle for Okinawa* (New York, 1969).

Belote, J.H. and W.M., *Titans of the Seas: The Development and Operations of American Carrier Task Forces During World War II* (New York, 1975).

Falk, S.L., *Decision at Leyte* (New York, 1966).

Reynolds, C.G. *The Fast Carriers: The Forging of An Air Navy* (New York, 1968).

Wheeler, R., *A Special Valor: The U.S. Marines and the Pacific War* (New York, 1983).

Y'Blood, W.T., *Red Sun Setting: The Battle of the Philippine Sea* (Annapolis, Md, 1980).

THE FIGHTING SERVICES

Armies and Land Warfare

Bartov, O., *The Eastern Front, 1941–45: German Troops and the Barbarisation of Warfare* (London, 1986).

Bidwel, S. and Graham, D., *Fire-power: British Army Weapons and Theories of War, 1904–45* (London, 1982).

Guderian, H., *Panzer Leader* (London and New York, 1952).

Holmes, R., *The Firing Line* (London, 1985).

Postan, M.M. *et al., Design and Development of Weapons: Studies in Government and Industrial Organisation* (London, 1964).

Navies and Sea Warfare

Hara, T. *et al., Japanese Destroyer Captain* (New York, 1961).

MacIntyre, D., *Battle of the Atlantic 1939–1945* (London, 1970).

Morison, S.E., *The Two-Ocean War: A Short History of the US Navy in the Second World War* (Boston, Mass., 1963).

Rohwer, J., *The Critical Convoy Battles of March 1943: The Battle for HX.229, SC.122* (London, 1977).

Roskill, S.W., *The War at Sea, 1939–45*, 3 vols. in 4 parts (London, 1954–61).

Till, G., *Air Power and the Royal Navy, 1914–1945: A Historical Survey* (London, 1979).

Air Forces and Air Warfare

Dean, Sir Maurice, *The Royal Air Force and Two World Wars* (London, 1979).

Galland, A., *The First and the Last: The German Fighter Force in World War II* (London, 1988).

Higham, R., *Air Power: A Concise History* (London, 1978).

Hogg, I.V., *Anti-aircraft: A History of Air Defence* (London, 1978).

Webster, Sir Charles and Frankland, N., *The Strategic Air Offensive against Germany, 1939–45* (London, 1961).

Prisoners of War

Best, G., *Humanity and Warfare: The Modern History of the International Law of Armed Conflicts* (London, 1980).

Hirschfeld, G. (ed.), *The Policies of Genocide: Jews and Soviet Prisoners of War in Nazi Germany* (London, 1986).

Reid, P.R., *The Colditz Story* (London, 1952).

Intelligence Services

Andrew, C., and Noakes, J. (eds.), *Intelligence and International Relations 1900–1945* (Exeter, 1987).

Deakin, F.W. and Storry, G.R., *The Case of Richard Sorge* (London, 1966).

Jones, R.V., *Most Secret War* (London, 1978).

Kahn, D., *The Code Breakers* (New York, 1967).

THE MOBILIZATION OF PEOPLES

The Impact of Total War

Costello, J., *Love, Sex and War: Changing Values 1939–1945* (London, 1985).

Marrus, M. R., *The Unwanted: European Refugees in the Twentieth Century* (New York, 1985).

Marwick, A., *War and Social Change in the Twentieth Century: A Comparative Study of Britain, France, Germany, Russia and the United States* (London, 1974).

Milward, A.S., *War, Economy and Society 1939–1945* (London, 1977).

Japan: Challenge and Response

Butow, R.J.C., *Tojo and the Coming of the War* (Princeton, N.J., 1961).

Cohen, J.B., *Japan's Economy in War and Reconstruction* (Minneapolis, MN, 1949).

Havens, T.R.H., *The Valley of Darkness: the Japanese People and World War Two* (New York, 1978).

Germany: Hitler's Home Front

Balfour, M., *Withstanding Hitler in Germany 1933–1945* (London and New York, 1988).

Beck, E. R., *Under the Bombs: The German Home Front 1942–45* (Lexington, KY, 1986).

Hoffmann, P., *German Resistance to Hitler* (Cambridge, Mass. and London, 1988).

Homze, E.L., *Foreign Labor in Nazi Germany* (Princeton, NJ, 1967).

Kershaw, I., *The "Hitler Myth": Image and Reality in the the Third Reich* (Oxford, 1987).

Kershaw, I., *Popular Opinion and Political Dissent in the Third Reich: Bavaria 1933–1945* (Oxford, 1983).

Milward, A.S., *The German Economy at War* (London, 1965).

Steinert, M.G., *Hitler's War and the Germans* (Athens, OH, 1977).

Stephenson, J., *The Nazi Organisation of Women* (London and New York, 1981).

The UK: A Democracy at War

Addison, P., *The Road to 1945: British Politics and the Second World War* (London, 1975).

Barnett, C., *The Audit of War* (London, 1986).

Calder, A., *The people's War: Britain 1941 45* (London, 1969).

Marwick, A., *The Home Front* (London, 1976).

Pelling, H., *Britain and the Second World War* (London, 1970).

The Soviet Great Patriotic War

Harrison, M., *Soviet Planning in Peace and War 1938–45* (Cambridge, 1985).

Linz, S.J. (ed.), *The Impact of World War II on the Soviet Union* (Princeton, NJ, 1985).

Porter, C. and Jones, M., *Moscow in World War II* (London, 1987).

Salisbury, H.F., *The Siege of Leningrad* (London, 1969).

Werth, A., *Russia at War 1941–45* (London, 1964).

Werth, A. *The Year of Stalingrad* (London, 1946).

The American Way of War

Costello, J., *Love, Sex and War: Changing Values 1939–1945* (London, 1985).

Perrett, G. *Days of Sadness, Years of Triumph: The American People, 1939–1945* (New York, 1973).

Vatter, H.G., *The U.S. Economy in World War II* (New York, 1985).

THE FRONTLINE CIVILIANS

The Bombing of Cities

Chisholm, A., *Faces of Hiroshima* (London, 1985).

Harrisson, T., *Living Through the Blitz* (London, 1985).

Middlebrook, M., *The Bomber Command Diaries: An Operational Reference Book 1939–45* (Harmondsworth and New York, 1985).

The New Yorker Book of War Pieces (London, 1989).

Occupation and Resistance

Dallin, A., *German Rule in Russia* (London, 1981).

Foot, M.R.D., *Resistance* (London, 1976).

Gillingham, J., *Belgian Business in the Nazi New Order* (Ghent, 1977).

Gordon, B.N., *Collaboration in France during the Second World War*, Ithaca, NY, 1981).

Gross, J.T., *Polish Society under German Occupation: The Generalgouvernement 1939–44* (Princeton, NJ, 1981).

Harterdorp, A. W. K., *Japanese Occupation of the Philippines* (Manila, 1967).

Hawes, S., and White, R. (eds.), *Resistance in Europe: 1939–1945* (London, 1975).

Herzstein, R.E., *When Nazi Dreams Come True* (London, 1982).

Hirschfeld, G., *Nazi Rule and Dutch Collaboration* (Oxford, 1988).

Jones, F. C., *Japan's New Order in east Asia, 1937–45* (London, 1954).

Kedward, H.R., *Occupied France: Collaboration and Resistance, 1940–44* (Oxford, 1985).

Kedward, H.R., and Austin, R. (eds.), *Vichy France and the Resistance: Culture and Ideology* (London, 1985).

Littlejohn, D., *The Patriotic Traitors: A History of Collaboration in German-Occupied Europe, 1940–45* (London, 1972).

Lukas, R.C., *Forgotten Holocaust: The Poles under German Occupation, 1939–44* (Lexington, KY, 1986).

Milward, A.S., *The Fascist Economy in Norway* (London, 1972).

Milward, A.S., *The New Order and the French Economy* (London, 1970).

Noakes, J., and Pridham, G. (eds.), *Nazism 1919–1945, Vol. 3, Foreign Policy, War and Racial Extermination* (Exeter, 1988).

Nu, U, *Burma Under the japanese* (New York, 1954).

Stafford, D., *Britain and the European Resistance 1940–45: A Survey of the Special Operations Executive, with Documents* (London, 1983).

Warmbrunn, W., *The Dutch under German Occupation 1940–45* (Stanford, Calif., and London, 1963).

Zee, H.A. van der, *The Hungry Winter: Occupied Holland 1944–45* (London, 1982).

Propaganda and the Arts

Balfour, M., *Propaganda in War 1939–1945* (London, 1979).

Briggs, A., *The History of Broadcasting in the United Kingdom* Vol. 3, *The War of Words* (London, 1970).

Kershaw, I., *The "Hitler Myth": Image and Reality in the Third Reich* (Oxford, 1987).

Klein, H. (ed.), *The Second World War in Fiction* (London, 1984).

Rhodes, A., *Propaganda: The Art of Persuasion: World War II* (Secaucus, NJ, 1987).

Welch, D. (ed.), *Nazi Propaganda: The power and the Limitations* (London, 1983).

War Crimes

Arendt, H., *Eichmann in Jerusalem* (Magnolia, Mass., 1983).

Bartov, O., *The Eastern Front, 1941–45: German Troops and the Barbarisation of Warfare* (London, 1985).

Gilbert, M., *The Holocaust* (London and New York, 1986).

Ienaga, S., *Japan's Last War* (Princeton, NJ, 1978).

Minear, R.H., *Victor's Justice* (Tokyo, 1971).

Tusa, A. and J., *The Nuremberg Trial* (London, 1983).

THE AFTERMATH OF WAR

The World of the Superpowers

Gaddis, J.L., *Strategies of Containment: A Critical Appraisal of Postwar American National Security Policy* (New York, 1982).

Hoffmann, E.P., and Fleron, F.J. Jr. (eds.), *The Conduct of Soviet Foreign Policy* (New York, 1980).

Hogan, M.J., *The Marshall Plan: American, Britain and the Reconstruction of Western Europe 1947–1952* (Cambridge, 1987).

Rubinstein, A.Z., *Soviet Foreign Policy Since World War II: Imperial and Global* (Cambridge, Mass., 1981).

Asia Reborn

James, D. Clayton, *The Years of MacArthur*, Vol. 3, *Triumph and Disaster, 1945–64* (Boston, Mass., 1985).

Pepper, S., *Civil War in China 1945–49* (Berkeley, Calif., 1978).

Pluvier, J.M., *South-East Asia from Colonialism to Independence* (Oxford, 1974).

Europe Divided

Bloomfield, J., *Passive Revolution: Politics and the Czechoslovak Working Class, 1945–8* (London, 1979).

Garlinski, J., *Poland in the Second World War* (London, 1985).

Rubinstein, A. Z., *Soviet Foreign Policy Since World War II: Imperial and Global* (Cambridge, Mass. 1981).

ACKNOWLEDGMENTS

Picture credits

1	Indian soldiers: India Office Library, London
2–3	Pilots run to Hurricane fighters: MARS, Braceborough, Lincs.
4	Clydebank, Scotland, after the Blitz of 1941: Glasgow Herald
6	Families in the Mayakovsky Metro station, Moscow: Alexander Meledin, Moscow
16–17	US forces unloading supplies, Iwo Jima, February 1945: PF
88–89	Troops boarding plane, Ipswich, Australia, September 1942: RHL
90–91	Japanese soldiers hail the rising sun: Pictorial Press, London
132–133	Shipyard workers, Bethlehem-Fairfield shipyards, Baltimore, Md, May 1943: LC
172–173	People looking for their dead, Kerch, USSR, 1942: Alexander Meledin, Moscow
220–221	Bomb damage, Cologne: USNA

9 FSP **13** AP **20–21** USNA **21t** PF **21c** HDC **21b** Mainichi Newspapers, Tokyo **22** AP **23t** USNA **23c** CP **23b** HPC **24tl** AP **24br** PF **26–27** CP **26b** PF **27** LC **28–29** AP **29r, 30bl, 30tr** ADNZ **30br** BPK **33t** PF **33b** UB **34t, 34b** IWM **35t** BPK **35c** ADNZ **35b** CP/IWM **36–37** RF **37t** Mainichi Newspapers, Tokyo **37b** PF **39t, 39b, 40t** USNA **40b** TPL **41t** USNA **41b** US Air Force, Washington, D.C. **42b, 42–43** USNA **45** HDC **46** Bison Picture Library, London **48** Royal Australian Air Force Association Aviation Museum, Bullcreek, Western Australia **49t** India Office Library, London **49b** IWM **50t** PF **50b** TPL **51t** IWM **51b** TPL **52–53** Harris Collection, Bampton, UK **52b** BK Singapore **53r** PF **55** UB **57tl** John Erickson, Edinburgh **57tr** Anatoli Garanin **57cl** ADNZ **57bl** APN, Novosti, Moscow **58** Georgi Zelma **59t** UB **59b** RHL/IWM **60–61** Alexander Meledin, Moscow **61b** IWM **62** FSP **63t** John Erickson, Edinburgh **63c** UB **63b** RHL **64–65** CP **65r** RHL **67t** CP **67c** PF **67b** TPL **68–69, 68tl** RHL **68bl** ADNZ **69** RHL **71tl, 71bl** IWM **71r** USNA **73tr, 73br** IWM **73c** RHL **73bl** PF **75** BPK **77t, 77b** USNA **78t** Bison Picture Library, London **78b** IWM **79** US Army/USNA **80t** IWM **80c** RHL **80b** US Army/USNA **81l** BPK **81r** IWM **82–83** PF **83, 84tl** USNA **84cl** US Naval Historical Center, Washington, D.C. **84bl** PF **84br** RHL **86t** USNA **86b** IWM **87** US Army/USNA **93, 94, 94–95** RHL **95t** Bison Picture Library, London **95b** BPK **96** RHL **96–97** David King Collection, London **98** RHL **98–99** CP **99bl** RHL **99tr** TPL **99br** BPK **100t** TPL **100b** HDC **101l** RHL **101r** PF **102–103, 102t, 103br** RHL **103tl** IWM **103tr** Pictorial Press Ltd, London **105** USNA **106** RHL **106–107** IWM **107** PF **108, 109b** USNA **109t** PF **110** MARS/US Navy, Braceborough, Lincs., UK **110–111, 111t, 111b** IWM **113** PF **114–115** MARS/IWM **115t, 115b** PF **116l** USNA **116r** RHL **116–117** IWM **117** PF **118t, 118–119** IWM **118b** AP **119l** BPK **119r** USNA **120t** Institute of Electrical Engineers, London **120b, 121tl, 121tr, 121br, 123, 123 inset, 124b, 124–125** IWM **124t** HDC **125b** AP **127t** G. Lacoste 127b David King Collection, London **128t** USNA **128b** Polish Institute and Sikorski Museum, London **129t** Nationalmuseet, Denmark **129b** PF **130t** TPL **130b** HDC **131l, 131r** IWM **135t** HPC **135b** RHL **136** PF **136–137** HPC **137l** Reginald Mount 137r Alexander Meledin, Moscow **138** RDZ-Bilddokument, Zurich **139t** RHL **139b** Pictorial Press Ltd **141t** RHL **141b** PF **142** Mainichi Newspapers, Tokyo **143t** PF **144t, 144b, 145b** Mainichi Newspapers, Tokyo **145t** Pictorial Press Ltd. **147** BPK **148t** EPA **148b** UB **149**BPK **150–151,151** UB **153** University of Reading, UK **154t** PF **154b** HDC **155t** Futile Press/IWM **155b** IWM **156t, 156b** PF **156–157** HDC **157** The Labour Party, London **159, 160t, 160–161** Alexander Meledin, Moscow **161t** John Erickson, Edinburgh **162** Alexander Meledin, Moscow **163, 163 inset** John Erickson, Edinburgh **164** Boris Kudoyarov **165tl** Mikhail Trakhman **165tr, 165br** Alexander Meledin, Moscow **167** LC **168t** IWM **168b, 169t, 169b, 170, 171t, 171b** LC **175 l** RHL **175r** Alexander Meledin, Moscow **176t** Glasgow Herald **176b** AP **177t** PF **177b** IWM **178–179** BPK **178b, 179t** UB **179b** LC **180l** IWM **180r** EPA **181** RHL **183** PF **184t, 184b, 185t, 185b** Pictorial Press, London **186–7** Magnum/Zucca Tallandier, London **186b, 187** BPK **188t** Magnum/Zucca Tallandier, London **188b** Roger-Viollet, Paris **189** RHL **190t** Nationalmuseet, Denmark **190b** ADNZ **191t** IWM **190–191** Roger-Viollet, Paris **192tl, 192cl** John Erickson, Edinburgh **192br** RHL **193** IWM **194** J-L Charmet, Paris **194–195t** Roger-Viollet, Paris **195t** RHL **195r** IWM **194–195** Bundesarchiv, Koblenz **197t** TPL **197b** PF **198** BPK **198–199** John Erickson, Edinburgh **199r** RHL **200t** Bundesarchiv, Koblenz **200b** IWM **201** Lee Miller Archives, Chiddingly, UK **202b, 202–203, 203t** Kobal Collection, London **203b** Aquarius, Brighton **204t, 204b, 205t** Kobal Collection, London **205b** Aquarius, Brighton **206t** Bridgeman Art Library, London **206b** Pictorial Press Ltd, London **207tl** Lee Miller Archives, Chiddingly, UK **207tr** Archiv für Kunst und Geschichte, Berlin (West) **207br, 208** IWM **208–209** The Tate Gallery, London **209b** IWM **211** HDC **212t, 212b** IWM **213** David King Collection, London **214t** Roger-Viollet, Paris **214B** Bundesarchiv, Koblenz **215t** HPC **215b** AP **216–217** IWM **217b** HPC/Bettmann Archive **218tl** Wiener Library, London **218–219** (main picture) Pictor International, London **218–219b** EPA **218b** EPA **219l** Wiener Library, London **225b** Alexander Meledin, Moscow **226** US Airforce/USNA **227l** Pictorial Press Ltd, London **227r, 228** PF **228–229** USNA **229** Saturday Evening Post, London **230t** TPL **230b** BPK **230–231** **USNA** **231t** FSP **231b, 232t, 232b** HDC **232–233** Zefa Picture Library, London **233tr, 233cr** TPL **233bl** David Low Centre for the study of cartoons and caricature, University of Kent at Canterbury **235t** Mainichi Newspapers, Tokyo **235b** TPL **236t** Government of India Information Bureau **236b** PF **234tr, 237bl** TPL **237cr** IWM **238bl, 238tr, 238br** TPL **239** Magnum, London **241** Pictorial Press Ltd, London **242t, 242c** TPL **242–243** BPK **243** Pictorial Press Ltd, London **244t, 244b, 245b** TPL **245t** PF **246** USNA **247tr** PF **247cr** TPL **247bl** Bundesbildstelle, Bonn **247br** Haus am Checkpoint Charlie, Berlin (West)

Abbreviations

ADNZ	Allgemeiner Deutscher Nachrichtendienst, Zentralbild, Berlin (East)
AP	Associated Press, London
BPK	Bildarchiv Preussischer Kulturbesitz, Berlin (West)
CP	Camera Press, London
EPA	Equinox Picture Archive, Oxford
FSP	Frank Spooner Pictures, London
HDC	Hulton Deutsch Collection, London
HPC	Hulton Picture Company, London
IWM	Imperial War Museum, London
LC	Library of Congress, Washington DC
PF	Popperfoto, London
RF	Rex Features, London
RHL	Robert Hunt Library, London
TPL	Topham Picture Library, Kent, UK
UB	Ullstein Bilderdienst, Berlin (West)
USNA	United States National Archive, Washington, D.C.

t = top, tl = top left, tr = top right, c = center, b = bottom etc.

Dr John Campbell and Equinox (Oxford) Ltd wish to thank the following people and institutions for their help in making this book: Paul Addison, Kathie Brown, Jane Higgins, Louise Jones, John Ridgeway, Graham Speake, Michelle VonAhn; Bodleian Library, Oxford; Imperial War Museum, London; London Library

Cartographic advisers
Duncan Anderson, Royal Military Academy, Sandhurst; David Parry, Imperial War Museum, London

Editorial and research assistance
Matthew Jones, William Philpott, Mike Pincombe, Elaine Welsh

Artists
Chris Forsey, Alan Hollingbery, Colin Salmon, Mick Saunders, Dave Smith

Photographs
Shirley Jamieson, Alison Renney

Cartography
Maps drafted by Alan Mais, Hornchurch; Euromap, Pangbourne; Lovell Johns, Oxford

Typesetting
Brian Blackmore, Anita Wright; OPUS Ltd, Oxford

Production
Clive Sparling

Color origination
Scantrans, Singapore

Index
Ann Barrett

INDEX

Page numbers in *italics* refer to illustrations or their captions. **Bold** page numbers refer to the subjects of special or ancillary text features.

M

N

O

P